Chemistry of Lignocellulosics: Current Trends

T0253443

Editor

Tatjana Stevanovic

Department of Wood and Forest Sciences
Laval University
Quebec City, Quebec, Canada

CRC Press
Taylor & Francis Group
Boca Raton London New York

CRC Press is an imprint of the
Taylor & Francis Group, an **informa** business

A SCIENCE PUBLISHERS BOOK

CRC Press
Taylor & Francis Group
6000 Broken Sound Parkway NW, Suite 300
Boca Raton, FL 33487-2742

First issued in paperback 2021

Version Date: 20180626

ISBN-13: 978-0-367-78086-9 (pbk)
ISBN-13: 978-1-4987-7569-4 (hbk)

Library of Cogress Cataloging-in-Publication Data

Names: Stevanovic, Tatjana, editor.
Title: Chemistry of lignocellulosics : current trends / editor, Tatjana
 Stevanovic (Department of Wood and Forest Sciences, Laval University,
 Quebec City, Quebec, Canada).
Description: Boca Raton, FL : CRC Press, [2018] | "A science publishers
 book." | Includes bibliographical references and index.
Identifiers: LCCN 2018027437 | ISBN 9781498775694 (hardback)
Subjects: LCSH: Lignocellulose. | Botanical chemistry. | Green chemistry. |
 Hemp. | Wood--Chemistry.
Classification: LCC TP248.65.L54 C4575 2018 | DDC 660.028/6--dc23
LC record available at https://lccn.loc.gov/2018027437

Visit the Taylor & Francis Web site at
http://www.taylorandfrancis.com

and the CRC Press Web site at
http://www.crcpress.com

Preface

This book is a result of a collective effort of experts working in the field of conversion of renewable sources using sustainable methods to obtain innovative products. It is reflecting the current trends of replacing the fossil-sources by renewable ones. Transformation of lignocellulosic biomass in the future will be performed within the biorefineries, which are designed to separate wood polymers, constructing the plant cell walls, from "free" molecules. These are commonly designated as extractives, since accessible by simple solvent extraction.

The first section of the book discusses the new applications of lignocellulosic fibers from alternative sources (other than wood). The transformation of hemp and hollow floss fibers will provide the access to innovative applications, while contributing to the sustainable transformation of forest resources by replacement of wood fibers.

Although this book does not provide an exhaustive overview of such alternative resources, it indicates how much the knowledge of their chemistry is important for creation of original applications. The non-destructive technique of examining the properties of lignocellulosic materials, the Near Infrared Spectroscopy, represents an interesting tool for estimating the potential of an oak wood for the specific application in wine industry.

The thermal transformation of wood and other biomass is not new, but interest in the development of biofuels has prompted the return to research on pyrolysis and a chapter of the second book section presents an original approach of combining industrial waste with the forest biomass for pyrolytic conversion into fuels. Another chapter discusses the application of thermal conversion in yet another original way, which is the wood welding. The gluing of solid wood pieces together without any adhesives is possible by simply inducing local pyrolysis by friction of wooden dowels against the wood panels. This technique is sustainable as it allows for the wood panel production without any petroleum-derived adhesives, exploiting the thermal conversion of wood constituents. The replacement of petroleum derived chemicals by those from renewable resources is explored in the chapter which describes the impregnation techniques used to densify wood, thus leading to new products with improved properties.

Finally, the third section of this book presents the current trends in application of major wood polymers: cellulose and lignins. The new applications of nanocellulose and related materials are accessible by their chemical modification. These modified nanocelluloses are designed to improve the compatibility with various polymer matrices for production of new composite materials. In a chapter presenting an overview of organosolv pulping, a new catalytic process is described which allows for the production of a highly pure lignin which is directly transformable into carbon fiber.

The two classes of wood extractives are presented: terpenoids and polyphenols. One chapter describes the historical uses of conifer resins, along with the current therapeutical and perfumer's palette applications of volatile (essential) oils. The great potential of application of extractable polyphenols resides principally in their bioactivities, strongly related to their antioxidant capacity. The application of flavonoid-rich extracts of Mexican oaks bark is illustrating such an application for treatments of skin inflammation. Finally, the last chapter of the book describes the transformation of condensed tannins into industrial foams.

The chosen examples and discussions presented in this book confirm the need for understanding the molecular level of lignocellulosics. The good knowledge of chemical structures of both structural biopolymers and extractives is therefore essential for the design of sustainable processes leading to innovative products, a result of a multidisciplinary team work.

Contents

Preface   iii

Part I: New Sources of Lignocellulosic Fibres and Their Properties

1. Waste Hemp (*Cannabis sativa*) Fibers as a Biosorbent and a   3
   Precursor for Biocarbon Sorbents: Influence of their Chemical
   Composition on Pb(II) Removal
   Mirjana Kostic, Biljana Pejic and Marija Vukcevic

2. Hollow Floss Fibers—Characterization and Potential   22
   Industrial Applications
   M. Robert, P. Ovlaque and M. Reza Foruzanmehr

3. Near Infrared Spectroscopy, A New Tool to Characterize Wood   42
   for Use by the Cooperage Industry
   Gilles Chaix, Thomas Giordanengo, Vincent Segura, Nicolas Mourey,
   Bertrand Charrier and Jean Paul Charpentier

Part II: Innovative Products of Chemical and Thermochemical
Treatment and Conversion of Lignocellulosics

4. Energetic Valorization of Lignocellulosic and Industrial Wastes   69
   by Thermal Gasification
   Paulo Brito, Octávio Alves, Luís Calado, Bruno Garcia and Eliseu
   Monteiro

5. Effect of Wood Welding Treatment on Chemical Constituents of   99
   Australian Eucalyptus Species
   Benoit Belleville, Georges Koumba-Yoya and Tatjana Stevanovic

6. Impregnation of Wood Products: Past, Present and Future Work   116
   Diane Schorr, Stéphanie Sabrina Vanslambrouck and Véronic Landry

Part III: Structural Polymers and Extractives From Lignocellulosic Sources—Properties and New Applications

7. **Latest Applications of TEMPO-Oxidized Cellulose Nanofibres Obtained from an Ultrasound-assisted Process**   147
Éric Loranger and *Claude Daneault*

8. **Processing and Characterization of Modified Nanocellulose/ Polyester Composites**   167
Jelena Rusmirović, Milica Rančić and *Aleksandar Marinković*

9. **Organosolv Processes: New Opportunities for Development of High Value Products from Lignins**   214
Georges Koumba Yoya and *Tatjana Stevanovic*

10. **Conifer Resins and Essential Oils: Chemical Composition and Applications**   231
Nellie Francezon and *Tatjana Stevanovic*

11. **Flavonoids from *Quercus* Genus: Applications in Melasma and Psoriasis**   253
Esquivel-García Roberto, Velázquez-Hernández María-Elena, Valentín-Escalera Josué, Valencia-Avilés Eréndira, Rodríguez-Orozco Alain-Raimundo and *Martha-Estrella García-Pérez*

12. **The Newer Chemistry of Condensed Tannins and Its Foams Application**   280
Antonio Pizzi

Index   307

Part I

New Sources of Lignocellulosic Fibres and Their Properties

Waste Hemp (*Cannabis sativa*) Fibers as a Biosorbent and a Precursor for Biocarbon Sorbents

Influence of their Chemical Composition on Pb(II) Removal

Mirjana Kostic, *Biljana Pejic* and *Marija Vukcevic*

Introduction

Nowadays, the mobilization of pollutants in the environment has become a serious concern due to the increased industrial activities and effluents discharged into the environment. Most of these effluents contain different organic and inorganic toxic substances (especially heavy metals), whose presence in the environment is a major concern. Heavy metals are permanent environmental pollutants due to their toxicity and bioaccumulative tendency. Toxicity of heavy metals represents a great danger to different forms of life and especially for human health. Also, due to their bioaccumulative tendency, heavy metals are easily incorporated into the food chain and thereby cause multiple diseases, such as renal dysfunction as well

Faculty of Technology and Metallurgy, University of Belgrade, Karnegijeva 4, 11 000 Belgrade, Serbia.
  Emails: biljanap@tmf.bg.ac.rs; marijab@tmf.bg.ac.rs
* Corresponding author: kostic@tmf.bg.ac.rs

as chronic alterations in the nervous system, gastrointestinal tract, brain and bones (Crini 2005, Demirbas 2008, Pejic et al. 2009).

Therefore, the pursuit of the optimal method for heavy metals removal from contaminated water represents one of the major ongoing topics. A number of methods have been developed for water purification: photocatalytic degradation, combined photo-Fenton and biological oxidation, aerobic degradation, ozonation, chemical precipitation, chemical oxidation or reduction, electrochemical treatment, evaporative recovery, filtration, reverse osmosis, ion exchange and membrane technology. Among them, adsorption on activated carbon has been proven to be an efficient technology for purification of drinking water and wastewater, though its large-scale application is limited by the high cost of the carbon adsorbent. Although adsorption may be a costly method, utilization of the inexpensive and renewable materials as alternative for conventional activated carbon could make the adsorption process cost effective. Therefore, in recent years, attention has been devoted to the development of different kinds of biosorbents which are very cost-effective adsorbents for purification of waste water and drinking water. The application of biosorbents derived from agricultural wastes for water purification is the equivalent of using one waste for cleaning up another (Volesky 2007). The biomass obtained as the byproducts or waste from industrial and agricultural production are increasingly used as environmentally- and economically-friendly natural materials for heavy metals removal. For example, different lignocellulosic waste materials, unmodified and chemically modified fibers (sisal, coir, jute, kenaf, ramie) and some other parts of plants such as corncobs, rice husk, coconut shells and sawdust have been used as biosorbents for adsorption of heavy metals (Kostic et al. 2014, Pejic et al. 2009, Sousa et al. 2010, Tan et al. 2010). Lignocellulosic fibers have excellent physico-mechanical and sorption properties that are the result of their heterogenous chemical composition and specific structure. The main constituents of lignocellulosic fibers are cellulose, hemicelluloses and lignin, while the content of pectin, fats and waxes is significantly lower, but their role in the fiber construction is also very important. In the middle lamella, elementary fibers are connected by pectins and encrusted by lignin, while inter-fibrillar regions in fibers are filled with hemicelluloses. Each elementary fiber can be considered as a network of ultrafine cellulose fibrils embedded in a matrix of hemicelluloses and lignin. The presence of cellulose, hemicelluloses and lignin functional groups make lignocellulosic fibers very effective biosorbents, due to the ability of carboxylic (primarily present in hemicelluloses, pectin and lignin), phenolic (lignin and extractives) and to a certain extent hydroxylic (cellulose, hemicelluloses, lignin, extractives and pectin) and carbonyl groups (lignin) to bind heavy metal ions from polluted waters. Strong bonding of metal ions by the hydroxylic, carboxylic and phenolic groups often involves complexation and ion exchange (Crini 2005, Pejic et al. 2009, Shukla and Pai 2005).

Despite the satisfactory results obtained using low-cost lignocellulosic waste materials as biosorbents, activated carbons are known to be more efficient in adsorbing a greater amount of pollutants. Adsorption on activated carbon has been proven to be an efficient technology for purification of drinking water and wastewater, while its large-scale application is limited by the high cost of the carbon adsorbent. Therefore, utilization of inexpensive and renewable materials as carbon precursors could make the adsorption process cost effective. Following the general trend of finding inexpensive adsorbents, different kinds of inexhaustible, sustainable and annually renewable industrial and agricultural lignocellulosic wastes were utilized as precursors for carbon adsorbents production (Nor et al. 2013, Vukcevic et al. 2012, Vukčević et al. 2014). The usage of the lignocellulosic wastes for production of carbon materials is very attractive from the point of decreasing the waste disposal costs and improving environment protection through the waste recycling and producing useful products. In that way, different carbon sorbents produced from lignocellulosic waste were used for water purification by removal of specific pollutants like dyes (Hameed et al. 2008, Hernández-Montoya et al. 2011), heavy metals (Hernández-Montoya et al. 2011, Rahman et al. 2014, Vukčević et al. 2014), pesticides (Salman and Hameed 2010, Vukčević et al. 2015) and phenols (Aber et al. 2009). Different carbon sorbents with high specific surface area and good sorption characteristics were obtained by using lignocellulosic fibers (Dumanh and Windle 2012, Reed and Williams 2004, Vukčević et al. 2014). Lignocellulosic fibers represent a good precursor for carbon production since they contain cellulose, hemicelluloses and lignin, which are rich in carbon. It was suggested that the complex structure and heterogeneous chemical composition of lignocellulosic fibers are one of the crucial factors that affect the specific surface area, amount of surface oxygen groups and morphology of resulting carbon sorbent (Suhas et al. 2007, Vukcevic et al. 2012). Additional factors that define the final properties of carbon materials are related to production conditions: carbonization and activation temperature, heating rate, type of activating agent and activating agent/carbon material ratio (Chunlan et al. 2005).

Different kinds of lignocellulosic materials of specific chemical composition and structure are already utilized as low-cost, renewable and easily available biosorbents and carbon sorbents, precursors for heavy metals removal from polluted water. Among other biomass types, hemp fibers were used as a biosorbent for heavy metals adsorption, as well as a precursor for biocarbon sorbents production (Pejic et al. 2011, Rosas et al. 2009, Vukcevic et al. 2014).

Hemp (*Cannabis sativa*), an old and controversial plant, has been planted agriculturally for many centuries to get its bast fibers and hempseed oil. Nowadays, due to increased demand for comfortable, biodegradable and ecological fibers, production of hemp fiber is experiencing a renaissance. The increased production of hemp fibers in the textile industry brings the considerable amount of waste in the form of short and entangled fibers. The hemp fibers are characterized by cross section complexity and specific surface morphology. It can be considered as composites of hollow cellulose fibrils held together by lignin and a hemicelluloses matrix. The primary fibrils are assembled into microfibrils with the hemicelluloses decorating the outside. Hemicelluloses act as the connection between the microfibrils, creating the primary structural network. Lignin can be deposited within this network in two ways: either as isolated lumps (when it acts to limit the movement of the microfibers, thus increasing stiffness by steric hindrance) or as a continuous matrix, which then supplements and presumably replaces the hemicelluloses in importance as a linker of the cellulose microfibrils. The hydrophobic lignin network affects the properties of other networks in a way that it acts as a coupling agent and increases the stiffness of the cellulose/hemicelluloses composite (Kostic et al. 2008, Kostic et al. 2014, Pejic et al. 2008, Thomas et al. 2011).

It was suggested that the complex structure and heterogeneous chemical composition of hemp fibers are one of the crucial factors that affect the specific surface area, amount of surface oxygen groups and morphology of biosorbents as well as biocarbon sorbents (Vukcevic et al. 2012, Vukcevic et al. 2014). The aim of this investigation was to attain the appropriate specific structure of hemp fibers which will be used as biosorbents and biocarbon precursors, in order to promote the formation of the required porosity and surface chemistry that facilitates heavy metals adsorption. For that reason, different biosorbents and biocarbon precursors were obtained by chemical modification of pristine short hemp fibers followed by carbonization. Equilibrium and kinetic studies of heavy metals adsorption on biosorbents and biocarbon, based on unmodified and modified hemp fibers, were performed using lead ions as a model of heavy metal ion. Additionally, lead is chosen due to it high toxicity and acute and chronic toxic effects in animal and human health.

Materials and Methods

Preparation of biosorbent and biocarbon samples

Short and entangled hemp fibers used, were obtained from ITES Odzaci, Serbia. Biosorbents and biocarbon precursors of different chemical composition were obtained by alkali and oxidative treatment of short hemp fibers, as it is described in the literature (Kostic et al. 2010, Pejic et al. 2011). The progressive removal of the hemicelluloses was brought about by treating the fibers with 17.5% NaOH solution, while the lignin was progressively removed by treating hemp fibers with 0.7% $NaClO_2$. Biosorbents and biocarbon precursors Hox5 and Hox60 were obtained after 5 and 60 minutes of oxidative treatment, respectively, while Hal5 and Hal45 were obtained by alkali treatments (5 and 45 minutes, respectively). Chemical composition of pristine (H) and modified hemp fibers was determined according to the scheme of Soutar and Bryden (Garner 1967) by successive removal of water solubles, fats and waxes, pectins, lignin and hemicelluloses. Variation of hemicelluloses and lignin content is shown in Fig. 1. After the alkaline treatment, the level of hemicelluloses content was lower than in the unmodified fibers, while lignin content was almost unchanged. On the other hand, the sodium chlorite treatment of hemp fibers progressively removed lignin and slightly changed the hemicelluloses content.

The samples obtained by chemical modification, along with the pristine short hemp fibers, were used as biosorbents and as biocarbons precursors. Biocarbons

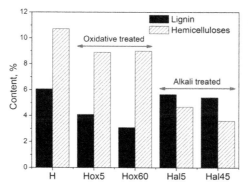

Fig. 1. Chemical composition of hemp and modified hemp fibers used as a biosorbents and biocarbon precursors.

were obtained by carbonization of unmodified and chemically modified hemp fibers in an electrical furnace at temperature 1000°C under constant nitrogen flow, with a heating rate of 5°C/min. After carbonization, five samples of hemp fiber-based biocarbons, denoted CH, CHox5, CHox60, CHal5 and CHal45, were obtained. The scheme of biosorbents and biocarbons preparation is given in Fig. 2.

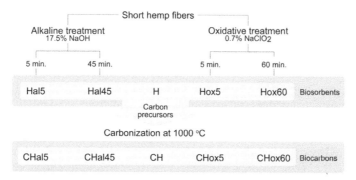

Fig. 2. Scheme of biosorbent and biocarbon samples preparation.

Materials characterization

Surface structure and morphology were studied by scanning electron microscopy (SEM JEOL JSM-6610LV).

Textural characteristics of biosorbent and biocarbon samples were obtained from the N_2 adsorption and desorption isotherms measured at the temperature of liquid nitrogen using Micromeritics ASAP 2020, Surface and Porosity Analyzer (Micromeritics Instrument Corporation, U.S.A.).

Temperature-programmed desorption (TPD) in combination with mass spectrometry was used to investigate the nature and thermal stability of biocarbon surface oxygen groups (Kalijadis et al. 2015, Vukčević et al. 2008). The TPD profiles were obtained using a custom-built set-up, consisting of a quartz tube placed inside an electrical furnace. Biocarbon samples were outgassed in the quartz tube and subjected to TPD at a constant rate of heating of 10°C/min to 900°C under high vacuum. The amounts of CO and CO_2 released from the carbon sample (0.1 g) were monitored using an Extorr 3000 quadrupole mass spectrometer (Extorr Inc, USA).

The amount of carbonyl and carboxyl groups, which acts as active sites for adsorption on the surface of biosorbents, was determined by calcium-acetate method described in literature (Parks and Hebert 1972, Praskalo et al. 2009). Additionally, the oxygen groups of biocarbon, that have acidic or basic properties, were determined by acid-base titration method (Kalijadis et al. 2011). The amount of acidic and basic sites were determined by mixing 0.1 g of biocarbon samples with 25 ml of 0.1 M NaOH and 0.1 M HCl, respectively, in 50 mL beakers. The beakers were sealed and shaken for 24 h. The solutions were then filtered and titrated with 0.1 M HCl and 0.1 M NaOH.

The point of zero charge (PZC) of the carbonized hemp fibers was determined by mass titration (Vukčević et al. 2008) by placing various amounts (0.05, 0.1, 0.5, 1 and 10% by weight) of biosorbent and biocarbon samples in 10 cm³ of 0.1 M KCl solution (prepared using preboiled water to eliminate CO_2). The beakers were held in an N_2 atmosphere to eliminate any contact with air. They were then placed in a thermostat shaker overnight. The limiting pH value was taken as the PZC.

Metal adsorption experiments

Aqueous solution of lead ions was prepared by using $Pb(NO_3)_2$. Adsorption of lead ions by different biosorbent and biocarbon samples was performed in the batch system at 25°C, from lead ions initial concentrations of 50 mg/dm³. 0.1 g of biosorbent and biocarbon samples were immersed in 50 ml of lead ions water solution and constantly shaken for two hours. Also, the effect of pH on the biosorption was studied. For this, the initial pH of 5 of lead ion aqueous solution was adjusted by stepwise addition of HNO_3 or NH_4OH. Lead ions concentration in the solution during the adsorption was measured by atomic absorption spectrometry (PYE UNICAM SP9, Pye Unicam, Ltd., U.K.).

Adsorption kinetics

The kinetic data, obtained from adsorption experiment, are examined by two kinetic models, Lagergren pseudo-first order and pseudo-second order (Vukcevic et al. 2014), as given in Eqs. (1) and (2), respectively. Lagergren proposed a method for adsorption analysis which is a pseudo-first-order kinetic equation in the linear form:

$$\log(q_e - q_t) = \log q_e - \left(\frac{k_1}{2.303}\right) \cdot t \tag{1}$$

where q_e and q_t are the amounts of lead ions adsorbed at equilibrium (mg/g), and at time t (min), respectively, and k_1 is the pseudo first order rate constant (min⁻¹).

A linear plot of log (q_e-q_t) against time allows one to obtain the rate constant from the slope.

The pseudo second order kinetics may be expressed as:

$$\frac{t}{q_t} = \frac{1}{k_2 \cdot q_e^2} + \left(\frac{1}{q_e}\right) \cdot t \qquad (2)$$

where k_2 (g mg^{-1} min^{-1}) is the rate constant of adsorption, q_e (mg/g) is the amount of lead ions adsorbed at equilibrium and q_1 (mg/g) is the amount of lead ions adsorbed at time t. The equilibrium adsorption capacity (q_e), and the second order constants k_2 (g mg^{-1} min^{-1}) can be determined from the slope and intercept of plot t/q_t vs. t.

Characterization of Biosorbents and Biocarbons

Surface structure and morphology

Changes in chemical composition, fiber surface and structure, which are the result of chemical treatments, can be ascribed to the decrease of lignin or hemicelluloses content as well as to the location of these components in the hemp fiber structure. As it can be seen from the SEM photographs of unmodified (sample H) and modified (samples Hox5 and Hox60) hemp fibers (Fig. 3.), removal of lignin from the middle lamellae and secondary wall of hemp fibers, through oxidative treatment, results in elimination of micro-pores and the less rigid cell wall, accompanied by formation of new capillary spaces in the inter-surfacial layer between partially separated elementary fibers. On the other hand, hemicelluloses removal from inter-fibrillar regions by alkaline treatment leads to fiber fibrillation as observed for samples Hal5 and Hal45. Both treatments lead to peeling of fiber surface (Fig. 3).

The morphology of biocarbon samples obtained by carbonization of unmodified and modified hemp fibers is also shown in Fig. 3.

Structure of lignocellulosic-derived carbon materials is directly dependent on the lignin and hemicelluloses content in the precursor material (Mohamed et al. 2010). Chemical modification induces liberation of fibrils in hemp fiber structure, which is noticeable in the morphology of carbonized samples based on modified hemp fibers. For both alkaline and oxidative precursor treatments, increasing the modification time increases the fibrillation rate of the carbonized samples. This fibrils liberation is more pronounced for carbonized samples obtained from alkali-treated hemp fibers, while in the case of sample CHox60 (Fig. 3), the presence of a few completely liberated nanofibrils (with a diameter less than 100 nm) is noticed. Also, some granular matters are dispersed on the surface of samples CHox5, CHal45 and in the large quantity on the sample CHox60. This granular matter may be attributed to the carbonization of hemicelluloses, which, as a volatile fraction of the biomass (Mohamed et al. 2010), can diffuse to the surface of the pyrolyzing material, during the pyrolytic process, and form granulated carbonized phase. Large amounts of granular matter, observed at the surface of CHox60, may

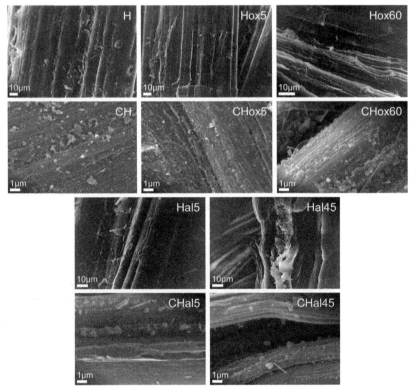

Fig. 3. SEM images of biosorbents and biocarbons.

be the consequence of high hemicelluloses content, uncovered by lignin removal from the middle lamellae. In the case of samples CHal5 and CHal45, despite the lower hemicelluloses content, increased precursor fibrillation may facilitate the diffusion of volatile pyrolytic products to the surface of pyrolyzing material to form granulated carbonized phase.

Textural characteristics

Values for textural characteristics (specific surface area (S_{BET})) of biosorbent and biocarbon samples, calculated from the N_2 adsorption and desorption isotherms, are summarized in the Table 1.

As can be seen from Table 1, the alteration of short hemp fibers' chemical composition affects specific surface area of both biosorbents and biocarbons. Although, biosorbent samples do not have a developed specific surface area, pronounced fibrillation induced by longer oxidative and alkali treatments leads to specific surface area increase. On the other hand, biocarbon samples are microporous with a developed BET specific surface area, ranging from 389 to 573 m²/g. The chemical composition of the lignocellulosic precursor has a significant effect on

Table 1. Surface characteristics of biosorbent and biocarbon samples.

Samples		S_{BET} (m²/g)	Amount of surface groups (mmol/g)*	pH$_{PZC}$
Biosorbents	H	0.236	0.587	4.4
	Hox5	0.217	0.637	3.8
	Hox60	0.312	0.546	4.6
	Hal5	0.185	0.619	4.1
	Hal45	0.258	0.598	4.3
Biocarbons	CH	518	0.810	10.58
	CHox5	429	0.854	11.05
	CHox60	389	1.459	10.37
	CHal5	426	1.045	10.72
	CHal45	574	1.060	10.95

* amount of carbonyl and carboxyl groups for biosorbents determined as described in the literature (Praskalo et al. 2009, Parks and Hebert 1979) and amount of surface oxygen groups for biocarbon obtained by acid-base titration (Kalijadis et al. 2011).

the textural characteristics of biocarbons. Lignin is one of the major components that affect specific surface area of biocarbons (Kennedy et al. 2004, Reed and Williams 2004). As shown in Fig. 1, for oxidatively-treated precursors (Hox5 and Hox60), lignin content is lower than for unmodified hemp fibers (H), and decreases with duration of modification. From Table 1, it is obvious that lower lignin content in oxidatively-treated precursors, leads to decrease in the specific surface area of obtained biocarbon samples (Hox5 and Hox60). However, specific surface area is not dependent only on the lignin content. Khezami et al. (2005) have suggested that microporosity of carbonized material is affected by the content of the cellulose component. Also, polymorphic form of cellulose may affect the specific surface area of carbonized material, as it was observed by Vukcevic et al. (2012). Experimental conditions used during the alkali treatment of carbon precursors may induce polymorphic transformation of cellulose I to more reactive cellulose II (Borysiak and Garbarczik 2003). This transition of cellulose I into cellulose II may be responsible for high specific surface area of sample CHal45.

Surface groups

The amount and accessibility of functional groups present on the surface of biosorbents and biocarbons may also be influenced by chemical modifications. Both chemical treatments used, at first, remove the accompanying components from the fiber surfaces, which leads to the liberation of the functional groups, increasing their amount (Table 1). Therefore, samples Hox5 and Hal5, obtained by shorter treatments, have increased amounts of functional groups. On the contrary, as shown in Table 1, samples Hox60 and Hal45 contain reduced amounts of functional groups, compared to the samples modified for a shorter time. This is a consequence of increased modification time and pronounced removal of hemicelluloses and lignin, which contain a considerable amount of functional groups.

TPD provides information on the nature of surface oxygen groups that decompose upon heating by releasing CO and CO_2. TPD profiles of CO and CO_2 evolution are presented in Fig. 4.

The TPD spectra of CO_2 desorption profiles (Fig. 4b) of all tested samples show an intensive peak at around 650°C, which coincides with the maximum in CO desorption profiles (Fig. 4a) indicating the existence of anhydride groups that decompose upon heating by releasing both CO and CO_2. Less intensive peaks at the lower temperature, observed in CO desorption profiles, may be due to thermal decomposition of carbonyl groups in α-substituted ketones and aldehydes (Vukčević et al. 2015), while increased CO evolution above 900°C can be attributed to phenols, ethers, carbonyls and quinines (Vukčević et al. 2008). Also, slight evolution of CO_2 at temperatures below 400°C, observed in CO_2 profiles, can originate from carboxylic

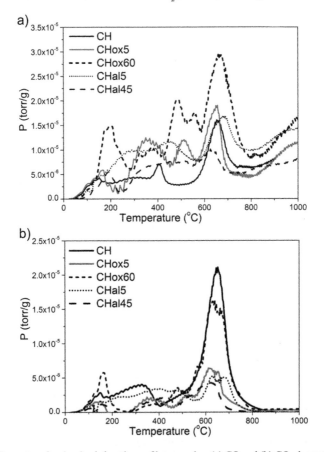

Fig. 4. TPD spectra of carbonized short hemp fiber samples: (a) CO and (b) CO_2 desorption profile.

groups. It is observed that modification of the pristine short hemp fibers prior to carbonization affects the content of oxygen groups at the surface of carbonized samples by increasing the CO evolving groups and decreasing the CO_2 evolving groups. In order to obtain quantitative information about the oxygen groups present on the surface of biocarbons, acid/base titration method was used. The results obtained by acid/base titration (Table 1) show that modification of carbon precursor increases the amount of surface oxygen groups. Also extension of oxidation and alkali treatment time, leads to the increased amount of the surface oxygen groups.

Additionally, the acid-base behavior of biosorbent and biocarbon samples was examined through the point of zero charge (PZC) determination. The PZC of the tested biosorbents and biocarbons was determined by mass titration (Vukčević et al. 2008). Point of zero charge pH (pH_{PZC}) is a pH of the solution at which the overall observed charge on the sorbent surface is zero. At the solution pH less then pH_{PZC}, functional groups are protonated, sorbent behaves as positively charged and repel the positive metal ions. On the other hand, increase in pH above pH_{PZC} makes the functional groups deprotonated, they act as negative species and attract and bind positive metal ions.

The values of pH_{PZC}, ranging from 3.8 to 4.6 (Table 1), indicate the acidic character of biosorbent surface. Obtained results are in agreement with the amount of surface functional groups; hence the increase in amount of acidic functional groups leads to the more acidic surface and lower values of pH_{PZC}. On the other hand, all biocarbon samples have a basic character with pH_{PZC} above 10 (Table 1), which is in compliance with the amounts of acidic and basic groups present on the biocarbon surface (Table 1). These results suggest that the surface of the tested biosorbents will be predominantly negatively-charged and attract the positive metal ions in the solution pH higher than 4.6, while biocarbons surface is positively-charged in the solution whose pH is less than 10.

Adsorption Experiments

Adsorption of heavy metal ions by biosorbents and biocarbon samples may be affected by several factors such as initial pH, initial metal ion concentration, contact time, fiber pretreatment, etc. The effect of the solution pH on the lead ions adsorption by biosorbent and biocarbon is illustrated in Fig. 5, for metal ions concentration of 50 mg/dm^3. As seen in the figure, adsorption of lead ions is strongly dependent on the solution pH for both tested biosorbent and biocarbon samples. The amount of adsorbed Pb^{2+} ions increase from negligible or very low to the maximum values in the pH range of 2–6. This can be explained by the fact that the pH of the adsorption medium affects the solubility of metal ions and the ionization state of the functional groups of the sorbent surface. According to pH_{PZC} value, functional groups present at the surface of the tested biosorbent (sample H) are positively charged at the solution pH lower than 4.4. As the pH increases above this point, amount of lead adsorbed increases until it reaches the maximal biosorption at pH 5. Biocarbon sample (CH) has basic character with pH_{PZC} 10.58 (Table 1). At low pH,

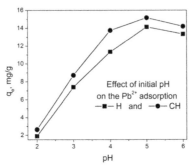

Fig. 5. Effect of initial pH on lead ions sorption on biosorbent and biocarbon samples (initial concentration of solution 50 mg/dm^3, RT, contact time 2 h).

of almost 2.0, all binding sites at biocarbon surface are protonated due to the high proton concentration and effective competition between the protons and the heavy metal ions for surface adsorption sites. With increasing pH, proton concentration decreases and metal ions compete more effectively for available binding sites, which increase adsorption. The high adsorption levels between pH 4.0 and 6.0 indicate that a high affinity for metal ions predominates in this pH region. Therefore, in order to decrease competition between protons and heavy metal ions for surface adsorption sites, and at the same time to prevent ions precipitation, adsorption of heavy metal ions on the biosorbent and biocarbon samples were performed in an aqueous solution at pH 5.

The effect of the contact time on biosorbents and biocarbons adsorption properties was studied up to the contact time of 2 h using Pb(NO$_3$)$_2$ aqueous solutions (initial concentration 50 mg/dm^3) in the batch system at 25°C (Fig. 6). Rapid adsorption of Pb^{2+} ions was observed for all tested samples, since in the first 5 min approximately more than 80% of the total uptake capacity of lead ions was adsorbed. Depending on the preparation parameters, the system attained

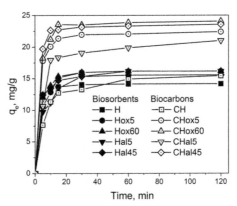

Fig. 6. Kinetics of lead ions sorption by biosorbent and biocarbon samples at initial conc. of 50 mg/dm^3 (pH 5, RT).

equilibrium in 30 min for biosorbents and CH sample, and in 15 min for other biocarbon samples. After this equilibrium period the amount of sorbed metal ions did not change significantly with an increase in contact time.

In order to gain insight into the adsorption kinetics, the experimental results obtained for lead initial concentration of 50 mg/dm³ were fitted with pseudo first and pseudo second order models. The plots for the pseudo first and the pseudo second order kinetics, along with appropriate correlation coefficients are given in Figs. 7a and 7b, respectively.

According to the values obtained for correlation coefficients, it can be concluded that adsorption of lead ions on biosorbent and biocarbon samples obey the pseudo second order kinetics. The adsorption rate constant, along with equilibrium uptakes, obtained by the pseudo second order kinetic model and experimentally, are listed in Table 2. The calculated values of the equilibrium uptake ($q_{e, cal}$) are very close to the ones obtained by the experiment ($q_{e, exp}$), confirming that adsorption of the lead by all tested biosorbents and biocarbons follows the pseudo second order

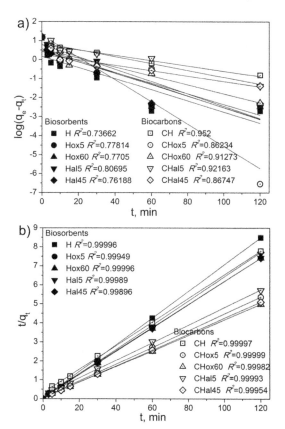

Fig. 7. Kinetic data and linear fit with (a) the pseudo first and (b) the pseudo second order kinetics for adsorption of lead ions on biosorbent and biocarbon samples.

kinetic. This fact indicates that the binding of lead ions on tested samples occurs mainly through chemical forces rather than physical forces of attraction.

The disadvantages of pseudo second order kinetic model are reflected in the inability to give information about the process that influences the adsorption rate. Adsorption process is very complex, and it takes place through three consecutive steps: external mass transport, transport through the fibers and adsorption at an interior active site. Therefore, the overall rate of lead ions adsorption will be controlled by the slowest step. Since the adsorption of lead ions was performed with a constant and rapid shaking, the effect of external mass transport on the adsorption rate is eliminated. Also, the last step in the adsorption process is considered to be an equilibrium reaction, assumed to be rapid and therefore considered negligible. On that base, the adsorption process primarily depends on the diffusion of lead ions through the porous structure of fibers. In order to describe the multi-scale phenomenon of lead ion transport through porous fiber matrices, effective diffusion coefficient of lead ions transport through biosorbent and biocarbon samples were calculated by mathematical models (Pejic et al. 2011, Vukčević et al. 2014) and presented in Table 2. The effective diffusion coefficients obtained for lead ion adsorption on biosorbents are much higher than the ones obtained for biocarbons. Since transport of lead ions may be affected by the materials' structure and porosity, lead ions much easily penetrate through the cracks and wide pores of biosorbent structures than through the microporous surface of biocarbons. The highest effective diffusion coefficient is obtained for lead ions transport through the structure of Hal45 sample, since pronounced fibrillation of this sample allows lead ions facilitate penetration into interfibrillar space in the fibers' structure.

According to the values of lead ions uptake capacities (Table 2), it can be noted that modification of hemp fibers increases the total uptake capacities for both

Table 2. The adsorption rate constant and equilibrium uptakes (calculated and experimental) obtained by the pseudo-second order kinetic model and effective diffusion coefficient of lead ions transport through the biosorbent and biocarbon samples.

Samples		k_2 (g/mg·min)	$q_{e,\,cal}$ (mg/g)	$q_{e,\,exp}$ (mg/g)	D_{eff} x 10^{12} (m²/s$^\beta$)
Biosorbents	H	0.2156	14.19	14.14	22.0
	Hox5	0.0549	16.35	16.16	22.8
	Hox60	0.0798	16.29	16.37	22.4
	Hal5	0.0286	16.50	16.16	22.8
	Hal45	0.0942	15.65	15.54	23.2
Biocarbons	CH	0.0129	16.03	15.57	0.91
	CHox5	0.0386	22.59	22.39	1.76
	CHox60	0.0351	24.31	24.06	1.70
	CHal5	0.0128	21.54	21.05	0.98
	CHal45	0.0578	23.69	23.60	1.83

biosorbent and biocarbon samples, especially for oxidatively-treated samples. Also, for biocarbon samples, longer treatment duration increases adsorption capacity, since the highest lead ions uptake was observed for the samples CHox60 and CHal45 (Fig. 6). Generally, adsorption of heavy metal ions can be affected by specific surface area, material structure and amount of surface oxygen groups that may act as active sites for adsorption. Taking into account the results obtained by surface characterization (Table 1), it can be observed that for the biosorbent samples, specific surface area and amount of functional groups has no decisive influence on the adsorption capacity. The highest biosorbent uptake capacities obtained for oxidized samples (lignin removed from middle lamellae), can be explained by the domination of sorption at outer surfaces of fibers and increased roughness of hemp fiber surfaces which induced new capillary spaces in inter-surfacial layers between completely or partially separated fibers (Pejic et al. 2009). On the other hand, for biocarbons, the clear dependence between the amount of surface oxygen groups and the adsorption can be noted: the highest adsorption capacity is obtained for sample CHox60 which contains the highest amount of surface oxygen groups. Also, the correlation between the adsorption (Fig. 6) and material structure (Figs. 1 and 3) shows that enhanced fibrillation increases the ability of biocarbons to remove lead ions, due to facilitated penetration of lead ions into the biocarbon structure. Obtained adsorption capacities of biosorbent and biocarbon samples are comparable or even higher than the adsorption capacities of different lignocellulosic biosorbents and derived activated carbons (Table 3). Notable differences between

Table 3. Uptake capacities of lead ions by different lignocellulosic biosorbents and biocarbons.

Samples		q_e (mg/g)	References
Biosorbents	Corncobs	43.4	Tan et al. 2010
	Rise husk	11.4	Roy et al. 1993
	Coconut shell	54.62	Sousa et al. 2010
	Sawdust	15.90	Yasemin and Zek 2007
	Sisal fibers	1.34	dos Santos et al. 2011
	Hemp fiber	8.05–18.20	Tofan et al. 2010
	Unmodified hemp fibers	14.14	This work
	Oxidatively treated hemp fibers	16.37	This work
Biocarbons	Palm kernel shells	74.63	Rahman et al. 2014
	Coconut shells	73.53	Rahman et al. 2014
	Palm pit	15.2	Giraldo 2008
	Cane sugar bagasse	13.7	Giraldo 2008
	Sawdust	17.5	Giraldo 2008
	Pericarp of pecans	75.4	Hernández-Montoya et al. 2011
	Unmodified hemp fibers	15.57	This work
	Oxidatively-treated hemp fibers	24.06	This work

biosorbent capacities may be the consequence of differences in the chemical composition (content of cellulose, hemicellulose, lignin, pectin and extractives), fiber structure (different thickness of cell walls and lumen, degree of crystallinity and fibrillar orientation) and specific surface area, i.e., presence of micropores and microcracks. As it can be observed from Table 3, biosorbent nature affects its sorption properties toward lead ions, which can be improved by utilization of an appropriate modification method. A remarkable feature is that the carbon material obtained by carbonization of lignocellulosic precursors retains some memory of the starting lignocellulosic structure. Therefore, specific chemical composition and nature of lignocellulosic precursor affect the adsorption properties of the resulting carbon sorbent, as shown in Table 3.

Conclusions and Outlooks

The production of innovative materials on the base of plant lignocellulosic fibers is becoming an important area of research since it offers a unique combination of high physical and specific chemical properties producing a variety of high-value products with low impact on the environment. Nowadays, using short and entangled lignocellulosic fibers, wastes from the textile industry, as low-cost and easily available adsorbents for purification of the contaminated water represents one of the major ongoing topics. Furthermore, short and entangled hemp fibers can be used as biosorbents and low-cost precursors for biocarbon production.

Distribution of the celluloses, lignin and hemicelluloses in the hemp fiber structure, as well as the changes in their amount induced by chemical modification, affects the surface properties, morphology and consequently, adsorption properties of the resulting hemp fibers-based biosorbents and biocarbons. Observed increase in sorption properties was ascribed not only to the decrease of lignin or hemicelluloses content, but also to the location of these components in the hemp fiber since lignin removal results in more homogenous middle lamella due to the gradual elimination of micro-pores and the less rigid cell wall, while hemicelluloses removal makes inter-fibrillar regions of fibers less dense and rigid, and fibrils more capable for rearrangement. Progressive removal of both hemicelluloses and lignin from the hemp fibers structure lead to liberation of elementary fibers, and that fibrous structure is preserved after carbonization. Lignin content and polymorphic transformation of cellulose I to a more reactive cellulose II primarily affects the porosity of biocarbon sorbents, while hemicelluloses induce more homogeneous distribution of adsorption active sites. Prolonged chemical modification of a carbon precursor greatly improved adsorption capacity of hemp fibers-based biocarbons, by increasing the amount of surface oxygen groups and the level of material fibrillation. Lead ions adsorption onto biosorbents and biocarbons follows the pseudo second order kinetic. This fact indicates that the binding of lead ions on tested samples occurs mainly through chemical forces rather than physical forces of attraction. Since transport of lead ions may be affected by materials structure and porosity, lead ions much easily penetrate through the cracks and wide pores of biosorbent structure than through the microporous surface of biocarbons. High adsorption capacities of tested biosorbent and biocarbon samples, enables the utilization of hemp fibers and hemp fiber-based biocarbons as filter materials for removal of lead ion from wastewater.

Acknowledgments

The authors wish to thank the Ministry of Education, Science and Technological Development of the Republic of Serbia for financial support through the projects of Basic Research, No. 172007 and No. 172029.

References

Aber, S., A. Khataee and M. Sheydaei. 2009. Optimization of activated carbon fiber preparation from Kenaf using K_2HPO_4 as chemical activator for adsorption of phenolic compounds. Bioresource Technol. 100: 6586–6591.

Borysiak, S. and J. Garbarczik. 2003. Applying the WAXS method to estimate the supermolecular structure of cellulose fibres after mercerisation. Fibres Text. East. Eur. 11: 104–106.

Buschle-Diller, G., C. Fanter and F. Loth. 1999. Structural changes in hemp fibers as a result of enzymatic hydrolysis with mixed enzyme systems. Text. Res. J. 69: 244–251.

Chunlan, L., X. Shaoping, G. Yixiong, L. Shuqin and L. Changhou. 2005. Effect of pre-carbonization of petroleum cokes on chemical activation process with KOH. Carbon 43: 2295–2301.

Crini, G. 2005. Recent developments in polysaccharide-based materials used as adsorbents in wastewater treatment. Prog. Polym. Sci. 30: 38–70.

Demirbas, A. 2008. Heavy metal adsorption onto agro-based waste materials: A review. J. Hazard. Mater. 157: 220–229.

dos Santos, W.N.L., D.D. Cavalcante, E.G.P. da Silva, C.F. das Virgens and F.S.D. Dias. 2011. Biosorption of Pb(II) and Cd(II) ions by Agave sisalana (sisal fiber). Microchem. J. 97: 269–273.

Dumanh, A.G. and A.H. Windle. 2012. Carbon fibres from celullosic precursors: a review. J. Mater. Sci. 47: 4236–4250.

Garner, W. 1967. Textile Laboratory Manual, Fibres. Heywood Books, London, UK.

Giraldo, L. and J. C. Moreno-Pirajan. 2008. Pb^{2+} adsorption from aqueous solutions on activated carbons obtained from lignocellulosic residues. Braz. J. Chem. Eng. 25: 143–151.

Hameed, B.H., I.A.W. Tan and A.L. Ahmad. 2008. Optimization of basic dye removal by oil palm fibre-based activated carbon using response surface methodology. J. Hazard. Mater. 158: 324–332.

Hernández-Montoya, V., D.I. Mendoza-Castillo, A. Bonilla-Petriciolet, M.A. Montes-Morán and M.A. Pérez-Cruz. 2011. Role of the pericarp of *Carya illinoinensis* as biosorbent and as precursor of activated carbon for the removal of lead and Acid blue 25 in aqueous solutions. J. Anal. Appl. Pyrol. 92: 143–151.

Kalijadis, A.M., M.M. Vukčević, Z.M. Jovanović, Z.V. Laušević and M.D. Laušević. 2011. Characterization of surface oxygen groups on different carbon materials by the Boehm method and temperature programmed desorption. J. Serb. Chem. Soc. 76: 757–768.

Kalijadis, A., J. Đorđević, T. Trtić-Petrović, M. Vukčević, M. Popović, V. Maksimović, Z. Rakočević and Z. Laušević. 2015. Preparation of boron-doped hydrothermal carbon from glucose for carbon paste electrode. Carbon. 95: 42–50.

Kennedy, L.J., J.J. Vijaya and G. Sekaran. 2004. Effect of two-stage process on the preparation and characterisation of porous carbon composite from rice husk by phosphoric acid activation. Ind. Eng. Chem. Res. 43: 1832–1838.

Khezami, L., A. Chetouani, B. Taouk and R. Capart. 2005. Production and characterisation of activated carbon from wood components in powder: cellulose, lignin, xylan. Powder Technol. 157: 48–56.

Kostic M., B. Pejic and P. Skundric. 2008. Quality of chemically modified hemp fibers. Bioresource Technol. 99: 94–99.

Kostic, M.M., B.M. Pejic, K.A. Asanovic, V.M. Aleksic and P.D. Skundric. 2010. Effect of hemicelluloses and lignin on the sorption and electric properties of hemp fibers. Ind. Crop. Prod. 32: 169–174.

Kostic, M., M. Vukcevic, B. Pejic and A. Kalijadis. 2014. Hemp fibers: Old fibers—New applications. pp. 399–446. *In*: Md. Ibrahim and M. Mondal (eds.). Textiles: History, Properties and Performance and Applications. Nova Science Publishers, Inc. New York.

Mohamed, A.R., M. Mohammadi and G.N. Darzi. 2010. Preparation of carbon molecular sieve from lignocellulosic biomass: A review. Renewable Sustainable Energy Rev. 14: 1591–1599.

Nor, N.M., L.L. Chung, L.K. Teong and A.R. Mohamed. 2013. Synthesis of activated carbon from lignocellulosic biomass and its applications in air pollution control—a review. J. Environ. Chem. Eng. 1: 658–666.

Parks, E.J. and R.L. Hebert. 1972. Thermal analysis of ion exchange reaction products of wood pulps with calcium and aluminium cations. Tappi J. 55: 1510–1514.

Pejic, M.B., M.M. Kostic, P.D. Skundric and J.Z. Praskalo. 2008. The effects of hemicelluloses and lignin removal on water uptake behavior of hemp fibers. Bioresource Technol. 99: 7152–7159.

Pejic, B., M. Vukcevic, M. Kostic and P. Skundric. 2009. Biosorption of heavy metal ions from aqueous solutions by short hemp fibers: Effect of chemical composition. J. Hazard. Mater. 164: 146–153.

Pejic, B., M. Vukcevic, I. Pajic-Lijakovic, M. Lausevic and M. Kostic. 2011. Mathematical modeling of heavy metal ions (Cd^{2+}, $Zn^{2+}$and Pb^{2+}) biosorption by chemically modified short hemp fibers. Chem. Eng. J. 172: 354–360.

Praskalo, J., M. Kostic, A. Potthast, G. Popov, B. Pejic and P. Skundric. 2009. Sorption properties of TEMPO-oxidized natural and man-made cellulose fibers. Carbohydr. Polym. 77: 791–798.

Rahman, M.M., M. Adil, A.M. Yusof, Y.B. Kamaruzzaman and R.H. Ansary. 2014. Removal of heavy metal ions with acid activated carbons derived from oil palm and coconut shells. Mater. 7: 3634–3650.

Reed, A.R. and P.T. Williams. 2004. Thermal processing of biomass natural fibre wastes by pyrolysis. Int. J. Energy Res. 28: 131–145.

Rosas, J.M., J. Bedia, J. Rodríguez-Mirasol and T. Cordero. 2009. HEMP-derived activated carbon fibers by chemical activation with phosphoric acid. Fuel 88: 19–26.

Roy, D., P.N. Greenlaw and B.S. Shane. 1993. Adsorption of heavy metals by green algae and ground rice hulls. J. Environ. Sci. Health, Part A 28: 37–50.

Salman, J.M. and B.H. Hameed. 2010. Effect of preparation conditions of oil palm fronds activated carbon on adsorption of bentazon from aqueous solutions. J. Hazard. Mater. 175: 133–137.

Shukla, S.R. and R.S. Pai. 2005. Adsorption of Cu(II), Ni(II) and Zn(II) on modified jute fibers. Bioresource Technol. 96: 1430–1438.

Sousa, F.W., A.G. Oliveira, J.P. Ribeiro, M.F. Rosa, D. Keukeleire and R.F. Nascimento. 2010. Green coconut shells applied as adsorbent for removal of toxic metal ions using fixed-bed column technology. J. Environ. Manage. 91: 1634–1640.

Suhas, P.J.M. Carrott and Ribeiro M.M.L. Carrott. 2007. Lignin—from natural adsorbent to activated carbon: A review. Bioresource Technol. 98: 2301–2312.

Tan, G., H. Yuan, Y. Liu and D. Xiao. 2010. Removal of lead from aqueous solution with native and chemically modified corncobs. J. Hazard. Mater. 174: 740–745.

Thomas, S., S.A. Paul, L.A. Pothan and B. Deepa. 2011. Natural fibres: Structure, properties and applications. pp. 3–42. *In*: S. Kalia, B.S. Kaith and I. Kaur (eds.). Cellulose Fibers: Bio- and Nano-Polymer Composites. Springer, Berlin.

Tofan, L., C. Paduraru, I. Volf and C. Balan. 2010. Kinetic and thermodynamic profile of Pb(II) sorption by untreated hemp fibers. Scientific Papers, Agronomy Series 53: 146–149.

Volesky, B. 2007. Biosorption and me. Water Res. 41: 4017–4029.

Vukcevic, M., A. Kalijadis, M. Radisic, B. Pejic, M. Kostic, Z. Lausevic and M. Lausevic. 2012. Application of carbonized hemp fibers as a new solid-phase extraction sorbent for analysis of pesticides in water samples. Chem. Eng. J. 211-212: 224–232.

Vukcevic, M., B. Pejic, M. Lausevic, I. Pajic-Lijakovic and M. Kostic. 2014. Influence of chemically modified short hemp fiber structure on biosorption process of Zn^{2+} ions from waste water. Fibers Polym. 15: 687–697.

Vukčević, M., A. Kalijadis, S. Dimitrijević-Branković, Z. Laušević and M. Laušević. 2008. Surface characteristics and antibacterial activity of a silver-doped carbon monolith. Sci. Technol. Adv. Mater. 9: ID 015006 (7pp). doi:10.1088/1468–6996/9/1/015006.

Vukčević, M.M., A.M. Kalijadis, T.M. Vasiljević, B.M. Babić, Z.V. Laušević and M.D. Laušević. 2015. Production of activated carbon derived from waste hemp (*Cannabis sativa*) fibers and its performance in pesticide adsorption. Microporous Mesoporous Mater. 214: 156–165.

Vukčević, M., B. Pejić, A. Kalijadis, I. Pajić-Lijaković, M. Kostić, Z. Laušević and M. Laušević. 2014. Carbon materials from waste short hemp fibers as a sorbent for heavy metal ions—Mathematical modeling of sorbent structure and ions transport. Chem. Eng. J. 235: 284–292.

Wang, H.M., R. Postle, R.W. Kessler and W. Kessler. 2003. Removing pectin and lignin during chemical processing of hemp for textile applications. Text. Res. J. 73: 664–669.

Yasemin, B. and T. Zek. 2007. Removal of heavy metals from aqueous solution by sawdust adsorption. J. Environ. Sci. 19: 160–166.

Hollow Floss Fibers

Characterization and Potential Industrial Applications

M. Robert,* P. Ovlaque and M. Reza Foruzanmehr

Introduction

The availability and the environmentally-friendly nature of renewable materials have caught the attention of both researchers and industries. Among the renewable materials, natural fibers not only have good properties (lightweight, high strength), but are also fairly cheap and available when they are extracted from the remnants of harvests. For instance, in western Canada oilseed flax is grown in over 600,000 to 800,000 ha. This can potentially produce 2,000 kg/ha of flax straw annually. Currently, more than 75% of this amount is usually burnt or thrown away (Reaney et al. 2006). The industrial use of these materials can also promote sustainable agriculture, which is the principal element of sustainable development (Pretty 2008, Chattopadhyay et al. 2011).

Another way to promote sustainable agriculture is promoting technologies and markets for local natural fibers. By doing so, local farmers find the motivation to breed these fibers. This can mitigate the social and economic problems of small agriculture, and reduce environmental impacts by providing renewable materials (Sharma et al. 2015).

Cellulosic natural fibers can be obtained from various parts of plants, such as leaves, fruits, basts, and seeds. Among them bast fibers (like flax) and seed

Center for Innovation in Technological Ecodesign (CITE), Faculty of Engineering, University of Sherbrooke, Sherbrooke, Canada, J1K 2R1.
* Corresponding author: Mathieu.Robert2@USherbrooke.ca

fibers (like cotton) are more conventional (Akil et al. 2011). The geometrical dimensions of these fibers as well as their properties depend on where the fibers were extracted. For example, fibers from seeds are shorter than fibers from stems and leaves (few centimeters to a few meters). Moreover, fiber extraction from stems is time consuming and leads to heterogeneous fibers in comparison with seed fibers (Smole et al. 2013).

Some seed fibers (floss fibers for instance) are hollow, as is the case with kapok, akund, and milkweed. These fibers have an extremely wide lumen and very thin wall. In fact, they are supposed to disperse grains for germination. This special hollow structure makes these fibers lightweight and therefore it provides plenty of opportunities for creation of low-density insulating materials.

Among the hollow floss fibers, milkweed (*Asclepias syriaca*) is a valuable fiber because it is easy to grow and it can be harvested twice a year (Reddy and Yang 2009). Milkweed is native to northwest North America. This territory includes the area within 35°–50° latitude north (N) and 60°–103° longitude west (W). In other words, the region is encircled by a line which passes from New Brunswick to Virginia to Tennessee, then through Kansas to Manitoba (Rahman and Wilcock 1991).

Scientists have been interested in using milkweed for more than a century. Milkweed has potentials of use in a variety of applications such as textile industry, composite manufacturing, paper production, oil absorbent, and thermal and sound insulation.

History of Hollow Seed Fibers

Hollow seed fibers such as those produced by members of the Apocynaceae family have had numerous well-known applications for centuries. These plants have been used in interesting applications including textiles and medicine (Unknown 1900).

Calotropis procera fibers were used during the 11th century in the former Bilad al Sudan empire in western Africa for manufacture of fireproof textiles (Cuoq 1975). According to the *Bulletin of Miscellaneous Information of the Royal Botanic Gardens, Calotropis gigantea* fibers were used as stuffing material for pillows and cushions in India. There are also some evidences in the report of exploration of the Ganges river by Fitch (1585), which indicate the use of hollow fibers for clothing. Europeans also used to spin *Calotropis gigantea* fibers, though the spinning difficulties and the insufficient source of fibers caused the complete decline of these fibers' usage (Unknown 1900). In the meantime, *Asclepias syriaca* hollow fibers were used in North America by the Amerindian tribes in numerous applications (Leprince 2014). The Amerindians transferred the knowledge of using the hollow fibers to French settlers after they started to colonize North America. The book Traités des asclepiades described the interest of the former French monarchy for developing an industry to exploit these fibers (Sonnini 1810). Even the King of France, Louis XV, started to use the fibers for his personal winter clothings (Leprince

2014). However, these activities were discontinued by late 18th century when the British Empire conquered the French colony in North America. In that period, the British Empire strongly dominated the businesses of cotton and silk. Hence, they considered *Asclepias syriaca* as a threat to their business and choked the rising economy based on these fibers.

The Second World War is an important milestone in the history of hollow floss fibers. Since there was no technology to fabricate low-density synthetic fibers, hollow kapok fibers became ideal filler materials for life jackets (Netravali and Pastore 2014). During the Second World War Japan invaded the countries in southern Asia which supplied kapok fibers and consequently and, the US navy was forced to find an alternative for kapok fibers. Therefore, the neglected milkweed floss fibers again started to be harvested in large quantities to compensate for the lack of kapok supply and sustain the life jacket industry (Patterson 2012). The change in demand for milkweed fibers due to the end of the war and the advent of new synthetic fibers diminished the demand, and the application of milkweed fibers again fell into oblivion.

Nowadays, hollow fiber properties and functions are progressively being rediscovered. In North America and especially in Canada the crop growing of *Asclepias syriaca* has started and has led to the formation of a private sector which is able to process fibers for various applications (Leprince 2015). This on the one hand encourages the local farmers to take advantage of growing more economic crops with higher benefits, and on the other hand, it paves the way for communities to achieve sustainable development (Bernstien 2016).

Plant Species

A huge variety of plants produce hollow fibers with very similar morphologies. Although they all exhibit hollow morphologies, they have slightly different structures and are produced by different species. Moreover, as it was previously mentioned, these plants grow all around the world. If the members of the Apocynaceae family are classified by the plant morphological characteristics, it will lead to the formation of a huge list consisting of 280 genera and 2000 species (Ashori and Bahreini 2009). It is not possible to describe all the species and thus, the most common ones that are found in the literature will be discussed in this chapter.

Kapok

Generally, wild weeds or bushes produce hollow fibers. However, kapok fibers is the only one that is produced by trees. Kapok trees (*Ceiba pentandra*) produce kapok hollow fibers (Fig. 1). The trees are variously known as silk-cotton tree, Ceiba and by many other indigenous names. Kapok-growing areas are found in the sub-tropical zones of America, and in Africa and Asia (Netravali and Pastore 2014). The trees can reach a height of upto 45 meters and a diameter of 2 meters. There is no need to cut down the tree in order to harvest the fibers. The fibers are

Fig. 1. Kapok fibers.

contained in pods and the pods can be harvested directly from the tree branches. Kapok is the most readily available hollow fiber in the world; almost 99,000 tons of kapok fibers were harvested in 2013 (Netravali and Pastore 2014).

Milkweed

Hollow fibers produced by weeds are often known as milkweed fibers. However, other names such as akund or estabragh are frequently used, for them, according to the location where these plants grow. Figure 2 shows the common names and corresponding regions at milkweeds. For instance, Estabragh is a Persian name for a hollow fibre producing plant which grows in Iran. A similar plant also grows in China, which is called akund.

Figure 3 depicts different subcategories of hollow fibers with corresponding family, genus, and species names. Among them, the aak fibers belong exclusively to the *Calotropis procera* genus. Akund and Mudar belong to the *C. procera* and *C. gigantea* genus in literature and milkweed is a common name for any species.

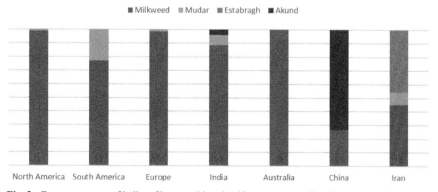

Fig. 2. Common names of hollow fibers used in scientific papers regarding the country of origin and the number of publications.

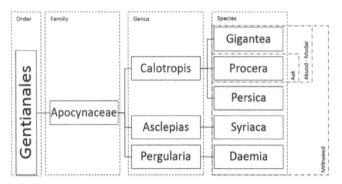

Fig. 3. Apocynaceae family and common names.

Concerning the Persian term 'Estabragh', only Gharehaghaji and Davoodi have indicated a species (*Asclepias procerais*) which has also been used to designate the *C. procera* species (Gharehaghaji and Davoodi 2008).

Asclepias syriaca

This species is also called common milkweed, as it is similar to a small wild weed producing a white sap. The plant can reach up 2 meters in height. It grows mainly in the eastern part of North America, according to the USDA plant database. The *Asclepias syriaca* species has small branches with opposite oval leaves. The flowering of this species occurs during summers and the floss fibers are ready to be harvested at the end of September (Stevens 2000, Holdrege 2010).

Calotropis procera and Calotropis gigantea

The species *Calotropis procera* can be found in North Africa and the Middle East. *C. procera* has been successfully introduced in semi-arid regions of California and

Fig. 4. *Asclepias syriaca* plant and fibers.

Fig. 5. Mudar shrub (Ashori and Bahreini 2009).

Australia (Rahman and Wilcock 1991). These species are frequently found in the region surrounded by India, China, and the Philippines. Unlike *Asclepias syriaca*, *C. procera* and *C. gigantea* can develop large shrubs with 3 meters high stems and 25 cm diameter width (Ashori and Bahreini 2009). Both *Calotropis* species produce fruits that contain the floss fibers. Fruit development on shrubs mainly occurs during the hot season.

Hollow Floss Fibers

Morphology

Unlike common natural fibers such as flax fibers, which have no lumen (a central hollow cavity in the fibers) or a tiny one, hollow floss fibers possess an enormously wide one. The lumen can form 77% of the volume of the fiber and creates a particular morphology in comparison with the common natural fibers (Lim and Huang 2007).

Another particular structural feature of hollow floss fibers is the thin cellulosic cell wall of these fibers (see Fig. 6). Normally, natural fibers are plant cells that grow together and form the fiber bundles. The plant cells consist of a large central vacuole (i.e., a micro-organelle filled with water) which is engulfed by a cell wall (composed of cellulose, hemicelluloses, pectin and lignin). When the cell dies the membrane of the vacuole is broken down so the cell loses its content except for its skeletal part, which is the cell wall. In fact, the lumen is the empty space of the lost cell content and the fiber wall is the remaining skeletal part of the plant cell. The cell wall thickness has been reported to be 1.27 μm for *Asclepias syriaca* (Drean et al. 1993) as compared to 2 μm for kapok fibers (Lim and Huang 2007). Regarding the diameter of the fiber, an average fiber diameter is around 20 μm for all types and species (see Table 1).

Hollow fibers have a conic morphology that was reported for several species such as *Asclepias syriaca*, *Calotropis gigantea* and kapok (Louis and Andrews 1987,

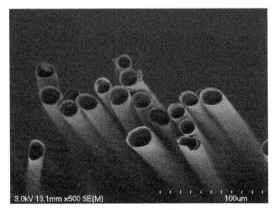

Fig. 6. *Asclepias syriaca* floss fibers.

Table 1. Length and diameter of hollow fibers.

Species	Length (mm)	Diameter (µm)	Reference
A. syriaca	20,1	22,4	(Drean et al. 1993)
C. procera	31	10–21	(Sakthivel 2005)
C. gigantea	17,1	30,4	(Ashori and Bahreini 2009)
C. persica	18,6	15,9	(Nourbakhsh et al. 2009)
Milkweed	14,2	-	(Reddy and Yang 2010)
Akund	31,24	20,63	(Yang, Huang et al. 2012)
Estabragh	26,4	25,3	(Kish and Najar 1998)
Estabragh	27,7	22,14	(Hassanzadeh et al. 2014)
Estabragh	27,89	-	(Bahari et al. 2015)
Estabragh	34,23	25,31	(Hasani et al. 2014)
Estabragh	33,5	-	(Gharehaghaji and Davoodi 2008)
Estabragh	26,6	28,93	(Merati 2014)
Kapok	19,08	19,28	(Yang, Huang et al. 2012)
Kapok	25	16,5	(Lim and Huang 2007)
Kapok	27.87	-	(Sunmonu and Abdullahi 1992)

Kish and Najar 1998, Merati 2014). The fibers can be considered as a truncated cone with a base-to-top diameter ratio equal to 3 (Kish and Najar 1998). Hollow fibers are shorter (20 mm) than other conventional fibers such as hemp or flax (70–80 mm) (Mohanty et al. 2005). This is because the natural bast fibers consist of a bundle of individual fibers, which are intertwound, and thus have longer dimensions. However, hollow floss fibers are single ultimate fibers and consequently are very short. Unlike most of the natural fibers, hollow fibers are directly harvested at the end of the fiber's life cycle with no need for retting.

Chemical composition of different species of hollow fibers

The chemical composition of the different species are shown in Table 2. The values show a certain homogeneity despite the variability of the plant species and

Table 2. Chemical composition of various species of hollow fibers.

Species	Cellulose (%)	Hemicellulose (%)	Lignin (%)	Wax (%)	References
C. gigantea	49	20	18	-	(Ashori and Bahreini 2009)
C. procera	55	24	18	3	(Sakthivel 2005)
A. syriaca	54,9	19,3	19,3	-	(Spiridon 2007)
A. syriaca	39,6	27,3	15,1	-	(Barth and Timell 1958)
C. persica	53,4	25,2	17,4	-	(Nourbakhsh et al. 2009)
Daemia	53	26	15	4	(Karthik and Murugan 2013)
Milkweed	55	24	18	-	(Gu et al. 1992)
Milkweed	52	-	21,3	-	(Reddy and Yang 2010)
Akund	55,45	21,91	16,15	-	(Yang, Huang et al. 2012)
Kapok	43	32	15	-	(Netravali and Pastore 2014)
Kapok	64	-	13	0.8	(Meiwu et al. 2010)

the techniques used for the characterization of the chemical composition. Hollow fibers exhibit a relatively lower amount of cellulose and a high amount of lignin and waxes in comparison with other conventional fibers.

Hollow floss fibers tend to have a low percentage of cellulose (around 55%) whereas flax or hemp fibers are usually reported having 20% higher of this amount (Mohanty et al. 2005). The content of lignin in the hollow fibers varies ranging from 13 to 21.3, while, bast fibers normally possess a very low percentage of lignin in their composition. For instance, the percentage of lignin in flax and hemp are 2.2 and 5.7 respectively (Mohanty et al. 2005). Only few researches were performed on the quantification of wax in the hollow floss fibers. The only works available on the wax content show a higher amount of wax compared to other natural fibers. As the hollow floss fibers are directly in contact with the environment, an epicuticular layer of wax may be there on the surface of the fiber (Koch and Ensikat 2008).

Structure of floss fibers

It is well known that the natural fibers cell walls are composed of a multilayered material. Based on the origin of the fibers the number of layers as well as their

Fig. 7. Structure of common seed hair (left) (Müssig 2010). Structure of cotton fibers (right) (Anderson and Kerr 1938).

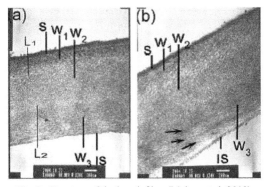

Fig. 8. Structure of the kapok fiber (Meiwu et al. 2010).

thickness may vary. Unlike the structure of bast fibers which has been well investigated, hollow floss fibers cell wall structure has not yet been characterized. It is possible to depict a universal structure for hollow floss fibers. However, it is better to consider the structure of seed fibers (like cotton seed fibers grow from seeds) rather than floss fibers. The structure of cotton fibers has been intensively studied and is described as the result of several concentric layers induced by the diurnal cycles (Fig. 7) (Müssig 2010). Moreover, according to Anderson and Kerr, the layers are the results of a difference in the density of the layer induced by the activity of the cell during the development of the fiber (Kerr 1937, Anderson and Kerr 1938). Kapok and cotton do not have the same structure and despite being seed fibers, Kapok fibers are produced by a tree and it may be irrelevant to generalize this model for weeds hollow fibers (Asclepias or Calotropis). Meiwu et al. (2010) studied the structure of kapok fiber walls with the help of transmission electron microscopy. Their study revealed clearly a 5-layer structure within the walls of the kapok fiber (Fig. 8). The fiber structure is composed of an external cuticle, a first interlaced structure , two closely packed layers and a final inner skin (Meiwu et al. 2010). Although proposing a universal structure for floss fibers and especially milkweed fibers may be irrelevant, both models can be applied to the Apocynaceae family fibers for various reasons.

Physical Properties of Floss Fibers

Moisture absorption

Moisture absorption and release of kapok and akund fibers were investigated by Yang et al. (Yang et al. 2015). The study showed that the hollow fibers tended to absorb more water than cotton fibers. The values showed that akund fibers reached an equilibrium state of 10.44% after a relatively long time. Kapok fibers reached a slightly lower state than akund (9.9%) but still higher state than cotton (7.5%). Moisture absorption of the common milkweed (*Asclepias syriaca*) was studied by Drean et al. (Drean et al. 1993). They used ASTM D2654 to characterize the moisture absorption of milkweed fibers and the results showed a correlation with the results found for kapok and akund. Milkweed fibers showed 9.9 wt% of moisture absorption. As the methods used in the two studies were different, the absorption time was also different from one study to another. Karthik and Murugan have studied the moisture regain of the *Pergularia daemia*. Their results showed once again that hollow fibers had a high rate of moisture absorption (10.0%) (Karthik and Murugan 2013). They also proved that the removal of wax on the fiber led to a higher amount of moisture absorption (12.0%). Moreover, Sakthivel et al. found a value of 11% regarding the moisture regain of the Mudar fibers (*Calotropis procera*) (Sakthivel 2005).

Overall, hollow fibers tend to absorb moisture around 10% of their weight regardless of the species.

Density

Several methods have been used to characterize the density of hollow fibers. Among them gas pycnometry has also been used. However, this method always shows a value around 1.5 which is very close to the density of cellulose (Karthik 2015). In fact, this method can only measure the density of the solid part of the fibers and not the whole hollow fiber.

It is very important to take the lumen and its corresponding volume into account when the density of a hollow fiber is being calculated. As previously explained, hollow fibers have two main components: the cellulosic wall and the lumen. Considering the lumen as a part of the fiber has a strong influence on the density of the fiber.

Literature show very different results for the density of hollow fibers. Reddy and Yang determined a density of 0.893 on the milkweed fiber without mentioning the species of fibers (Reddy and Yang 2010). The result was found using the sink-float method in xylene and carbon tetrachloride. The same method was used by Sakthivel for measuring the density of Mudar fibers (*Calotropis procera*) (Sakthivel 2005). According to the results, the *C. procera* fibers have a density of 0.97. Ashori and Bahraini also determined the density of Mudar fibers (*Calotropis gigantea*) and found a value of 0.68 while they did not indicate the method they used (Ashori and Bahreini 2009). Bahari et al. also found a density of 0.9 for Estabragh fibers without explaining the methodology (Bahari et al. 2015). Another study was conducted on kapok fibers by Meiwu et al. and determined the density of fiber close to 0.3 (Meiwu et al. 2010).

With density values ranging from 0.3 to 0.9 (see Table 3), it is not clear what the correct density of the hollow fibers is. However, it is possible to find a value based approximatively on the geometrical calculation. As hollow fibers are very short, it is possible to assume the fiber as a cylinder of a very short length (see Fig. 9) with an average diameter of 20 μm (Table 1), an average wall thickness of 2 μm (Lim and Huang 2007) and an average density of the wall of 1.48 measured using a gas pycnometer on *Asclepias syriaca*. Theoretically, the volume of fiber is equal to the volume of the cylinder and determined by the following equation:

$$V_{fiber} = \pi \times R^2 \times \delta x$$

where R is the external diameter and δx the length of the cylinder considered for the volume calculation. The mass of fiber is equal to the density of wall multiplied by its volume. Then the mass of fiber is given by the following equation:

$$M_{fiber} = V_{wall} \times d_{cellulose} = \pi \times (R^2 - (R-e)^2) \times \delta x \times d_{wall}$$

where e is the thickness of the fiber's wall.

Table 3. Density of hollow fibers.

Species	Method	Density	Reference
Gigantea	-	0,68	(Ashori and Bahreini 2009)
Daemia	Sink/float in xylene	0.8–0.9	(Karthik and Murugan 2013)
Procera	Sink/float in xylene and carbon tetrachloride	0,97	(Sakthivel 2005)
Estabragh	-	0,97	(Bahari et al. 2015)
Milkweed	Sink/float in xylene and carbon tetrachloride	0.893	(Reddy and Yang 2010)
Kapok	-	0,31–0.38	(Cicala et al. 2010)
Kapok	-	0.3	(Meiwu et al. 2010)
Kapok	-	0,3	(Zheng and Wang 2014)

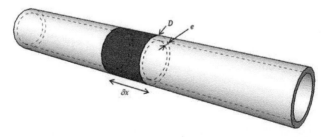

Fig. 9. Schematic diagram of the cylindrical model to calculate the volume and density of hollow fibers.

The theoretical density of the fiber can be calculated by the final equation below:

$$d_{fiber} = \frac{M_{fiber}}{V_{fiber}} \quad \frac{\pi \times (R^2-(R-e)^2) \times \delta x \times d_{wall}}{\pi \times R^2 \times \delta x} = \frac{R_2-(R-e)_2}{R^2} \times d_{wall}$$

Then by applying average values of hollow fibers, the theoretical hollow fiber density is close to 0.296 g/cm³. This value shows that the conventional methods of measuring the density of hollow fibers have yielded overestimations. A different value was found using the immersion method, which can be explained by the fact that the lumen was partially filled by liquid during the immersion test.

Tensile properties

Tensile properties of hollow fibers are presented in Table 4. Few publications were found dealing with the mechanical properties of natural hollow fibers. The majority of studies has characterized the mechanical properties of fibers for the textile industry. Accordingly, the ultimate tensile stress is often expressed in g/tex instead of Pascal. Only Ashori and Bahreini have published values for the Young modulus of hollow fibers (Ashori and Bahreini 2009). Other studies were manly focused on the textile industry, and only ultimate tensile stress, elongation and linear density were presented. Nevertheless, despite inhomogeneity in the results, hollow fibers exhibit quite low tensile properties. In fact, hollow fibers are not supposed to have high mechanical properties and their purpose is to suspend the seeds in air and then disperse them as far as they can fly. However, bast fibers grow to withstand gravity and lateral forces. In fact, the primary function of the fibers naturally determines if they need to have high or low mechanical properties. Values of UTS in italic in the table were recalculated when the density values were available (Chawla 2005).

Moreover, the growing conditions strongly influence the mechanical properties of fibers. Yang et al. (2012) showed in their study that the properties of Akund

Table 4. Mechanical properties of hollow fibers.

Species	Young's modulus (GPa)	Tensile strength (MPa)	Breaking strenght (g/tex)	Strain (%)	Linear density (tex)	References
Gigantea	8,2	296	-	1,6	-	(Ashori and Bahreini 2009)
Procera	-	*231*	24,3	2	0,26	(Sakthivel 2005)
Akund	-	-	29,6	2,45	0,1	(Yang, Huang et al. 2012)
Estabragh	-	*362*	38	2,6	0,37	(Bahari et al. 2015)
Estabragh	-	-	29,9	2,2	0,17	(Gharehaghaji and Davoodi 2008)
Kapok	-	-	16.7	3.6	0.68	(Yang, Huang et al. 2012)
Kapok	4	93.3	-	1.2	-	(Cicala et al. 2010)
Kapok	-		-	1,83–4,23	-	(Zheng and Wang 2014)

were influenced dramatically by the environment and the growing conditions (Yang, Cheng et al. 2012). In that study, the effect of parameters like the growing altitude, the pod size, and the location of fibers in the pod on the mechanical properties were investigated. They showed that an increase in the altitude tends to decrease the ultimate tensile strength of the fiber. They also noticed a strong correlation between the size of the pods and the properties of the fibers. The smaller the pods, the weaker the fibers. Eventually, they found that the fibers located in the upper part of the pods are more resistant to tensile stress than those in the lower part.

Thermal properties

Yang et al. (2012) characterized the thermal properties of Akund and kapok fibers (Yang, Huang et al. 2012). Their results were compared with the cotton fibers. The results showed that both fibers had poor thermal stability. In both cases, the main onset temperature occurred few degrees before the one of cotton. Regarding the residual mass, cotton had higher values in comparison with these fibers. However, the two hollow fibers presented lower degradation speed than cotton.

The lower thermal stability of hollow fibers compared to cotton fibers can be explained by two reason—the content of cellulose and the degree of crystallinity. As it was shown in Table 2, hollow fibers possess higher amounts of lignin and hemicelluloses and lower amounts of cellulose, compared with cotton fibers (92.56% of cotton is cellulose) (Yang, Huang et al. 2012). Hemicellulose and lignin have lower molecular weight in comparison with cellulose and thus their thermal degradation starts well before that of cellulose (Gu et al. 1992). In addition, hollow fibers' cellulose presents a degree of crystallinity of about 28.92% while cotton fibers' cellulose generally presents a degree of crystallinity close to 78.4% (Yang, Huang et al. 2012). The higher crystallinity of cotton fibers naturally lead to a higher degradation temperature (Morgado and Frollini 2011). That is why cotton exhibits superior thermal properties compared with akund and kapok fibers.

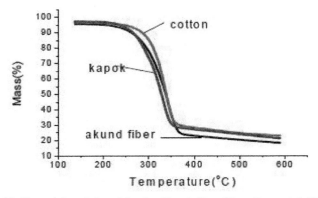

Fig. 10. Thermal degradation of akund and kapok fibers (Yang, Huang et al. 2012).

Alkaline resistance

Alkaline resistance of hollow fibers and specially estabragh fiber, was studied by Merati (Merati 2014). The alkaline resistance of Estabragh was evaluated in cementitious materials. Merati found that estabragh fibers had very poor alkaline resistance. The strength of the fiber decreased from 3.65 CN to 2.17 CN after 28 days in caustic soda (40 g/l). According to the research done by Merati, adding the fibers to the cement did not improve the long-term strength of cement. However, the study showed that hollow fibers decreased the number of cracks appearing during the early stages of cement curing.

Alkaline resistance on the Mudar fibers (*Calotropis procera*) was also studied by Sakthivel et al. (Sakthivel 2005). The study placed focus on the spinnability of hollow fibers. As hollow fibers are very smooth, it is hard to manage the spinnability. Therefore, to improve the friction, a short alkaline treatment was applied on Mudar fibers. The treatment involved soaking the fibers in a 5% NaOH solution for 5 minutes. The fiber morphology was studied by scanning electron microscopy and showed that the fiber's geometry is strongly deteriorated by the treatment. Alkali-treated fibers exhibited a convoluted-like morphology same was with cotton fibers. Moreover, they noticed that fibers tend to agglomerate after the treatment.

Acoustic properties

A huge amount of void in hollow fibers provides the opportunity for these fibers to be used in sound-proofing applications. Hollow fibers' acoustic properties were studied by Hasani et al. for automotive application (Hasani et al. 2014). Their study showed that Estabragh presented a good acoustic performance. The noise absorption coefficient of a conventional hollow polyester web was improved by the addition of Estabragh fiber. This study mainly attributed the high damping properties of Estabragh to the presence of lumen in the fiber. The results showed that a pure Estabragh web has a noise absorption coefficient almost double of the one of pure polyester.

Industrial Applications of Floss Fibers

Composites manufacturing

Apocynaceae fibers have been used as reinforcements in both thermoset and thermoplastic polymers. Although, the number of studies on thermosets is higher than the one on thermoplastics, researchers and industries are interested to use hollow fibers in both resins and even in cementitious matrices.

One study was done on natural hollow fibers using polypropylene (PP) by Reddy and Yang at the University of Nebraska (Reddy and Yang 2010). However, the authors did not specify the species of the fibers used but it is likely *Asclepias syriaca*. The composites were fabricated using 35% and 50% fibers by thermal compression. First, the fibers were carded four times to make mats and then the composites were made by compression molding at 190°C for 140 seconds. The results showed that the PP/hollow fibers composites had superior mechanical properties compared to conventional PP/kenaf composites. Using 50 wt% hollow fibers as reinforcement increased the flexural strength two times and elastic modulus four times in comparison with PP/kenaf composites.

Srinivas et al. studied the mechanical properties of *Calotropis gigantea* fibers' reinforced with unsaturated polyester (UP) composites (Srinivas and Babu 2013). The authors studied the effect of increasing the contents of the fiber on the mechanical properties of the composites. The composites were produced by hand layup molding with fiber contents ranging from 0% to 30%. The article does not provide an indication of the state of the fibers or even their distribution in the specimens. The results showed that the addition of *Calotropis gigantea* fiber of up to 30% in the polyester resulted in an increase of 100% in the tensile strength and an increase of 47% in the elastic modulus. The study also showed that the composites with 30 wt% fibers exhibited 1.54 times higher flexural modulus and 1.3 times higher impact compared with the reference. Moreover, the study could successfully decrease the density of UP by 30% for the composite with 30 wt% fibers.

Another well-known hollow fiber which has been used in composite materials is kapok. Like the other hollow natural fibers, very few publications have focused on the study of kapok fibers-reinforced composites (Mwaikambo et al. 2000, Venkata Reddy et al. 2008a, 2008b, Venkata Reddy, Shobha Rani et al. 2009, Venkata Reddy, Venkata Naidu, et al. 2009). To the best of our knowledge, there is still no study which shows the effect of kapok fibers on the mechanical properties of thermoplastics. However, Venkata Reddy et al. conducted a research on the mechanical properties of kapok/UP composites. The authors found a significant increase in the mechanical properties of the composites by adding 9% fibers to the polymer. The composites were processed by reinforcing the UP using kapok fabrics. They also studied the effect of fibers' pre-treatments on the mechanical properties of the composites. The alkaline treatment of the fibers (30 minutes in a 2% solution of NaOH) slightly improved the properties. Although they have claimed for the amelioration in the mechanical properties of composites, they did not show any microscopy on the treated fibers. Therefore, it is not possible to check the state of the fiber to discuss the real reason for the improvements (Venkata Reddy et al. 2008a, 2008b, Venkata Reddy, Shobha Rani et al. 2009, Venkata Reddy, Venkata Naidu et al. 2009).

A handful of studies have been done on hollow fibers' reinforced composites. Although, the studies are very different and limited, it seems that Apocynaceae fibers can improve the mechanical properties of polymeric matrices. Table 5 summarizes the results of these studies.

Oil absorbents

Oil sorption capacities of natural fibers are generally much higher than those of polypropylene (Choi 1996). According to Choi and Cloud, milkweed fibers are the

Table 5. Mechanical properties of hollow fiber composites.

	Asclepias syriaca[a]	*Calotropis gigantea*[b]	Kapok[c]	Alkali treated Kapok[d]	Kapok[e]	Alkali treated Kapok[f]
Fiber content	50%	30%	9%	9%	5%	5%
Polymeric matrices	PP	UP	UP	UP	UP	UP
Process	Thermocompression	Manual	Manual	Manual	Manual	Manual
Elastic modulus (Tensile mode)	+ 392%	+ 57%	−18%	+ 19%	-	-
Elastic modulus (Flexural mode)	+ 65%	+ 154%	+ 723%	+ 731%	+ 303%	+ 327,1%
Tensile strength	+ 57%	+ 100%	+ 182%	+ 232%	-	-
Flexural strength	+ 195%	+ 80%	+ 125%	+ 160%	+ 50%	+ 56%
Improve in impact resistance	+ 33%	+ 130%	+ 133%	+ 145%	-	-

[a] (Reddy and Yang 2010)
[b] (Srinivas and Babu 2013)
[c] (Venkata Reddy et al. 2008a, 2008b, Venkata Reddy, Shobha Rani et al. 2009)
[d] (Venkata Reddy et al. 2008a, 2008b, Venkata Reddy, Shobha Rani et al. 2009)
[e] (Venkata Reddy, Venkata Naidu et al. 2009)

best natural oil absorbents followed by Kapok fibers (Choi and Cloud 1992, Choi 1996). Lim and Huang and Choi and Cloud have explained this high oil absorption capacity due to two phenomena (Choi and Cloud 1992, Lim and Huang 2007). First, hollow fibers create interaction with oil due to their waxy surface. This interaction is induced by a chemical compatibility with oil as both are hydrophobic (Lim and Huang 2007). Second, absorption mechanism is related to capillarity in the lumen of the fibers. Therefore, the small radius of lumen increases the capillarity pressure which promotes oil absorption.

Kobayashi et al. found that kapok fibers are able to absorb more than twice the amount of oil absorbed by polypropylene which is considered an excellent oil absorbent (Kobayashi et al. 1976). Kapok fibers are able to absorb almost 40 g of oil per gram of fiber (Hori et al. 2000, Lim and Huang 2007). Same trend was also found for milkweed fibers (Choi 1996).

Oil absorbents made of hollow fibers can be reused even though the oil sorption capacity will be lowered with increasing number of cycles they are used. However, the sorption capacity remains still higher than that of new polypropylene (Lim and Huang 2007).

Thermal insulation

Down feather is one of the most famous materials which has been used for thermal insulation applications. However, there are always some ethical and economic concerns with using this material. Hollow fibers can find applications in the thermal insulation with fewer concerns. The thermal insulation capacity of milkweed fibers (*A. syriaca* and *A. speciosa* blended) was studied by Crews et al. (Crews et al. 1991). Winter coats isolated with milkweed fibers, synthetic fibers or down feather were tested to measure their insulating efficiency. Results showed surprisingly that a coat insulated with 100% milkweed fibers had a very low insulation value. Moreover, they also noticed that standard commercial drycleaning procedures affected strongly the fiber and their insulating efficiency. Nevertheless, the 50/50 milkweed/down coat happened to be as effective as the pure down feather coat even after the cleaning process. Their conclusion showed that milkweed fibers are very useful to decrease the use of down feather with no loss of thermal efficiency. In addition, as the milkweed fiber costs half the price of goose down, the use of milkweed fibers may decrease the price of the insulating materials.

Parmar et al. found a similar result regarding the use of milkweed for thermal insulation (Parmar et al. 2015). Their results showed that the addition of milkweed fibers to common cotton or polyester fabric resulted in an increase in the thermal insulation value. Thermal insulation properties of fabrics were improved by 14% for cotton and 10% for polyester when half of the fibers are replaced by milkweed fibers. Furthermore, they also noticed that milkweed fibers greatly induced ultraviolet protection properties to the fabrics. Based on literature, Parmar et al. have explained this property due to the high percentage of lignin in milkweed fibers.

Kapok also exhibits similar behaviors when it is used in thermal insulating applications. Cui et al. have shown that the thermal conductivity of wadding decreased when the share of kapok in fabrics increased (Cui and Wang 2009, Cui et al. 2010). Few patents have also been published on hollow fibers for different applications. For instance, in 1949 the Milkweed Products Dev. Corp. patented an insulated life preserver for below 0° Celsius temperatures (Boris 1952). A building insulation material made from hollow fibers has also been patented recently (Ali 2013).

Conclusions

Hollow fibers are a truly sustainable product that can be harvested in almost any part of the world with constant properties regardless of the type or species. Hollow fibers can be recognized by the following properties:

- Hollow fibers are shorter than common natural fibers.
- Hollow fibers' composition is characterized by a lower amount of cellulose and a high amount of lignin, hemicellulose and wax than common natural fibers.
- Hollow fibers show higher moisture regain than common natural fibers.
- The density of hollow fibers is almost five times lower than common natural fibers.
- The tensile properties of hollow fibers are lower compared to liber fibers.
- The thermal stability is lower than common natural fibers.

- Hollow fibers are very sensitive to alkaline environment.
- The acoustic properties of hollow fibers are excellent.

In this chapter, different species of hollow fibers and their corresponding mechanical properties were described. It is quite clear that the properties of fibers do not depend directly on the species of hollow fibers. Even though there is a strong relationship between the wall structure and the tensile properties of the fibers, there is no clear link between the structure of a fibers' wall and the species. Hollow fibers should not be used or considered in the same way as conventional natural fibers. These fibers possess lower tensile properties in comparison with conventional fibers. However, hollow fibers have the potential of numerous applications such as oil absorbence or thermal insulation, which are not available in other natural fibers.

References

Akil, H.M., M.F. Omar, A.A.M. Mazuki, S. Safiee, Z.A.M. Ishak and A. Abu Bakar. 2011. Kenaf fiber reinforced composites: a review. Materials & Design 32(8-9): 4107–4121.

Ali, M.E.-S. 2013. Natural fiber insulation material and method for making the same.

Anderson, D.B. and T. Kerr. 1938. Growth and structure of cotton fiber. Industrial & Engineering Chemistry 30(1): 48–54.

Ashori, A. and Z. Bahreini. 2009. Evaluation of *Calotropis gigantea* as a promising raw material for fiber-reinforced composite. Journal of Composite Materials 43(11): 1297–1304.

Bahari, N., H. Hasani, M. Zarrebini and S. Hassanzadeh. 2015. Investigating the effects of material and process variables on the mechanical properties of low-density thermally bonded nonwovens produced from Estabragh (milkweed) natural fibers. Journal of Industrial Textiles.

Barth, F.W. and T.E. Timell. 1958. The Constitution of a Hemicellulose from Milkweed (Asclepias syriaca) Floss, 80.

Bernstien, J. 2016. How a Quebec company used milkweed to create a one-of-a-kind winter coat. CBC News, 13 Oct.

Boris, B. 1952. Insulated Life Preserver Garment.

Chattopadhyay, S.K., S. Singh, N. Pramanik, U.K. Niyogi, R.K. Khandal, R. Uppaluri and A.K. Ghoshal. 2011. Biodegradability studies on natural fibers reinforced polypropylene composites. Journal of Applied Polymer Science 121(4): 2226–2232.

Chawla, K.K. 2005. Fibrous Materials. Cambridge University Press.

Choi, H.M. and R.M. Cloud. 1992. Natural sorbents in oil spill cleanup. Environmental Science & Technology 26(4): 772–776.

Choi, H.-M. 1996. Needlepunched cotton nonwovens and other natural fibers as oil cleanup sorbents. Journal of Environmental Science and Health. Part A: Environmental Science and Engineering and Toxicology 31(6): 1441–1457.

Cicala, G., G. Cristaldi, G. Recca and A. Latteri. 2010. Composites based on natural fibre fabrics. *In*: Woven Fabric Engineering. Polona Dobnik Dubrovski.

Crews, P.C., S.A. Sievert, L.T. Woeppel and E.A. Mccullough. 1991. Evaluation of milkweed floss as an insulative fill material. Textile Research Journal 61(4): 203–210.

Cui, P. and F. Wang. 2009. An investigation of heat flow through kapok insulating material 19: 88–92.

Cui, P., F.-M. Wang, A. Wei and K. Zhao. 2010. The performance of kapok/down blended wadding. Textile Research Journal 80(6): 516–523.

Cuoq, J. 1975. Recueil des sources arabes concernant l'Afrique occidentale du 8 au 16 siècle. -Bilad Al-Sudan. Centre national de la recherche scientifique.

Drean, J.-Y.F., J.J. Patry, G.F. Lombard and M. Weltrowski. 1993. Mechanical characterization and behavior in spinning processing of milkweed fibers. Textile Research Journal 63(8): 443–450.

Gharehaghaji, A.A. and S.H. Davoodi. 2008. Mechanical damage to estabragh fibers in the production of thermobonded layers. Journal of Applied Polymer Science 109(5): 3062–3069.

Gu, P., R.K. Hessley and W.-P. Pan. 1992. Thermal characterization analysis of milkweed flos. Journal of Analytical and Applied Pyrolysis 24(2): 147–161.

Hasani, H., M. Zarrebini, M. Zare and S. Hassanzadeh. 2014. Evaluating the acoustic properties of estabragh (milkweed)/hollow- polyester nonwovens for automotive applications. Journal of Textile Science & Engineering 4(3).

Hassanzadeh, S., H. Hasani and M. Zarrebini. 2014. Analysis and prediction of the noise reduction coefficient of lightly-needled Estabragh/polypropylene nonwovens using simplex lattice design. The Journal of The Textile Institute 105(3): 256–263.

Holdrege, C. 2010. The story of an organism: Common milkweed. The Nature Institute http://www.natureinstitute.org/txt/ch/Milkweed.pdf.

Hori, K., M.E. Flavier, S. Kuga, T.B.T. Lam and K. Iiyama. 2000. Excellent oil absorbent kapok [*Ceiba pentandra* (L.) Gaertn.] fiber: fiber structure, chemical characteristics, and application. Journal of Wood Science 46(5): 401–404.

Karthik, T. and R. Murugan. 2013. Characterization and analysis of ligno-cellulosic seed fiber from *Pergularia Daemia* plant for textile applications. Fibers and Polymers 14(3): 465–472.

Karthik, T. 2015. Studies on the Spinnability of Milkweed Fibre Blends and its Influence on Ring Compact and Rotor Yarn Characteristics.

Kerr, T. 1937. The structure of the growth rings in the secondary wall of the cotton hair. Protoplasma 27(1): 229–241.

Kish, M.H. and S.S. Najar. 1998. Structure and properties of a natural celulosic hollow fiber. International Journal of Engineering 11(2): 101–108.

Kobayashi, Y., R. Matuo and M. Nishiyama. 1976. Method for Adsorption of Oils: US4061567A.

Koch, K. and H.-J. Ensikat. 2008. The hydrophobic coatings of plant surfaces: Epicuticular wax crystals and their morphologies, crystallinity and molecular self-assembly. Micron. 39(7): 759–772.

Leprince, J.M. 2014. Le Québec deviendra-t-il l'eldorado de la soie d'Amérique? Radio Canada 20 Oct.

Leprince, J.M. 2015. La soie d'Amérique passe en production industrielle. Radio Canada 26 Oct.

Lim, T.-T. and X. Huang. 2007. Evaluation of kapok (*Ceiba pentandra* (L.) Gaertn.) as a natural hollow hydrophobic–oleophilic fibrous sorbent for oil spill cleanup. Chemosphere 66(5): 955–963.

Louis, G.L. and B.A.K. Andrews. 1987. Cotton/milkweed blends: a novel textile product. Textile Research Journal 57(6): 339–345.

Meiwu, S., X. Hong and Y. Weidong. 2010. The fine structure of the kapok fiber. Textile Research Journal 80(2): 159–165.

Merati, A.A. 2014. Reinforcing of cement composites by estabragh fibres. Journal of The Institution of Engineers (India): Series E 95(1): 27–32.

Mohanty, A.K., M. Misra and L.T. Drzal. 2005. Natural Fibers, Biopolymers, and Biocomposites. London: CRC Press.

Morgado, D.L. and E. Frollini. 2011. Thermal decomposition of mercerized linter cellulose and its acetates obtained from a homogeneous reaction. Polímeros 21(2): 111–117.

Müssig, J. (ed.). 2010. Industrial Application of Natural Fibres: Structure, Properties, and Technical Applications. Chichester, West Sussex, U.K.; Hoboken, N.J: Wiley.

Mwaikambo, L.Y., E. Martuscelli and M. Avella. 2000. Kapok/cotton fabric–polypropylene composites. Polymer Testing 19(8): 905–918.

Netravali, A.N. and C.M. Pastore. 2014. Sustainable Composites: Fibers, Resins and Applications. Pennsylvania: DEStech Publications, Inc.

Nourbakhsh, A., A. Ashori and M. Kouhpayehzadeh. 2009. Giant milkweed (*Calotropis persica*) fibers—a potential reinforcement agent for thermoplastics composites. Journal of Reinforced Plastics and Composites 28(17): 2143–2149.

Parmar, M.S., M. Bahl and J.V. Rao. 2016. Milkweed blended fabrics and their thermal insulation and UV protection properties. Indian Journal of Fibre & Textile Research (IJFTR) 40(4): 351–355.

Patterson, C. 2012. Milkweed fruits: Pods of plenty. The Washington Post, 25 Sep.

Pretty, J. 2008. Agricultural sustainability: concepts, principles and evidence. Philosophical Transactions of the Royal Society B: Biological Sciences 363(1491): 447–465.

Rahman, M.A. and C.C. Wilcock. 1991. A taxonomic revision of Calotropis (Asclepiadaceae). Nordic Journal of Botany 11(3): 301–308.

Reaney, M.J.T., W. Hartley Furtan and P. Loutas. 2006. A Critical Cost Benefit Analysis of Oilseed Biodiesel in Canada: a BIOCAP Research Integration Program Synthesis Paper.

Reddy, N. and Y. Yang. 2009. Extraction and characterization of natural cellulose fibers from common milkweed stems. Polymer Engineering & Science 49(11): 2212–2217.

Reddy, N. and Y. Yang. 2010. Non-traditional lightweight polypropylene composites reinforced with milkweed floss. Polymer International 59(7): 884–890.

Sakthivel, J.C. 2005. Some studies on Mudar fibers. Journal of Industrial Textiles 35(1): 63–76.

Sharma, V., H.K. Vinayak and B.M. Marwaha. 2015. Enhancing sustainability of rural adobe houses of hills by addition of vernacular fiber reinforcement. International Journal of Sustainable Built Environment 4(2): 348–358.

Smole, M., S. Hribernik, K. Stana-Kleinschek and T. Kreze. 2013. Plant fibres for textile and technical applications. *In*: Advances in Agrophysical Research. Stanislaw Grundas.

Sonnini, C.S. 1810. Traité des Asclépiades particuliérement de l'Asclépiade de Syrie: précédé de quelques observations sur la culture du coton en France. Paris.

Spiridon, I. 2007. Modifications of *Asclepias syriaca* fibers for paper production. Industrial Crops and Products 26(3): 265–269.

Srinivas, C.A. and G.D. Babu. 2013. Mechanical and machining characteristics of calotropis gigentea fruit fiber reinforced plastics. Inter. J. Eng. Res. Tech. 2: 1524–1530.

Stevens, M. 2000. USDA-NRCS Plant Guide - Common Milkweed (*Asclepias Syriaca* L.). https://plants.usda.gov/plantguide/pdf/cs_assy.pdf.

Sunmonu, O.K. and D. Abdullahi. 1992. Characterization of fibres from the plant *Ceiba pentandra*. The Journal of The Textile Institute 83(2): 273–274.

Unknown, 1900. Madar. (*Calotropis gigantea*, R. Br.). Bulletin of Miscellaneous Information (Royal Botanic Gardens, Kew), 1900 (157/168): 8–12.

Venkata Reddy, G., S. Venkata Naidu and T. Shobha Rani. 2008a. Impact properties of kapok based unsaturated polyester hybrid composites. Journal of Reinforced Plastics and Composites 27(16-17): 1789–1804.

Venkata Reddy, G., S. Venkata Naidu and T. Shobha Rani. 2008b. Kapok/glass polyester hybrid composites: Tensile and hardness properties. Journal of Reinforced Plastics and Composites 27(16-17): 1775–1787.

Venkata Reddy, G., T. Shobha Rani, K. Chowdoji Rao and S. Venkata Naidu. 2009. Flexural, compressive, and interlaminar shear strength properties of kapok/glass composites. Journal of Reinforced Plastics and Composites 28(14): 1665–1677.

Venkata Reddy, G., S. Venkata Naidu and T. Shobha Rani. 2009. A study on hardness and flexural properties of kapok/sisal composites. Journal of Reinforced Plastics and Composites 28(16): 2035–2044.

Yang, X., L.D. Cheng, L.Q. Huang and W.H. Fan. 2012. Study on the correlation between the property of akund fiber and its growing conditions. Advanced Materials Research 476-478: 1934–1938.

Yang, X., L.Q. Huang and L.D. Cheng. 2012. Study on the structure and the properties of akund fiber. Applied Mechanics and Materials 217-219: 617–621.

Yang, X., L. Huang, L. Cheng and J. Yu. 2015. Studies of moisture absorption and release behaviour of akund fiber. Advances in Mechanical Engineering 4(0): 356548–356548.

Zheng, Y. and A. Wang. 2014. Kapok fiber: Structure and properties. pp. 101–110. *In*: K.R. Hakeem, M. Jawaid and U. Rashid (eds.). Biomass and Bioenergy. Springer International Publishing.

Near Infrared Spectroscopy, A New Tool to Characterize Wood for Use by the Cooperage Industry

Gilles Chaix,[1,2,3] *Thomas Giordanengo,*[4] *Vincent Segura,*[5]
Nicolas Mourey,[4] *Bertrand Charrier*[6] and *Jean Paul Charpentier**[*,5]

General Introduction

It is important for the industry to be able to quickly measure and/or evaluate the mechanical, physical and chemical properties of wood. Whether it is upstream in the field, on a living tree, when transformed into timber or in industrial product, during and after the material is put into operation. Among the non-destructive measurement techniques that have been developed since the 1980s, spectroscopy based on emission or absorption in the near infrared (NIR spectroscopy), i.e., in the

[1] CIRAD, UMR AGAP, 34398 Montpellier, France.
  Email: gilles.chaix@cirad.fr
[2] AGAP, Univ Montpellier, CIRAD, INRA, Montpellier SupAgro, Montpellier, France.
[3] ESALQ-USP, Wood Anatomy & Tree-Ring Lab, Piracicaba, Brazil.
[4] R&D Tonnellerie Radoux – Pronektar, 10 Avenue Faidherbe, 17500 Jonzac, France.
  Emails: t-giordanengo@sciage-du-berry.fr; n-mourey@radoux.fr
[5] BioForA, UMR INRA-ONF and Wood Analysis Technical Platform GénoBois, INRA Val de Loire, 2163 avenue de la Pomme de Pin 45075 Orléans, France.
  Emails: vincent.segura@ inra.fr; jean-paul.charpentier@inra.fr
[6] Université de Pau et des Pays de l'Adour, Xylomat, IUT, 40004 Mont de Marsan, France.
* Corresponding author: jean-paul.charpentier@inra.fr

frequencies between 800 and 2500 nm, offers a new field of possibilities for analysis and evaluation of the quality of materials. While NIR spectroscopy is still not widely used in the wood industry, it is more developed in other industrial sectors such as petrochemicals, pharmaceuticals, agriculture, agrotechnology, food and textiles.

Since the early 1990s, numerous research studies have explored the value of using NIR spectroscopy to estimate the properties of wood material. This represents more than 500 scientific articles published in this sector with an average since the end of the 2000s of about 40 articles published each year (Tsuchikawa 2007, Tsuchikawa and Kobori 2015). This chapter attempts to take stock of the theoretical and practical aspects of SPIR technology and highlights through multiple examples, the wide potential of this measurement tool in the field of wood quality assessment and in the use and processing of wood material.

In the first part, is discussed the general elements of the NIR spectroscopy methodology and present the significant results of the scientific literature. In the second part, we report the first industrial application of NIR spectroscopy in the field of the chemical quality of oak wood for cooperage. We do not seek to be exhaustive but rather to describe some of the most iconic applications, and to make the reader want to go further, such as the examples of durability conferred and cooperage. For an exhaustive view, we recommend reading bibliographic reviews (Tsuchikawa 2007, Tsuchikawa and Schwanninger 2013, Tsuchikawa and Kobori 2015) and specialized works (Osborne et al. 1993, Bertrand and Dufour 2006).

NIR Spectroscopy and Wood Chemistry: Theoretical Approach and Fields of use of the Technology

Principles and theoretical approach of NIR spectroscopy

Spectroscopy can be defined as the study of the interaction of electromagnetic waves with matter. The electromagnetic spectrum is generally divided into various regions as a function of the wavelength of the radiation where the gamma rays are found which are the most energetic, the X-rays, the ultraviolet, the visible, the infrared (IR), microwaves and radio frequency waves. Each region is associated with a type of atomic or molecular transition involving different energies.

The infrared (IR) domain, characterized by vibrational-type energy transitions and comprises three spectral ranges: the near-IR (NIR) between 800 and 2500 nm, the mid IR (MIR) between 2500 and 25000 nm and far IR (FIR), beyond 25000 nm.

NIR spectroscopy is the spectral region of harmonic bands and combination bands (Fig. 1). This region is dominated by the absorption of the bonds of the functional group X-H, where X corresponds to carbon, oxygen or nitrogen atoms, and H denotes the hydrogen atom. An absorption spectrum is most frequently represented by the decimal logarithm of the inverse of the reflectance as a function of wavelengths in nanometers or of the wave number in cm^{-1}. Spectral information is repeated in the regions of the harmonic and combination bands, which facilitates

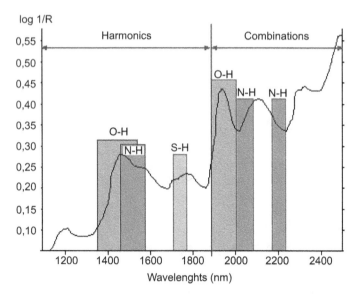

Fig. 1. Near-infrared absorption bands, particularly OH, NH and SH bonds, the C-H bonds are at the origin of the absorption bands throughout the NIR region (after Bertrand 2002).

the construction of the quantitative models. Schwanninger et al. (2011) conducted a review of the absorbance bands of the main wood compounds in the NIR domain.

When light radiation comes into contact with matter, it can be absorbed, transmitted or reflected. The transmittance T is defined as the fraction of light energy passing through the sample from one side to the other: $T = IT/I0$ (IT being the transmitted light intensity, $I0$ the luminous intensity emitted by the source). The reflectance R represents the part of the reflected light. It is defined by the ratio of the intensity reflected on the incident intensity: $R = IR/I0$ (IR being the reflected light intensity).

Diffuse reflection contains rich information on the chemical composition of the material. Absorbance A is defined as the logarithm of the inverse of the transmittance T or the reflectance R: $A = \log(1/T)$ or $A = \log(1/R)$

Most molecules are *IR* active and are represented by diverse bands due to different types of molecule vibrations. Band intensities depend on the concentration of the compound on its molar absorption coefficient and on individual properties of the functional group. For complex materials like wood, bands are broad and overlapping and are often not assignable. However, *IR* spectrum could be considered like a "fingerprint" of the analyzed material.

The equipment

Many commercial instruments are specifically dedicated to industrial analytical applications in the NIR region. Their principles differ according to the technical options chosen by the builders; one can classify them into three main technologies:

- **Sequential spectrometers**
  These devices are characterized by recording the light absorptions one after the other, scanning the spectral region in a sequential order. Typical instruments in this group are optical filters or monochromatic spectrometers.

- **Multiplexed spectrometers**
  These spectrometers have a single photosensitive sensor that simultaneously receives information corresponding to several wavelengths. Among the multiplexed devices, Fourier transform (FTNIR) spectrometers are widely used in research and quality control.

- **Multichannel spectrometers**
  Unlike multiplexed spectrometers, multichannel spectrometers have multiple photosensitive sensors that can independently record the absorption of light at specific wavelengths. Examples of devices belonging to this group include diode array systems and hyperspectral cameras.

Chemometrics for NIR spectroscopy analysis

Chemometrics is "the science of using mathematical, statistical and computer methods to improve the extraction of information obtained from analytical data" (Geladi 1995).

NIR spectra provide information on the chemical and physical properties of the samples studied. Obtaining this information is not immediate and requires a mathematical treatment that can be relatively complex.

Chemometric methods can be classified according to their linear or non-linear nature (Bertrand 2002). Linear methods assume that useful information can be extracted from linear combinations of predictor variables. In this approach, a spectrum is considered as the sum of pure spectra or underlying signals. Non-linear methods, which are essentially connectionist, are part of the field of artificial intelligence and mainly involve neural networks. Chemometric methods can also be classified according to their supervised (predictive) or unsupervised (descriptive) nature. In supervised methods, NIR spectral data are used to predict a qualitative or quantitative variable. With regard to unsupervised methods (for example, principal component analysis named PCA), spectral data are available only without further information on the nature of the data. The vast majority of NIR spectroscopy applications are based on supervised and linear methods for forecasting quantitative variables.

To calibrate a property by NIR spectroscopy, it is necessary to measure on the same lot of samples the NIR absorbance spectra as well as the values of this property by a reference method. Thanks to chemometrics, a model linking the spectrum and the reference measurement is obtained. If the resulting model is efficient, it can be used to estimate the property of new samples only from the NIR absorbance spectrum. Many methods are available but partial least squares regression (PLS-Regression) is the most used. Via et al. (2014) proposed to assess and compare the performance of PLS and PCR (principal components regression). Their conclusion was that PLS was found to provide better predictive diagnostics but PCR was better for model interpretation and wavenumber selection. Many software programs allow the development and application of these mathematical algorithms and recently a free and shared internet tool was developed by a French team and made available to chemometricians (http://chemproject.org).

Calibration optimization

Limitations and recommendations for a good calibration

The limit detection level of a component on complex products is of the order of 0.5 to 1% of the mass. There is not much hope to use NIR spectroscopy for components in very low concentrations. Moreover, the minerals do not absorb in the NIR, so only the organic matter can be detected.

The penetration depth of IR radiation in solid wood does not generally exceed 1 to 2 mm and only a few millimeters for powdered wood. These penetration depths depend on the characteristics of the emission source, the density and the compaction of the powders.

Attention must also be paid to the quality of the reference measurement. This measurement is the basis of calibration and determines its quality. It is important to perform a repeatability and accuracy analysis of the reference method measurement to determine the Standard Error of Laboratory (SEL). SEL has to be compared to the error of the NIR spectroscopy calibrations. The quality of calibration will depend very much on the quality of the reference measurement. The more noise it contains, the less effective modeling will be. Similarly, the quality of the spectral data and thus the state of the spectrometer will have an influence on the quality of the calibration.

The calibration samples must be numerous (generally between 100 and 300 samples depending on the property to be calibrated), representative of those that will be analyzed later (the selection can be made on the basis of the spectral information measured in advance on the samples) and have a sufficiently large variability of the character of interest to cover the range of variation thereof. Random selection of samples is not desired for calibration: a Gaussian type distribution is usually observed with many values close to the mean and extreme values is infrequent. A selection of samples that balance their distribution well along the gradient of variation is preferable.

Sample preparation must be well defined and repeatable. Sampling conditions, grinding, particle size (for powders), drying, storage, filling (for powders) and presentation for solid wood (ligneous orientation, temperature, surface quality, etc.), must be fixed according to a protocol established by the operator. This protocol will be maintained in the future for the prediction of new samples. In fact, keeping the measuring device in good condition, thanks to the internal tests but also thanks to the regular tests performed on reference samples, is a decisive element for a quality operationality.

Simple calibration does not determine the reliability of the models. In other words, the calibration error (RMSEC: Root Mean Square of Standard Error Calibration) is not sufficient to evaluate the performance of a model. However, it is imperative to validate the models to achieve this goal of reliability. The validation step consists in applying the predictive models to a second set of samples that were not used for calibration. It is considered that the validation is independent (named external validation). It is generally accepted to use a cross-validation, especially when the number of samples is low (100–150), or more simply to check the feasibility of a calibration. The latter consists of successively placing each of the observations available in the validation set (named full cross-validation), or randomly dividing the samples into several validation groups (few is better), and using them successively to validate the model (see Fig. 2). The latter is

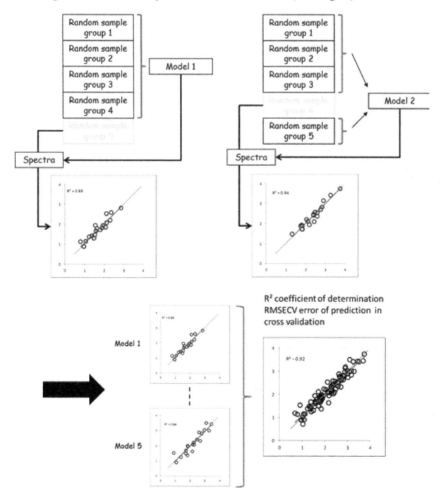

Fig. 2. Cross-validation process for NIR spectroscopy calibration and validation.

recommended and increasingly used. Another option is to divide the sample set into two group and use one for calibration and the other for validation and repeat it several times (minimum of ten times). After this, it is better to use an independent sample set to validate the model.

This set of constraints and recommendations is however essential, as is the dexterity of the operator. Failure to follow them means that costly NIR devices are no longer used, or that the NIR technology is abandoned because it is judged ineffective although many laboratories have shown its efficiency.

Spectral data pretreatments commonly used

Raw spectral data are often spoiled with defects related to surface differences, grain size variation and presence of random noise resulting in deformation of baseline of the NIR spectrum. Spectral pre-treatments can be used before applying a chemometric method to improve the quality of the spectral signal. These pre-treatments are numerous and of several types, and the most common are:

i) *Vector Normalization (SNV)*: The average and the standard deviation are calculated for each spectrum. Then each absorbance value of the spectrum is subtracted with the mean value and the result is divided by the standard deviation of the spectrum. This standardization is very effective in reducing the effect of particle size on absorbance, for example, and any other uncontrollable environmental effect.

ii) *The derivative*: The calculation of a derivative type pre-treatment is characterized by three criteria: the order of the derivative, the segment on which the derivative is calculated and the number of points taken into account for the smoothing preceding the operation of derivation. The derivative is effective in reducing peak overlap problems and large baseline variations (Fig. 3).

iii) *Multiplicative Scatter Correction (MSC)*: It is based on the linear modeling of a spectrum according to the average spectrum of a sample set. The "error" represents effects that cannot be modeled by a constant and multiplier. It therefore reflects spectral variations from "useful" chemical information.

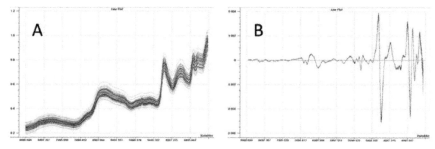

Fig. 3. Example of pretreatment of spectral data of oak wood, (A) raw spectrum, (B) spectra after a second derivative (smoothing on 25 points of polynomial type of order 3 before derivation).

Calibration evaluation criteria

Optimizing a calibration requires knowing how to use different statistical criteria to ensure the reliability and quality of the model.

The standard errors of calibration (SEC or RMSEC) corresponds to the standard deviation of the residuals of the modeling. It is representative of the fit quality of the model to the calibration batch.

The standard errors of cross-validation (SECV or RMSECV) is the standard deviation of the residues obtained with cross-validation groups. RMSECV is calculated in the cross-validation process (Wold 1978). The spectra are separated into sets (or groups or segments) of the same number of samples, chosen randomly or systematically. As many models as groups are built and validated by the removed samples.

The standard errors of prediction (SEP or RMSEP) is used to evaluate the quality of the prediction model. A validation test is performed on a group of samples completely independent of that used to create the calibration model. SEP is the standard deviation of the residues obtained with the validation batch.

In addition, the ratio RMSEP/RMSECV was sometime adopted to evaluate the robustness of the model; a model with RMSEP/RMSECV lower than 1.2 was usually considered robust (Alves et al. 2012).

The rank (number of factors, number of latent values) is the number of PLS factors introduced to develop the equation of a model. It is determined during the calibration process by comparing the SEC and SECV curves. Increasing the rank improves the fit of the calibration. However, using too high a rank may result in overfitting.

The ratio of performance to deviation (RPD) is the relationship between population variability (standard deviation) and prediction error (Williams and Sobering 1993, Williams 2014). It is used to judge the quality of the calibration and its potential applications. If the prediction error is close to the standard deviation, calibration can only predict the population mean. There is RPD degrees adapted for wood, forages, feeds, soils (complex materials). For a value less than or equal to 1.9, it is not recommended to use the model. A value between 1.9 and 2.4 generally makes it possible to sort and screen the samples in a few classes. Between 2.5 and 2.9, the sorting will be better and can be used, for example, for breeding programs. Between 3.0 and 3.4, the prediction is satisfactory and can be used in quality control. Between 3.5 and 4.0, the calibration is considered very satisfactory and can be used in process control. Beyond 4.1, the model is considered as excellent and valid for all analytical applications for this type of material (Williams 2014).

The coefficient of determination in calibration (R^2c), in cross-validation: (R^2cv), in validation (r^2) provide information between the NIR model prediction and the reference measurements for the same samples. Its value is between 0 and 1, a value close to 1 indicates a very good fit.

Recently, Williams et al. (2017) report in an exhaustive tutorial almost 40 items essential to show the results on NIR spectroscopy model. These must be reported and explained in a NIR spectroscopy project to allow duplication of the application and methods reported. These items are very important if the final project is to extend it to the level of industrial application.

Application development outside the laboratory

NIR absorbance is sensitive to many parameters such as temperature (Chauchard et al. 2004) or wood moisture content (Giordanengo et al. 2008), and spectrometers are not necessarily suitable for all industrial atmospheres (thermal amplitudes, vibrations, dust, vapors, etc.). When a measurement methodology is developed and validated under controlled laboratory conditions, specific developments are usually necessary in order to obtain a robust measurement methodology for applications in industrial conditions. Sensor design and measurement methodology are critical steps in optimizing application performance. Furthermore, it is also common to use chemometric methods to suppress the influence of NIR signal disturbance factors (Roger et al. 2003, Chauchard et al. 2004, Zeaiter 2004) as well as for calibration transfer between devices.

The development of new miniaturized and portable equipment makes it possible to envisage for the future the deported use on the log parks or the timber stocks or even directly on the field on standing trees.

The main fields of application of NIR spectroscopy to evaluate wood chemical properties and its products

Quantification of wood macromolecules

Cellulose content

The cellulose content of wood is strongly correlated with the yield of kraft pulp (Kube and Raymond 2002). Schimleck et al. (2000) obtained excellent calibration of cellulose content with an r^2 greater than 0.90 for the two eucalyptus species *E. globulus* and *E. nitens*. Raymond and Schimleck 2002 showed the reliability of the NIR spectroscopy to predict the cellulose content for *Eucalyptus globulus*, as did Downes et al. 2010 with *Eucaluptus nitens*. Similar work has been done by Wright et al. 1990 on a series of pine species where a coefficient of determination in cross validation of 0.73 was obtained between the measured cellulose content and that predicted by the NIR spectroscopy. Clarke and Wessels (1995) also measured the cellulose content of a series of eucalyptus species by NIR spectroscopy and the calibrations gave a coefficient of determination equal to 0.73 between the measured value and the predicted value. Schimleck et al. 2000, demonstrated the ability of NIR spectroscopy both in the control of the pulp quality of eucalyptus plantations and in the selection of individuals with the highest pulp yield.

Lignin content and lignin quality

NIR spectroscopy was used to control the delignification process during kraft cooking of Scots pine (*Pinus sylvestris*) and Norway spruce (*Picea abies*) samples (Lindgren et al. 1998). It has been shown that measurements of NIR spectra made during the various delignification steps can be used in the prediction of pulp yield, kappa index and Klason lignin.

Usually, the quantification of lignin is difficult not only because of the variability of its monomeric composition but also because lignin may be covalently linked with carbohydrates, proteins, phenols and other compounds in the cell wall. These products may interfere with lignin determination leading to over- or under-estimation (Monties 1989). In the works of Brinkmann et al. (2002), the lignin content of 5 different species (*Nicotiana tabacum, Populus* x *canescens, Fagus sylvatica, Quercus robur* and *Picea abies*) was determined using two chemical methods: first a gravimetric method based on acid hydrolysis of polysaccharides (acid detergent fiber technique), and second a spectrophotometric method measuring UV absorbance (thioglycolic acid process). The coefficient of determination between the data obtained by these two methods of analysis and those by the NIR spectroscopy were very high ($R^2 > 0.98$). Yeh et al. (2004) calibrated the lignin level using PIR spectra measured from chips and *Pinus taeda* wood powder. The results obtained in validation between the measured and predicted values were 0.81 for the R^2_{cv} with SECV = 0.73 in crossvalidation for the powders and $r^2 = 0.71$ with a SEP = 0.87 in external validation for the chips.

He and Hu (2013) investigated the feasibility of using FT-NIR spectroscopy to rapidly determine the lignin content of 147 different species of Chinese wood trees species, including softwoods and hardwoods. The automatic selection of relevant wavenumbers combined with the appropriate data pre-processing methods produced satisfactory prediction models (obtained values of R^2 of prediction between 0.9775 and 0.9521).

Hodge and Woodbridge (2004) developed multispecific lignin content calibration for 5 tropical and subtropical pine species (*Pinus caribaea, Pinus maximinoi, Pinus oocarpa, Pinus patula* and *Pinus tecunumanii*) from Brazil and Columbia. Calibrations, developed from the reflectance spectra, gave values of r^2 around 0.90 in validation. Denis et al. (2013) and Chaix et al. (2012) developed calibrations for the main wood chemical compounds (extractives, lignins, holocellulose, cellulose, syringyl/guaiacyl (S/G ratio) with several species of *Eucalyptus*, and obtained relevant results (r^2 of 0.74 to 0.96 and RPD of 2.0 to 4.8). Alves et al. (2011) also proposed a determination of an S/G ratio of *Eucalyptus globulus* wood lignin using analytical pyrolysis as reference method.

To establish the best PLS-R model for prediction of lignin content in wood of Norway spruce, Schwanninger et al. (2011a, 2011b) took care of the influence of the particle size of the wood powder. They also verified several data-processing methods to select wavenumber range in spectra taken with a fiberoptic probe.

Baillères et al. (2002) and Denis et al. (2013) worked on the calibration of the S/G ratio and obtained calibrations for efficient prediction ($r^2 = 0.90$, SEP = 0.28, RPD = 2.4 for Baillères et al. (2002) and $R^2_{cv} = 0.92$, SECV = 0.13, RPD = 2.5 for Denis et al. (2013). Alves et al. (2006) have effectively developed calibrations for another criterion: the ratio H/G in maritime pine.

Holocellulose content

Holocellulose represents all the cellulose and hemicellulose polysaccharides constituting wood. Thus, the hemicellulose and cellulose content was calibrated by NIR spectroscopy in *Liquidambar styraciflua* and *Pinus taeda* (Schultz and Burns 1990). Denis et al. (2013) and Chaix et al. (2012) also investigated the possibility of calibrating holocellulose levels in eucalyptus. They obtained, in cross-validation for holocellulose, $RPD_{cv} = 3.2$, SECV = 2.07% and $R^2_{cv} = 0.90$. Kothiyal et al. (2015) showed multispecific models (*Eucalyptus*, *Dalbergia*, *Leucaena*, and *Populus*) with RMSEP from 0.64 to 1.05% and an r^2 around 0.90

Hemicellulose content and individual sugars

Yao et al. (2010) calibrated hemicelluloses in acacia and observed SEP = 0.55% and an $r^2 = 0.95$. For *Acacia mangium*, Karlinasari et al. (2014) validated a model with an SEP = 1.9, $r^2 = 0.82$ and RPD = 2.38. Chaix et al. (2012), calibrated hemicellulose levels for various eucalyptus species and cross-validation, obtained an $RPD_{cv} = 2.4$, SECV = 1.61% and $R^2_{cv} = 0.82$. All of this work tends to show that one can develop a calibration for hemicelluloses that can be used in the classification of samples.

As part of the *Tree4Joules* project (http://tfj.lrsv.ups-tlse.fr), the first sugar calibration results for eucalyptus and poplar showed the effectiveness of NIR spectroscopy for measuring arabinose, galactose, glucose and xylose.

A recent study by the INRA Orleans team on more than 6,000 samples has highlighted the contribution of NIR spectroscopy for the genetic study of European populations of black poplar to qualify wood quality criteria (lignins, lignin monomers, cellulose, hydrolysed and saccharifiable sugars, and extractives) related to its use for bioenergy (Gebreselassie et al. 2017).

Extractives wood compounds

Extractives content

The methods of analysis used to quantify extractives are complex and destructive. In the early 2000s, Kelley et al. (2004) used NIR spectroscopy to predict the extractive content of *Pinus taeda* solid wood. They obtained coefficients of determination between measured and predicted values between 0.62 and 0.88. The work of Gierlinger et al. (2002) also demonstrated the effectiveness of the NIR spectroscopy technique in assessing the natural durability of larch by determining the level of extractives from solid and powdered heartwood. However, calibrations based on solid wood spectra are less predictive than those constructed from powdered wood spectra. This is probably due to the addition of variable factors such as wood structure, sample geometry or surface properties (Brunner et al. 1996, Tsuchikawa et al. 1998, Zhang et al. 2015). Several studies were then developed to predict the extractive content from wood pulp (Meder et al. 1999, Baillères et al. 2002, Schimleck et al. 2003, Zhari et al. 2008, Alves et al. 2012). The correlations obtained between the quantity of the extracts and the value predicted by the NIR spectroscopy were each time high. These results confirm the effectiveness of the NIR spectroscopy technique in predicting the content of extractives whether from solid wood or sawdust.

The measurement of extractives compounds by NIR spectroscopy has also proved its effectiveness in tropical timber. Ribeiro da Silva et al. (2013) have shown that it is possible to quantify the presence of phenols in mahogany wood with an accuracy of 10% and a minimum concentration of 2.4%. Niamke et al. (2014) also showed it on teak for the total extractives compounds but also on some extractable molecules linked in particular to the natural durability of teak wood.

For Scots pinewood they are stilbene type molecules that give duramen its natural durability. The INRA team in Orléans associated with Finnish researchers has recently shown the feasibility of the indirect measurement of pinosylvine and its monomethyl ether by NIR spectroscopy applied on wood cores with models of sufficient quality to be used for genetic selection (Pulkka et al. 2016).

In the particular case of oak, wood extractives can be up to 10% of the dry wood mass (Scalbert et al. 1989). The work of Zahri et al. (2008) confirmed strong correlations ($R^2 = 0.93$) between chemical analyses of total oak phenols and NIR spectral data, highlighting the low percentage error in the prediction of total phenols (RMSEP = 0.54%). NIR spectroscopy is therefore an effective tool for predicting the amount of phenolic compounds in oakwood (RPD = 5.4). This prediction makes it possible to measure the amount of total phenols irrespective of the variability of the origin of the wood (provenance, species, tree). On solid wood, it can be seen that the prediction of the total phenol content from wood, whether at the axial or radial woody plane, is very satisfactory. From a practical point of view, these two measurement plans could be recommended for a routine implementation of the NIR spectroscopy technique to evaluate the phenol content in a fast, reliable and non-destructive way in industries related to oakwood.

Moreover, other studies have been conducted in France by several teams of cooperage Radoux, Ondalys, IRSTEA, INRA Loire Valley Orleans and CIRAD, and resulted in the establishment of the first European industrial tool for online measurement of the quantity of phenolic compounds in oakwood for the barrel production: *Oakscan®* process (Giordanengo et al. 2009). The approach and the tool are detailed in the next section of this chapter.

Evaluation of Cooperage Oakwood Quality by NIR Spectroscopy

Oak barrels were invented over 2000 years ago during the Gallo-Roman era. They quickly replaced amphora, which was heavier and more fragile. However, an oak barrel is not simply a container. It transmits certain compounds to the liquids it contains and modifies their sensorial properties. Nowadays, oak casks are not just used for the transport of beverages, but have become an essential tool in the production of wines and spirits. Indeed, "cask-matured" wines and spirits have their own complexity and originality.

Quality criteria for oak used in cooperage

In the 1990s, university research studies began to be interested in the use of oak casks in the making of wines and spirits (Chatonnet 1991, Masson 1996, Vivas 1997). Mainly, two physico-chemical phenomena were studied: the extraction of volatile and non-volatile compounds stemming from the oak and dissolution of the dioxygen from the air through the barrel.

The distinctive aroma resulting from oak barrel aging is defined by several descriptors, such as aromatic notes of "vanilla", "coconut", "spicy", "roasted" or "smoked". It is now considered that a hundred or so volatile compounds from wood have an effect on the bouquet of the wine (Boidron et al. 1988, Jarauta et al. 2005). Some compounds having a direct impact on the bouquet of the wine have been identified amongst these molecules: *cis* and trans β-methyl-γ-octalactone ("*boisé*/coconut" aroma), also named whisky lactones as their aroma is similar to that of a whisky, vanillin (aroma of "pastry vanilla") or eugenol and isoeugenol (clove aroma). It can also be remarked that notes of roasted and roasted coffee appear in the wine during aging. The wood's furaldehydes, in particular furfural, are extracted towards the wine and are at the origin of a very fragrant thiol compound, 2-furanemethanethiol (Tominaga et al. 2000).

Furthermore, non-volatile compounds are released during the aging process and contribute to the palate taste and structure of the wine. Amongst these compounds are ellagitannins which influence the astringency and bitterness of the wine (Glabasnia and Hofman 2006), and which are involved in many chemical reactions with wine compounds (Singleton 1987, Quideau et al. 2005, Chassaing et al. 2010). Other compounds, like some lignans, seem to have an impact on the bitterness of the wine. Recent works also highlight an increase in the sweetness of wines aged in barrels, which could be related to a family of compounds: the quercotriterpenosides (Marchal 2010).

Finally, a phenomenon of oxygenation of wines and spirits occurs during the barrel aging (Vivas and Glories 1996). Dissolved dioxygen is involved in the reduction of astringency and softening of red wines, and enables to avoid reduction aromas.

During manufacturing, three stages condition the sensorial typicity of a barrel: selection of the raw material, natural drying of the wood or maturation, and the heat treatment of the oak cask, or toast. Selecting the wood is primordial since it determines the initial chemical composition of the barrel. Natural drying in open yards with influence of weathering, or maturation, generally lasts 24 months and allows the wood to dry, lessens the ellagic tannins content and modifies the aromatic compounds content (Masson and Puech 2000). Traditionally, toasting was used to bend and assemble the casks. This step in manufacture has been widely studied, since thermal processing forms numerous aromatic compounds which give the wine the characteristic aromas of 'roasted', 'spicy' or 'smoked' (Chatonnet 1991). This toasting process is often considered to be the cooper's signature.

The chemical composition of oakwood is highly variable (Mosedale et al. 1996, Snakkers et al. 2000, Dousset et al. 2002, Guilley et al. 2004, Prida et al. 2006). Botanical species, geographic provenance, ecological conditions during growth, forestry, genetic factors or the position of wood in the log are so many parameters influencing this variability. The selection of the wood to be used in the production of the staves therefore plays an essential role in determining the quality of a barrel. For this reason, coopers place a great deal of importance on their supplies of oak.

Originally, geographic provenance was the first criterion for oak selection. Cooperage then turned to a macroscopic anatomical criterion, which was identifiable and which could, to some extent, be correlated with the composition of the wood: the width of annual growth or *grain*. Using grain as a criterion for selecting wood became commonplace within the profession. Nevertheless, wines aged in oak barrels from the same forests, or with the same grains, still present a significant sensorial heterogeneity. The best way to monitor the enological quality of oak wood is to analyze it chemically. However, such analyses are usually carried out on simulated extractions which are time-consuming and not readily compatible with wood selection on a large scale. Moreover, they are complex and costly to set up.

Then, we directed our studies towards the use of fast and non-destructive monitoring tools to assess the quality of oak wood. NIR spectroscopy was used to develop a specific sensor.

Development of NIR sensor to assess chemical quality of cooperage oakwood

In France, Tonnellerie Radoux developed a measurement process of oak polyphenols content, named Oakscan®, in partnership with INRA Val de Loire - Orléans, CEMAGREF, CIRAD and Ondalys Company (Giordanengo et al. 2009). It is the first time that NIR spectroscopy was introduced in the cooperage industry.

A collection of oak samples was prepared in order to build the NIR calibrations. Four hundred pieces of French oak were selected from the cooper's maturation stock with the aim of including the largest possible variation of wood composition. These samples were taken from 33 different batches of wood and were selected according to geographic provenance and their grain. First, the samples were characterized by chemical analysis, then measured by NIR spectroscopy.

Chemical analysis of the polyphenols was carried out in a laboratory by several 80% acetone extractions on wood powders. The amount of polyphenols present in the extracts was quantified by four methods:

- The extract rate is determined by measuring the loss of mass in the wood powders after extraction and drying.
- The optical density at 280 nm is measured in the final extract.
- The total phenols present in the extracts are dosed by a colorimetric method using the Folin-Ciocalteu reagent (Boizot and Charpentier 2006).
- The concentrations of eight oak ellagitannins in the extracts are determined by high performance liquid chromatography.

These reference analyses provide a rich information about the polyphenolic content of the oak samples and allow to build four NIR calibrations.

Figure 4 presents the correlations obtained for the four calibrations of the reference analyses characterizing the polyphenolic content of oakwood. These calibrations were built by Partial Least Squares regression, in controlled laboratory conditions. The NIR calibrations of optical density at 280 nm, total phenols measured by Folin-Ciocalteu and of ellagitannins are efficient (Table 1), and the NIR estimations are well correlated to the reference chemical analyses.

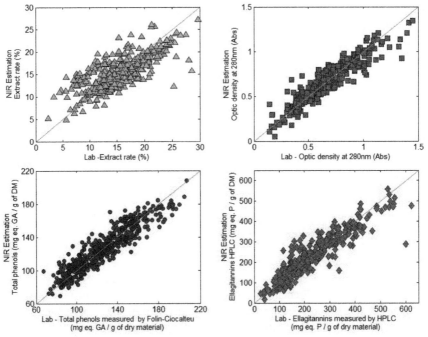

Fig. 4. Results of NIR calibrations for the extract rate, for the optical density at 280 nm, for the total phenols in the wood extracts measured by the Folin-Ciocalteu method and for the ellagitannins content measured by HPLC (cross validation).

Table 1. NIR calibrations results of the four reference chemical analyses of oakwood.

Analysis	Unit	Reference analyses			NIR Calibrations				
		Mean	Standard deviation	SEL	Rank	SECV	R^2_{CV}	bias $_{CV}$	SECV / SEL
Extract rate	%	15,4	4,9	2,1	6	3,3	0,56	-0,004	1,6
Optic density at 280nm	Absorbancy	0,64	0,24	0,05	5	0,09	0,85	-0,0008	1,9
Total phenols	mg equivalent gallic acid / g of dry material	122	24	6,9	4	10,7	0,86	-0,2	1,5
Ellagitannins	mg equivalent pyrogallol / g of dry material	223	116	21	10	49	0,82	-0,08	2,4

SEL: *Standard Error of Laboratory*; Rank: number of PLS latent variables used in the model; SECV: *Standard Error of Cross Validation*; R^2_{cv}: determination coefficient obtained in cross validation; bias$_{cv}$: bias of the results obtained in cross validation.

The prediction models of optical density at 280 nm, of total phenols and of ellagitannins content obtained are enough efficient to consider an industrial application aimed at measuring the polyphenols content of oak wood by NIR spectroscopy.

Industrialization of Oakscan® process

At the cooperage workshop, an NIR sensor was designed to function on the stave machining line. Adaptation on the line proved to be a difficult task, particularly to succeed in making the measurement reliable in the face of disrupting parameters such as temperature, variations in working conditions or the moisture content of the wood. Various methods were applied in order to obtain a robust spectral acquisition, by optimizing the physical conditions of the measurement and by processing the spectral information data with chemometric methods (Roger et al. 2003, Giordanengo et al. 2008).

In practice, the measurement process is used to create different selections of wood. The body of a barrel is composed of about 30 oak staves. Each machined stave is scanned in under one second. The three characteristics of polyphenolic content are estimated from the NIR calibrations, then an *Index of Polyphenols* between 0 and 100 is calculated and attributed to the piece. Figure 5 represents a control chart of the staves analyzed by Oakscan® process during one morning's production, and illustrates the variability of oak polyphenol content.

The oak staves are sorted into three classes with this *Index of Polyphenols* and then marked (Plate 1). Sorting wood according to this chemical criterion allow to improve the homogeneity of the raw material within each selection and thus favorize the reproducibility of wine aging in barrel over the years. Furthermore, it is possible to adapt the tannic potential of oak wood to the specifications of aging needed for each wine.

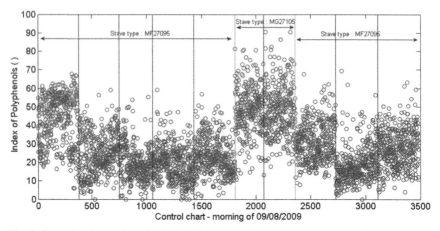

Fig. 5. Example of an Index of Polyphenols control chart, measurements realized on ten pallets of staves. Each point represents the Index of Polyphenols of one scanned stave. The vertical lines represent a change of pallet. Two types of staves are produced: MF27095–Fine Grain French oak stave for the production of 225 Li barrels and MG27105–Large Grain French oak stave for the production of 400 Li barrels.

Plate 1. Stock of Oakscan® sorted staves and barrel during the toasting process.

The first comparative trials on wine began with the 2006 vintage. They were carried out in several appellations of the French regions of Bordeaux, Burgundy, Rhône Valley and Languedoc-Roussillon, as well as in Spain, Chile and California. The influence of the oak selection process has thus been studied for the production of red and white wines of different specifications from these regions and from different grape varieties (Merlot, Cabernet Sauvignon, Pinot Noir, Syrah, Grenache, Sauvignon Blanc, etc.). These trials have confirmed the importance of selecting oak wood according to its polyphenol content, and its value in order to monitor oak impact on wines and spirits.

In order to study deeply the interactions between oak and wine, this novel criterion for wood selection was studied during a Ph.D. thesis (Institut des Sciences de la Vigne et du Vin, Bordeaux, Michel 2012). The conducted research validated the correlation between the richness in tannins of the oak selections and the concentration of ellagitannins released in wine, as well as the influence of Oakscan® selections on sensorial descriptors such as roundness, fullness, bitterness or astringency.

Following on this experience, the technology has been applied to other activities in the industry. A similar process has been applied on stave milling, when the oak logs are first transformed into "merrain" (rough barrel stave). An NIR sensor allows the measurement of the level of tannins of oak wood before the natural drying, and thus characterize the distinctive types of stocked wood. Furthermore, the process has also been introduced into the manufacture of *Oak for Enology* (Giordanengo et al. 2012). Oak chips or tank staves provide an economical solution for producing oaked wines. New selections of *Oak for Enology* that have been sorted according to their richness in polyphenols exist since 2012.

Conclusions and perspectives

The entire production of French oak barrels manufactured by Tonnellerie Radoux has been measured and sorted according to the NIR spectroscopy process Oakscan®, since the summer of 2009. This technology has been introduced into the first transformation of wood, when oak logs are transformed into *merrain*, as well as in production of *Oak for Enology*. NIR analysis of the richness in polyphenols of oakwood ensures a better monitoring of oak tannic impact than that offered by traditional selection methods. This new sorting criterion allows to adapt the oak to the desired enological objectives and also contributes to improving the reproducibility of wine aging in the barrel. Furthermore, Tonnellerie Radoux constitutes a large data base of analyses, thus leading to improved knowledge about the oak-supplying regions.

General Conclusion

Research on NIR spectroscopy technology since the 1980s has demonstrated the effectiveness of this method in the non-destructive testing of wood. Today, multiple parameters related to the chemical composition but also to the anatomical organization of wood as well as certain mechanical properties can be characterized using NIR spectroscopy methods. Following the example of the first industrial use of NIR spectroscopy in a company in the wood industry with Oakscan® process on oakwood of the French cooperage Radoux, other systems could have a significant benefit to this sector. Thus, the feasibility of measuring the natural or conferred durability by NIR spectroscopy has been possible too. The technique can be used for timber grading, identification of species or origin and quality control in processing units. Research shows that these results can be used for both temperate and tropical wood species.

The development of rapid, non-destructive and cost-effective tools such as NIR spectroscopy allowed the assessment of large populations needed to provide genetic parameter estimates such as chemical properties of wood, in a short time at lower costs (Perez et al. 2007, Gaspar et al. 2011). Today, NIR spectroscopy method can be used in genetic improvement programs and in genetic studies of forest species thanks to its capacity to phenotype a large number of samples in a short time (Lepoittevin et al. 2011, Gebresselassie et al. 2017). Thus, as for growth, genetic control of wood chemicals properties can be estimated (Raymond and Schimleck 2002, Hein et al. 2012) and when this is proved, certain traits can be included as the selection criteria in the breeding programs (Estopa et al. 2017). It is one of the few technologies that has made possible the production of selected varieties on the characteristics of wood interests, the main objective of production forests (Resende et al. 2012). With the increasing improvement in the quality of portable equipment of a smaller and smaller size, applications with spectra taken directly in the field on standing trees are conceivable and would bring an additional evolution towards high-throughput phenotyping for this type of genetic research. Similarly, NIR spectroscopy is used for the mass phenotyping of samples used in quantitative genetic studies (Raymond and Schimleck 2002, Hein et al. 2012), association genetics (Wegrzyn et al. 2010, Gebresselassie et al. 2017), genetic mapping (Gion et al. 2015) and genomic studies for various forest species (Denis et al. 2013). Studies conducted by the forest tree genetic team at the INRA VDL in Orleans (France) and other teams (Posada et al. 2009, O'Reilly-Wapstra et al. 2013) show the opportunity to use the NIR spectrum as a variable (biological marker) and usable as such to explore genetic diversity in the selection programs, it is called "phenome". Other authors showed the heritability map of NIR spectrum wavelength (Hein and Chaix 2014).

In view of all the work carried out since the late 1980s, NIR spectroscopy is therefore a non-destructive testing technology, which has shown its interest in wood analysis. With the evolution of technologies and in particular measuring devices and signal processing capacities, it should increase its level of efficiency in the years to come and thus be more present in the production chains of the companies in the wood industry.

References

Alves, A., M. Schwanninger, H. Pereira and J. Rodrigues. 2006. Calibration of NIR to assess lignin composition (H/G ratio) in maritime pinewood using analytical pyrolysis as the reference method. Holzforschung 60: 29–31.

Alves, A., R.F.S. Simões, J.S. Desmond, R.E. Vaillancourt, B.M. Potts, M. Schwanninger and J. Rodrigues. 2011. Determination of syringyl/guaiacyl ratio of *Eucalyptus globulus* wood lignin by near infrared-based partial least squares regression models using analytical pyrolysis as the reference method. Journal of Near Infrared Spectroscopy 19: 343–348.

Alves, A., A. Santos, P. Rozenberg, L.E. Pâques, J.P. Charpentier, M. Schwanninger and J.C. Rodrigues. 2012. A common near infrared based partial least squares regression model for the prediction of wood density of *Pinus pinaster* and *Larix* × *eurolepis*. Wood Science & Technology 46(13): 157–175.

Alves, A., A. Santos, R.F.S. Simões, C.A. Santos, B.M. Potts, J. Rodrigues and M. Schwanninger. 2012. Determination of *Eucalyptus globulus* wood extractives content by near infrared-based partial least squares regression models: comparison between extraction procedures. Journal of Near Infrared Spectroscopy 20(2): 275–285.

Baillères, H., F. Davrieux and F. Ham-Pichavant. 2002. Near infrared analysis as a tool for rapid screening of some major wood characteristics in a eucalyptus breeding program. Annals of Forest Sciences 59: 479–490.

Bertrand, D. 2002. La spectroscopie proche infra-rouge et ses applications dans les industries de l'alimentation animale. INRA Productions Animales 15(3): 209–220.

Bertrand, D. and E. Dufour. 2006. La spectroscopie infrarouge et ses applications analytiques. Collection Sciences et Techniques Agroalimentaires. Paris, FRA: Lavoisier, 660 p.

Boidron, J.N., P. Chatonnet and M. Pons. 1988. Influence du bois sur certaines substances odorantes des vins. Connaissance de la Vigne et du Vin 22: 275–294.

Boizot, N. and J.-P. Charpentier. 2006. Méthode rapide d'évaluation du contenu en composes phénoliques des organes d'un arbre forestier. Le Cahier des Techniques de l'INRA, n° spécial 2006, Méthodes et outils pour l'observation et l'évaluation des milieux forestiers, prairiaux et aquatiques, pp. 79–82.

Brinkmann, K., L. Blaschke and A. Polle. 2002. Comparison of different methods for lignin determination as a basis for calibration of near-infrared reflectance spectroscopy and implications of lignoproteins. Journal of Chemical Ecology 28(12): 2483–2501.

Brunner, M., E. Eugster, E. Trenka and L. Bergamin-strotz. 1996. FT-NIR spectroscopy and wood identification. Holzforschung 50(2): 130–13.

Chaix G., F.M. Tomazello, S. Nourissier and A. Diombokho. 2012. Global Nirs models to predict main chemical compounds of eucalyptus woods. *In*: Forests and forest products – International Union Of Forest Research Organizations (IUFRO) Division 5, Estoril, Portugal, 8–13 July 2012.

Chassaing, S., D. Lefeuvre, R. Jacquet, M. Jourdes, L. Ducasse, S. Galland, A. Grelard, C. Saucier, P.-L. Teissedre, O. Dangles and S. Quideau. 2010. Physicochemical studies of new anthocyano-ellagitannin hybrid pigments: about the origin of the influence of oak C-glycosidic ellagitannins on wine color. European Journal of Organic Chemistry 1: 55–63.

Chatonnet, P. 1991. Incidence du bois de chêne sur la composition chimique et les qualités organoleptiques des vins. Thèse DER. Université de Bordeaux II.

Chauchard, F., J.-M. Roger and V. Bellon-Maurel. 2004. Correction of the temperature effect on near infrared calibration—application to soluble solid content prediction. Journal of Near Infrared Spectroscopy 12-3: 199–205.

Clarke, C.R.E. and A.M. Wessels. 1995. Variation and measurement of pulp properties in eucalypts. pp. 93–100. *In*: Potts, B.M. and N.M.G. Vienna (eds.). Eucalypt Plantations: Improving Fibre Yield and Quality. Proceedings of the CRCTHF-IUFRO Conference, Feb. 1995, Hobart, Australia.

Denis, M., B. Favreau, S. Ueno, L. Camus-Kulandaivelu, G. Chaix, J.-M. Gion, S. Nourrisier-Montou, J. Polidori and J.-M. Bouvet. 2013. Genetic variation of wood chemical traits and association with underlying genes in *Eucalyptus urophylla*. Tree Genetics & Genomes 9: 927–942.

Dousset, F., B. De Jeso, S. Quideau and P. Pardon. 2002. Extractives content in cooperage oak wood during natural seasoning and toasting; influence of tree species, geographic location, and single-tree effects. Journal of Agricultural and Food Chemistry 50: 5955–5961.

Downes, G.M., R. Meder, N. Ebdon, H. Bond, R. Evans, K. Joyce and S.G. Southerton. 2010. Radial variation in cellulose content and Kraft pulp yield in *Eucalyptus nitens* using near infrared spectral analysis of air-dry wood surfaces. Journal of Near Infrared Spectroscopy 18(2): 147–155.

Estopa, R.A., F.R. Milagres, R.A. Oliveira and P.R.G. Hein. 2017. NIR spectroscopic models for phenotyping wood traits in breeding programs of *Eucalyptus benthamii*. Cerne 22: 367–375. https://doi.org/10.1590/0104776020172303231 9.

Gaspar, M.J., A. Alves, J.L. Louzada, Morais, A. Santos, C. Fernandes, M.H. Almeida and J.C. Rodrigues. 2011. Genetic variation of chemical and mechanical traits of maritime pine (*Pinus pinaster* Aiton). Correlations with wood density components. Annals of Forest Science 68: 255–265.

Gebreselassie, M.N., K. Ader, N. Boizot, F. Millier, J.P. Charpentier, A. Alves, R. Simões, J.C. Rodrigues, G. Bodineaud, F. Fabbrinie, M. Sabattie, C. Bastien and V. Segura. 2017. Near-infrared spectroscopy enables the genetic analysis of chemical properties in a large set of wood samples from *Populus nigra* (L.) natural populations. Industrial Crops & Products 107: 159–171.

Geladi, P. 1995. An Overview of Multivariate Spectral Data Analysis Spectral Data Analysis in Infra Red Spectroscopy: The Future Waves, edited by Davies A.M.C and Willamp. NIR Publications, Chichester, UK.

Gierlinger, N., M. Schwanninger, B. Hinterstoisser and R. Wimmer. 2002. Rapid determination of heartwood extractives in *Larix* sp. by means of Fourier transform near infrared spectroscopy. Journal of Near Infrared Spectroscopy 10: 203–214.

Gion, J.M., P. Chaumeil and C. Plomion. 2015. EucaMaps: linking genetic maps and associated QTLs to the *Eucalyptus grandis* genome. Tree Genetics & Genomes 11: 1–7.

Giordanengo, T., J.-P. Charpentier, J.-M. Roger, S. Roussel, L. Brancheriau, G. Chaix and H. Baillères. 2008. Correction of moisture effects on near infrared calibration for the analysis of phenol content in eucalyptus wood extracts. Annals of Forest Science 65(8): 803.

Giordanengo, T., J.-P. Charpentier, N. Boizot, S. Roussel, J.-M. Roger, G. Chaix, C. Robin and N. Mourey. 2009. *OAKSCAN™*: Procédé de mesure rapide et non destructif des polyphénols du bois de chêne de tonnellerie. Revue Française d'Œnologie 234: 10–15.

Giordanengo, T., J. Michel, P. Gauthier, J.-P. Charpentier, M. Jourdes, P.-L.Teissedre and N. Mourey. 2012. Application du procédé *OakScan®* aux Bois pour l'Œnologie : Sélection et influence de la teneur en polyphénols du chêne sur le profil aromatique et structurant du vin. Revue Française d'Œnologie 255 : 16–24.

Glasbania, A. and T. Hofmann. 2006. Sensory-directed identification of taste-active ellagitannins in ethanol/water extracts of American white oak wood (*Quercus alba* L.) and European oakwood (*Quercus robur* L.) and quantitative analysis in bourbon whiskey and oak matured red wines. Journal of Agricultural and Food Chemistry 54: 3380–3390.

Guilley, E., J.P. Charpentier, N. Ayadi, G. Snakkers, G. Nepveu and B. Charrier. 2004. Decay resistance against *Coriolus versicolor* in Sessile oak (*Quercus petraea* Liebl.): analysis of the between-tree variability and correlations with extractives, tree growth and other basic wood properties. Wood Science and Technology 38: 539–554.

He, W. and H. Hu. 2013. Rapid prediction of different wood species extractives and lignin content using near infrared spectroscopy. Journal of Wood Chemistry and Technology 33(1): 52–64. DOI: 10.1080/02773813.2012.731463.

Hein, P.R.G., J.M. Bouvet, E. Mandrou, B. Clair, P. Vigneron and G. Chaix. 2012. Age trends of microfibril angle inheritance and their genetic and environmental correlations with growth, density and chemical properties in *Eucalyptus urophylla* S.T. Blakewood. Annals of Forest Science 69: 681–691.

Hein, P.R.G. and G. Chaix. 2014. NIR spectral heritability: a promising tool for wood breeders? Journal of Near Infrared Spectroscopy 22: 141–147.

Hodge, G.R. and W.C. Woodbridge. 2004. Use of near infrared spectroscopy to predict lignin content in tropical and sub-tropical pines. Journal of Near Infrared Spectroscopy 12: 381–390.

Jarauta I., J. Cacho and V. Ferreira. 2005. Concurrent phenomena contributing to the formation of the aroma of wine during aging in oak wood: an analytical study. Journal of Agricultural and Food Chemistry 53: 4166–4177.

Karlinasari, L., M. Sabed, I.N.J. Wistara and Y.A. Purwanto. 2014. Near infrared (NIR) spectroscopy for estimating the chemical composition of (*Acacia mangium* Willd.) wood. J. Indian Acad. Wood Sci. 11(2): 162–167.

Kelley, S.S., T.G. Rials, L.R. Groom and C.L. So. 2004. Use of near infrared spectroscopy to predict the mechanical properties of six softwoods. Holzforschung 58: 252–260.

Kothiyal, V., Jaideep, S. Bhandari, H.S. Ginwal and S. Gupta. 2015. Multi-species NIR calibration for estimating holocellulose in plantation timber. Wood Sci. Technol. 49: 769–793.

Kube, P.D. and C.A. Raymond. 2002. Prediction of whole tree basic density and pulp yield using wood core samples in *Eucalyptus nitens*. APPITA J. 55: 43–48.

Lepoittevin, C., J.P. Rousseau, A. Guillemin, C. Gauvrit, F. Besson, F. Hubert, D. da Silva Perez, L. Harvengt and C. Plomion. 2011. Genetic parameters of growth straightness and wood chemistry traits in *Pinus pinaster*. Ann. For. Sci. 68: 873–884. https://doi.org/10.1007/s13595-011-0084-0.

Lindgren, T. and U. Edlund. 1998. Prediction of lignin content and pulp yield from black-liquor composition using near infrared spectroscopy and partial least squares. Nord. Pulp Pap. Res. J. 13: 76–80.

Marchal, A. 2010. Recherches sur les bases moléculaires de la saveur sucrée des vins secs, Approches analytique et sensorielle. Thèse DER. Université de Bordeaux II, 300 p.

Masson, G. 1996. Composition chimique du bois de chêne de tonnellerie. Essai de discrimination des espèces et des forêts. Thèse DER. Université de Montpellier II, 150 p.

Masson, E. and J.L. Puech. 2000. Les méthodes de séchage du bois: 1-influence du séchage en étuve sur les teneurs en ellagitannins et en composés volatils de merrains de chêne sessile (*Quercus petraea* (Matt.) Liebl.) Vᵉ Colloque des Sciences et Techniques de la Tonnellerie, Bordeaux, France.

Meder, R., S. Gallagher, K.L. Mackie, H. Bohler and R.R. Meglen. 1999. Rapid determination of the chemical composition and density of *Pinus radiata* by PLS modelling of transmission and diffuse reflectance FTIR spectra. Holzforschung 53: 261–266.

Michel, J. 2012. Classification et influences des polyphénols du bois de chêne sur la qualité sensorielle des vins (Application du procédé *Oakscan®*). Thèse DER. Université de Bordeaux II, 226 p.

Monties, B. 1989. Molecular structure and biochemical properties of lignins in relation to possible self-organization of lignin networks. Annales des sciences forestières. INRA/EDP Sciences 46 (Suppl), pp. 848s–855s.

Mosedale, J., B. Charrier, N. Crouch, G. Janin and P. Savill. 1996. Variation in the composition and content of ellagitannins in the heartwood of European oaks (*Quercus robur* and *Quercus petraea*). A comparison of two French forests and variation with heartwood age. Annals of Forest Science 53: 1005–1018.

Niamké, F.B., N. Amusant, A.A. Kadio, M.F. Thevenon, S. Nourissier, A.A. Adima, C. Jay-Allemand and G. Chaix. 2014. Rapid prediction of phenolic compounds as chemical markers for the natural durability of teak (*Tectona grandis* Linn f.) heartwood by near infrared spectroscopy. Journal of Near Infrared Spectroscopy 22(1): 35–43.

O'Reilly-Wapstra, J.M., J.S. Freeman, R. Barbour, R.E. Vaillancourt and B.M. Potts. 2013. Genetic analysis of the near-infrared spectral phenome of a global Eucalyptus species. Tree Genetics & Genomes 9: 943–959.

Osborne, B.G., T. Fearn and P.T. Hindle. 1993. Practical NIR spectroscopy with applications in food and beverage analysis. (2nd Ed.). Addison-Wesley Longman Ltd: Harlow UK.

Perez, D.D., A. Guillemain, P. Alazard, C. Plomion, P. Rozenberg, J.C. Rodrigues, A. Alves and G. Chantre. 2007. Improvement of Pinus pinaster Ait. elite trees selection by combining near infrared spectroscopy and genetic tools. Holzforschung 61: 611–622.

Posada, H., M. Ferrand, F. Davrieux, P. Lashermes and B. Bertrand. 2009. Stability across environments of the coffee variety near infrared spectral signature. Heredity 102: 113–119.

Prida, A., J.C. Boulet, A. Ducousso, G. Nepveu and J.L. Puech. 2006. Effect of species and ecological conitions on ellagitannin content in oak wood from an even-aged and mixed stand of *Quercus robur* L. and *Quercus petraea* Liebl. Annals of Forest Science 63: 415–424.

Pulkka, S., V. Segura, A. Harju, T. Tapanila, J. Tanner, L. Pâques and J.P. Charpentier. 2016. Prediction of stilbene content from heartwood increment cores of scots pine using NIRS methodology. Journal of Near Infrared Spectroscopy 24: 517 528.

Quideau, S., M. Jourdes, D. Lefeuvre, D. Montaudon, C. Saucier, Y. Glories, P. Pardon and P. Pourquier. 2005. The chemistry of wine polyphenolic C-glycosidic ellagitannins targeting human topoisomerase *II*. Chemistry—A European Journal 11: 6503–6513.

Raymond, C.A. and L.R. Schimleck. 2002. Development of near infrared reflectance analysis calibrations for estimating genetic parameters for cellulose content in *Eucalyptus globulus*. Can. J. For. Res. 32: 170–176.

Resende, M.D.V., M.F.R. Resende Jr., C.P. Sansaloni, C.D. Petroli, A.A. Missiaggia, A.M. Aguiar, J.M. Abad, E.K. Takahashi, A.M. Rosado, D.A. Faria, G.J. Pappas Jr., A. Kilian and D. Grattapaglia. 2012. Genomic selection for growth and wood quality in Eucalyptus: capturing the missing heritability and accelerating breeding for complex traits in forest trees. New Phytologist 194: 116–128.

Ribeiro da Silva, A.R., T.C.M. Pastore, J.W.B. Braga, F. Davrieux, E.Y.A. Okino, V.T.R. Coradin, J.A.A. Camargos and A.G.S. Soares do Prado. 2013. Assessment of total phenols and extractives of mahogany wood by near infrared spectroscopy (NIRS). Holzforschung 67: 1–8.

Roger, J.-M., F. Chauchard and V. Bellon-Maurel. 2003. EPO-PLS external parameter orthogonalisation of PLS application to temperature-independent measurement of sugar content of intact fruits. Chemometrics and Intelligent Laboratory Systems 66: 191–204.

Scalbert, A., B. Monties and G. Janin. 1989. Tannins in wood: comparison of different estimation methods. Journal of Agriculture and Food Chemistry 37: 1324–1329.

Schimleck, L.R., C.A. Raymond, C.L. Beadle, G.M. Downes, P.D. Kube and J. French. 2000. Applications of NIR spectroscopy to forest research. Appita Journal 53: 458–464.

Schimleck, L.R. and Y. Yazaki. 2003. Analysis of *Pinus radiata* D. don bark by near infrared spectroscopy. Holzforschung 57: 520–526.

Schultz, T.P. and D.A. Burns. 1990. Rapid secondary analysis of lignicellulose: comparison of near infrared and fourier transform infrared (FTIR). Tappi J. 73(5): 209–212.

Schwanninger, M., J.C. Rodrigues and K. Fackler. 2011. A review of band assignments in near infrared spectra of wood and wood components. Journal of Near Infrared Spectroscopy 19(5): 287–308.

Schwanninger, M., Rodrigues, N. Gierlinger and B. Hinterstoisser. 2011a. Determination of lignin content in Norway spruce wood by Fourier Transformed near infrared spectroscopy and partial least square regression. Part 1: Wavenumber selection and evaluation of selected range. Journal of Near Infrared Spectroscopy 19: 319–329.

Schwanninger, M., Rodrigues, N. Gierlinger and B. Hinterstoisser. 2011b. Determination of lignin content in Norway spruce wood by Fourier Transformed near infrared spectroscopy and partial least square regression. Part 2: Development and evaluation of the final model. Journal of Near Infrared Spectroscopy 19: 331–341.

Singleton, V.L. 1987. Oxygen with phenols and related reactions in musts, wines and model systems. Observations and practical implications. American Journal of Enology and Viticulture 38: 69–77.

Snakkers, G., G. Nepveu, E. Guilley and R. Cantagrel. 2000. Variabilité géographique, sylvicole et individuelle de la teneur en extractibles de chênes sessiles français (*Quercus petraea* Liebl.): polyphénols, octalactones et phénols volatils. Annals of Forest Science 57: 251–260.

Tominaga, T., L. Blanchard, P. Darrier and D. Dubourdieu. 2000. A powerful aromatic volatile thiol, 2-Furanmethanethiol, exhibiting roast coffee aroma in wines made from several *Vitis vinifera* grape varieties. Journal of Agricultural and Food Chemistry 48: 1799–1802.

Tsuchikawa, S., M. Torii and S. Tsustumi. 1998. Directional characteristics of near infrared light in the process of radiation and transmission from wood. Journal of Near Infrared Spectroscopy 6: 47–53.

Tsuchikawa, S. 2007. A review of recent near infrared research for wood and paper. Applied Spectroscopy Review 42: 43–71.

Tsuchikawa, S. and M. Schwanninger. 2013. A review of recent near infrared research for wood and paper-Part 2. Applied Spectroscopy Review 48: 560–587.

Tsuchikawa, S. and H. Kobori. 2015. A review of recent application of near infrared spectroscopy to wood science and technology. J. Wood Sci. 61: 213–220.

Via, B.K., C. Zhou, G. Acquah, W. Jiang and L. Eckhardt. 2014. Near infrared spectroscopy calibration for wood chemistry: Which chemometric technique is best for prediction and interpretation? Sensors 14: 13532–13547; doi:10.3390/s140813532.

Vivas, N. and Y. Glories. 1996. Role of oak wood ellagitannins in the oxidation process of red wines during aging. American Journal of Enology and Viticulture 47(1): 103–107.

Vivas, N. 1997. Recherches sur la qualité du chêne français de tonnellerie (*Q. Petraea* Liebl., *Q. robur* L.) et sur les mécanismes d'oxydoréduction des vins rouges au cours de leur élevage en barriques. Thèse DER. Université de Bordeaux II.

Wegrzyn, J.L., A.J. Eckert, M. Choi, J.M. Lee, B.J. Stanton, R. Sykes, M.F. Davis, C.-J. Tsai and D.B. Neale. 2010. Association genetics of traits controlling lignin and cellulose biosynthesis in black cottonwood (*Populus trichocarpa*, Salicaceae) secondary xylem. New Phytologist 188: 515–532.

Williams, P. and D. Sobering. 1993. Comparison of commercial near infrared transmittance and reflectance instruments for analysis of whole grains and seeds. Journal of Near Infrared Spectroscopy 1: 25–32.

Williams, P. 2014. The RPD statistic: a tutorial note. NIR News 25: 22–26.

Williams, P., P. Dardenne and P. Flinn. 2017. Tutorial: Items to be included in a report on a near infrared spectroscopy project. Journal of Near Infrared Spectroscopy 25(2): 85–90. DOI: 10.1177/0967033517702395.

Wold, S. 1978. Cross-validatory estimation of the number of components in factor and principal component model. Technometrics 20: 397–405.

Wright, J., M. Birkett and M. Gambino. 1990. Prediction of pulp yield and cellulose content from wood using near infrared reflectance spectroscopy. Tappi J. 73: 164–166.

Yeh, T.F., H.M. Chang and J.F. Kadla. 2004. Rapid prediction of solid wood lignin content using transmittance near-infrared spectroscopy. Journal of Agricultural and Food Chemistry 52(6): 1435–1439.

Zahri, S., A. Moubarik, F. Charrier, G. Chaix, H. Baillères, G. Nepveu and B. Charrier. 2008. Quantitative assessment of total phenols content of European oak (*Quercus petraea* and *Quercus robur*) by diffuse reflectance NIR spectroscopy on solid wood surfaces. Holzforschung 62(6): 679–687.

Zhang, M., Y. Liu and Z. Yang. 2015. Correlation of near infrared spectroscopy measurements with the surface roughness of wood. BioResources 10: 6953.

Zeaiter, M. 2004. Mesures robustes en ligne des solutés organiques par spectrométrie infrarouge et étalonnages multivariés. Ph.D. thesis, Montpellier 2 University, 158p.

Part II

Innovative Products of Chemical and Thermochemical Treatment and Conversion of Lignocellulosics

Energetic Valorization of Lignocellulosic and Industrial Wastes by Thermal Gasification

Paulo Brito,* Octávio Alves, Luís Calado, Bruno Garcia and
Eliseu Monteiro

Introduction

Currently, humanity is confronting two major environmental problems that demand scientific and innovative solutions: the rise in the average temperature of the planet and an increasing production of residues by the society. The first problem has resulted due to the rise of carbon dioxide emissions in the last 200 years due to the massive use of fossil fuels, and therefore it is imperative to develop new sources and ways for energy and fuel production that can be sustainable and simultaneously with neutral emissions of carbon dioxide. The second problem is the increasing production of residues and effluents resulting from a growing human population and from the way that consumer goods are commercialized. As an example, the average generation of solid residues in the EU is around 5 ton/year per capita. In developed countries, this value is greater and has a higher growth rate. Any of the previous problems require a persistent work in scientific research focused on the implementation of innovative solutions that can be sustainable for the environment, economy and society. Thus, the present work aims to valorize biomass resources with special attention to biomass residues in order to contribute to the development

Polytechnic Institute of Portalegre, Campus Politécnico, 10, 7300-555 Portalegre, Portugal.
* Corresponding author: pbrito@estgp.pt

of new scientific and applied approaches that are able to solve both problems in a simultaneous manner, i.e., to create alternative fuels through the valorization of residues. It is intended to evaluate the feasibility of energetic valorization of forest and industrial residues based on thermal co-gasification.

Biomass was one of the first sources of energy to be discovered and used by mankind. The use of coal and oil, sources that are based on the transformation of biomass over a very long time as a result of pressure, temperature and an absence of oxygen, resulting in compounds rich in hydrocarbon compounds with no oxygen that give them a high calorific value, leading to a gradual decline in the use of biomass and biofuels. However, during the latter 200 years humanity has used this resource in an intensive way, releasing to the atmosphere all this carbon accumulated during millions of years, changing the normal cycle of that element; at the present stage, the quantity of carbon dioxide in the atmosphere has doubled. As this gas contributes to the greenhouse effect, it is increasing the planet's temperature with serious consequences for the environmental equilibrium. The solution to this problem passes through a progressive diminishment of the use of fossil fuels and through the introduction of alternative fuels, namely those derived from biomass.

There are basically three types of biomass materials, which are classified according to their origin. Energy crops, which are grown primarily for the production of energy. Their function is to capture solar radiation and store it in the biomass. Examples include rapeseed (*Brassica napus*), the common sunflower (*Helianthus annuus*), Chinese silver grass or miscanthus (*Miscanthus sinensis*) and corn (*Zea mays*). Another type of biomass consists of agricultural and forest residues that are generated when grains are harvested, trees pruned and cut down, such as, for example, straw, tips from the pruning of vineyards and olive groves, waste materials resulting from the pruning, cleaning and lopping of forests. These waste materials are particularly suited for use as sources of energy for the farms themselves, in order to increase the yield of growing chains. Finally, we have the by-products and organic waste that is generated in the processing of biomass for the development of food products, from which energy can be recovered. These by-products include organic waste from the agri-food sector, waste from farming and cattle-raising and waste from the industrial processing of wood and plant fibres; organic waste includes a portion of household waste and sludge from domestic and industrial effluents, such as, for example, food production waste.

With the need to change our paradigm in energy terms with a view to reducing our dependence on oil and seeking environmental and economic sustainability, biomass emerges as a good alternative contribution to the energy mix. Currently, this source of energy occupies fourth place globally as an energy resource, supplying about 14% of the total energy required on the planet. Despite the fact that International Energy Agency estimates that the weight of bio-energy in the global energy paradigm will rise to 26% by 2020, the truth is that there is growing concern regarding the sustainability of its use, since the issue of energy needs for a growing world population come hand in hand with the availability of food for that same population.

Taking into account the availability of existing land around the globe and the fact that the supply of food to people in sufficient quantity is considered a priority, it can easily be seen that the amount of land available for biomass cultivation is limited and will become increasingly limited the greater the rate of growth in world population. Thus, alternative sources of biomass must be envisaged that do not involve the cultivation of raw materials that are also food crops or that are produced in rich soils which can be used for food production. Second generation biofuels, whose raw materials are non-food substances or agro-industrial and forestry waste, have been prioritized for development.

In Portugal, Directive 2008/0016/COD set a binding target that 20% of the energy consumption of member states should come from renewable energy sources, and also another binding target that, by 2020, 10% of the energy used in the transport sector should come from renewable sources. From that year, there are targets stating that 31% of final consumption of primary energy, 60% of electricity produced and 10% of energy consumed in road transport should come from renewable sources. In order to honour those commitments, some steps are necessary to estimate the energy potential of the most significant renewable energy sources and promote economic analysis of its application.

This work presents the scope, potential and technologies related to the use of biomass resources with a focus on thermal gasification of industrial, agricultural and forest waste. The section Biomass Residues gives an overview and a characterization of some waste biomass resources available with potential for energetic valorization. The section Thermal Gasification presents the thermal gasification technology. In section Wastes Gasification Results, some thermal gasification results obtained with pilot industrial plants are presented. Economic Considerations gives an economic analysis of a thermal gasification case study application.

Biomass Residues

Solid residues with energetic valorization can be divided into three basic classes: domestic residues (sewage sludges from wastewater treatment plants and municipal solid wastes), forest and agriculture residues, and industrial residues.

Domestic residues

Residual wastes originated by human activity represent a problem of public concern caused by their toxicity and by the higher amounts generated. Streams of sewage sludge (SS) originated in wastewater treatment plants and municipal solid waste (MSW) achieved quantities of around 9.9×10^6 and 245.1×10^6 ton during 2012 in EU, and are a significant source of pathogens, heavy metals and organic pollutants that must be treated properly (Han 2016, Manara 2012). To deal with these problems in a sustainable manner, the EU directive 2008/98/EC set a hierarchy of methodologies for managing the waste which include prevention in production, re-use, energy recovery and disposal, by this order of priority (Lombardi 2015). In the case of MSW, there is a fraction of refuse that remain after the recovery of materials that can be recycled with higher proportions of organic matter (e.g., paper, wood and plastic) that contribute for an appreciable calorific value; due to these characteristics, such fractions are commonly called refuse-derived fuels (RDF) (Di Lonardo 2016).

Essentially, RDF's come from solid and liquid residues with a higher calorific value and which are produced in several industrial processes (e.g., hydrocarbon residues and solvents). They can be used in dedicated energy valorization plants or in industrial plants as a replacement for the conventional fossil fuels. Residues with good combustible properties may follow three pathways: (a) usage in combustion processes immediately after the production process that generated them; (b) submission to a pre-treatment for improvement of their oxidizing properties and separation of fractions for material valorization; and (c) usage as secondary feedstocks in the generation of new fuels with superior qualities. Fuels originated from solid residues may include biomass, meat meal, wood wastes, fragments of tires and rubber wastes, chars (obtained from pyrolysis), domestic and industrial sludges after dehydration, paper and plastic, and textiles.

As an example, waste tires constitute a solid residue that became a great environmental challenge around the world. Usually, this residue is set to landfills, and during 2015 its production in Portugal was about 18×10^6 tonnes. Due to these larger productions, the possibility of implementation of new techniques based on thermal conversions like combustion and gasification have gained a strong attention by the society for the valorization of non-recyclable materials. In addition, recent scientific studies have proposed a new technology named hydrothermal gasification where residues are converted to a combustible gas in an aqueous medium and under high temperatures (350–750°C) and pressures (> 22.1 MPa). This process is especially appropriate for very wet residues like sludges where the conventional gasification and incineration may become inefficient (Basu 2013).

Conventional processes adopted in the treatment of RDF include application as agriculture fertilizers, incineration/combustion with energy recovery and landfilling. Incineration is viable for reduction of mass and volume of wastes enabling them to be landfilled with a lesser occupation of space, and also destroys the pathogens and makes available the heat produced for energy generation. However, problems related with the toxicity of ashes, the unviability of implementation of small-scale plants (e.g., for rural areas) and the higher costs of flue gas decontamination due to the emissions of toxic compounds (e.g., fly ash with heavy metals, NO, SO_x and HCl) have limited the use of this solution (Fytili 2008, Furness 2000, Wu 2014). Deposition in landfills as the last method in the hierarchy of waste management has created several issues to the environment due to the occupation of large areas and to the contamination of underground waters and food chains created by the polluted leachates over time (Han 2016, Di Lonardo 2016). The EU legislation, such as directive 1999/31/EC, has restricted the forwarding of wastes to the landfills and increased the costs associated with this operation, which forced waste operators to find new alternatives for the management of their wastes (Manara 2012).

Since RDF in its dry form possesses significant amounts of carbon that attribute to them good calorific values ranging from 14.8 to 21.4 MJ/kg (for comparison, coal has typical values among 14.6 and 26.7 MJ/kg), these wastes may be submitted in a gasification process to produce a synthesis gas (syngas) for combustion or to create new biofuels through, for instance, the Fischer-Tropsch method (Manara 2012, Gallardo 2014). This is also true for forest and agricultural wastes, like pine sawdust, rice husk, almond shells or olive kernels, which also produce a syngas with calorific values among 4–16 MJ/m^3 (Alauddin 2010).

In many occasions, it is imperative to implement pre-treatments such as grinding, drying and torrefaction with the aim of obtaining adequate particle sizes, good mechanic and hydrophobic properties, and moisture contents less than 7%, therefore preparing the feedstocks for an efficient gasification (Lee 2014, Park 2015). Some studies already stated the possibility of reducing the environmental impact and costs associated with gasification when compared with some of the conventional solutions currently adopted, such as incineration or landfilling (Lumley 2014, Di Gianfilippo 2016).

Some gasification plants in operation have processing capacities between 100 and 420 ton/day and are able to produce at most 10 MWe of electric power (Arena 2012, Judex 2012). Facilities directed to treat vegetable biomass residues may achieve typical powers ranging from 0.1 and 2.5 MW, while coal gasification plants continue to exhibit a much greater potential to produce energy in the range of 250–300 MW (Sansaniwal 2017).

Composition of syngas obtained from gasification includes a range of pollutants such as solid particles, tars, polycyclic aromatic hydrocarbons (PAH's), heavy metals, H_2S, NH_3 and HCl, that are toxic for the environment and health and may cause problems of fouling or corrosion in the equipment (Woolcock 2013). In spite of inhibiting the formation of components like dioxins and furans due to the lower levels of oxygen present in the reactor, a gasification process requires the implementation of strategies with a view of reducing the remaining pollutants during the conversion or the treatment of syngas, in order to ensure the compliance with legislation related to atmospheric emissions (e.g., EU directive 2000/76/EC) and to extend the service life of equipment, especially when turbine machines are used to produce electric or thermal energy (Arena 2012).

Forest and agriculture residues

Regarding forest residues generated from clean-up operations, their valorization is highly relevant in Southern European countries like Portugal, which exhibit lower humidity levels and high temperatures during Summer that promote the occurrence of fires inside green zones.

The National Directorate of Forestry indicates that, in Portugal, there is a fuel potential available in forest residues of approximately 3.6 Mton of dry matter/year, which corresponds to around 1.6 million ton of oil equivalent. By comparison, Greece has 1.2 Mton/year, England, 1.7 Mton/year, Switzerland, 1.8 Mton/year, Italy, 3.2 Mton/year and Norway, 3.5 Mton/year.

These availabilities of biomass from the forest residues are, in principle, insufficient to meet the energy needs of the 13 biomass plants up for tender in Portugal. There is therefore a major discrepancy between the potential and actual availability of residues in forests. Due to the mountainous terrains found in much of Portugal, which makes the extraction and transport of forest residues a high-cost activity, it will be economically viable to harness only a small portion of these residues for energy production—between 43 and 65% of total production, according to the latest estimates.

In a study that focused primarily on estimating the biomass production potential in nine of the 15 municipalities in the district of Portalegre, Portugal, residues from forest and agricultural crops were included in that estimate, and it was found that the agro-forestry industry produces a significant range of residues that could be used for producing electricity, amounting to around 43,700 MWh/year, equivalent to the electricity consumption of three municipalities of the region in 2010 (Gonçalo Lourinho 2015). The most significant sources of agricultural waste in this region are herbaceous plants and olive trees, with availabilities of around 1,500 and 1,200 dry tonnes/year; in relation to forest residues. Species such as *Q. suber*, *Q. ilex* and *E. globulus* are of particular importance, as it is estimated that around 15,800, 9,000 and 6,400 tonnes/year of biomass from these species, respectively, are available for energy purposes (Gonçalo Lourinho 2015).

One of the issues strongly affecting the evaluation of biomass is its transport, particularly the associated cost. The transport of biomass is often divided into primary and secondary transport. Primary transport consists of the first transport of the biomass forest residues along forest tracks, from the felling or harvesting site to the banking ground or first point of landing. This type of transport, also called extraction or logging, may be carried out using an agricultural tractor, a forwarder (machines with high productivity and efficiency), a lorry or, in the case of steeper slopes, cable cranes. It is usually performed by small-scale (transport) equipment whose main tasks are loading the biomass, moving the vehicle after loading and unloading operations at the banking ground. The material may be unloaded onto the ground, or directly into a secondary (end) transport vehicle as a means of optimizing resources (Adriano Guilhermino 2017).

If the exploitation of the biomass is considered an integrated process in forestry activity and included in a framework of operational and economic optimization, then it is assumed that the primary transport encompasses all the above-mentioned operations (collection and processing, grinding, sieving or baling) and that the amounts previously shown include all activities carried out in the forest until the secondary or end transport. Secondary transport starts from the first point of landing or banking ground to the energy production plants and can be carried out using a tractor with a trailer (for distances not exceeding 10 km, due to its cargo capacity and speed) or lorry (for distances of up to around 100 km). In terms of volume, lorries can carry loads of between 70 and 90 m³, corresponding to weights of 16 to 26 tonnes of biomass, depending on packaging and humidity. Within the biomass, cost structure for the Alto Alentejo region, secondary transport accounts for approximately 16 to 23% of the total cost of the raw material for an average distance of 26.4 km. Primary transport has the greatest weight that ranges from 77 to 84% of the total cost. This is not unconnected to the fact that the primary transport includes grinding or chipping operations. With this type of transport, it is important to note that loading and unloading operations are normally performed using a hydraulic crane already fitted to the lorries, making the use of other equipment for the purpose unnecessary (Adriano Guilhermino 2017).

Industrial residues

Industrial residues such as waste tires, plastics, oils and meat meal exhibit also a good potential for energetic valorization. During 2015, the generation of these residues was about 18 million tonnes in Portugal. Due to the higher quantities that were produced, the possibility of adoption of thermal conversion techniques like combustion and gasification have gained a high attention by the community for the valorization of such non-recyclable materials. Waste tires present an interesting higher heating value (HHV) which ranges from 29 to 39 MJ/kg and are composed by ca. 90% of organic matter, and therefore they may behave as a good fuel. In many applications, waste tires are used in combustion chambers as received, or in pulverized or granulated forms. Because the HHV is higher compared with coal, they are appropriate for use in cement plants or foundry furnaces. In terms of contaminant emissions, combustion of waste tires present lower values than those reported for coal. The high temperatures observed in cement plant furnaces allow the introduction of waste tires as a replacement for coal (Rowhani 2016). Common compounds found in tires include natural and synthetic rubber (both among 14–27%), fillings like coal and silica (26–28%) and plastics made of oils and resins (5–6%). The majority of studies based on gasification of tires uses steam as the oxidizing agent and consider the use of residues with particle sizes among 3–5 mm and temperatures in the range of 600 to 800°C.

Regarding plastic residues, they presented an amount of 36 Mton generated in 2015. The energetic valorization of such residues has been done mainly in incineration facilities and cement plant furnaces. In the last few years, it was seen good benefits in the management of plastic residues through the introduction of restrictions by the EU in the landfilling of organic wastes and the adoption of requirements to achieve high levels of material recovery by recycling, in which plastics must be recovered in the first place using mechanical or chemical processes. According to the Plastics Europe Foundation, 25.1 Mt of plastic residues were generated in 2011 and, from these, 6.3 Mt were recycled and 8.6 Mt were valorized for energy production purposes. The remaining part was sent to landfills. Plastic materials may achieve higher values of HHV in the order of 40 MJ/kg, due to the higher contents of carbon and hydrogen and lower amounts of ash. However, the possibility of gaseous contaminant emissions like chlorine, hydrochloric acid, phosgene, benzene and its derivatives, ammonia, hydrogen cyanide, formic acid, formaldehyde, phenols, polychlorinated dioxins and furans is a severe objection to the energetic valorization of plastic residues. Due to the toxic emissions caused by combustion operations, several polymers like polyvinyl chloride (PVC), polyurethane (PUR), polymethyl methacrylate (PMMA), polyamide (PA) and resin of phenol-formaldehyde (PF) require special attention during the selection of the technology employed in combustion or gasification, where investment costs are an important aspect for the final decision (Wang 2004). The energy content of plastics is comparable to other liquid fuels like gasoline and heating oil ($\approx$ 45 MJ/kg) and is higher than that of coal ($\approx$ 30 MJ/kg) (Ergut 2007), and in this manner their valorization for energy production is an interesting option which reduces the amount of residues destined to landfills in more than 90%. As the solutions for thermal conversion occur at temperatures higher than 900°C, an important part of toxic emissions is avoided (Wang 2004). Wasilewski (2013) made a comparative evaluation of the characteristics of several plastics, some of which are reported in Table 1. It is interesting to note the higher calorific values for plastics, especially the case of polyethylene (PE) that is higher than 40 MJ/kg.

Kukačka et al. (2010) refers that incineration and gasification seem to be the most advantageous processes for the valorization of mixed plastic wastes. Considering the higher costs supported for their transportation due to the lower density, the same study recommends a dispersed network of incineration and gasification plants. A significant amount of studies based on gasification of plastics involve the mix of these residues with other materials (i.e., co-gasification). Ahmed et al. (2011) analysed the properties of the syngas generated from gasification and pyrolysis of plastics and rubber, within temperatures between 800 and 900°C, and concluded that co-gasification with forest biomass originated a significant production of hydrogen that was independent of the temperature (Kaewluan 2011).

Another residue with an interesting potential for energy recovery is meat meal, which is a by-product from production, slaughter and transport of animals and is composed mainly of bones and viscera. With the appearance of bovine spongiform encephalopathy (BSE), a neurodegenerative disease that affects essentially bovine

Table 1. Ultimate analysis and calorific values for the main types of plastics (Wasilewski 2013).

Plastic	Ultimate analysis (%)						Ash (%)	Calorific value (MJ/kg)
	C	H	O	N	S	Cl		
PE	81.89	12.37	0.00	0.46	1.92	0.97	2.39	41.80
PP	68.89	9.13	14.61	1.82	1.29	1.24	2.93	30.90
PVC	37.56	4.94	44.00	0.42	0.71	4.43	7.94	13.69
PA	65.39	10.38	10.54	8.49	1.41	0.43	3.36	36.76
PS	88.48	8.36	0.00	0.50	1.12	0.16	1.38	38.97
PET	56.40	5.68	33.10	0.44	0.80	1.43	2.15	21.81
Average	59.18	7.94	23.68	1.05	1.16	2.37	4.66	26.41

domestic cattle, the use of meat meal pellets in animal food was forbidden by directive 94/381/EC. As a result, there is a high availability of meat meal that requires an efficient and economic valorization.

Several research works have been done in Europe about thermal conversion by using gasification and combustion processes. Some of them stated that co-combustion of meat meal with other biomass (e.g., coal) achieved good performances in terms of energy content and generation of stable by-products that do not harm the environment (McDonnell 2001). Although co-combustion of meat meal with other fuels in cement plant kilns is being practiced (Soni 2009), there are some studies that showed that emissions of NO_x depend on the nitrogen content present in the feedstock, and this fact may constitute an obstacle (Cummins 2006). Because of problems associated with NO_x emissions and of some social concerns regarding the environmental impact caused by the incineration of meat meal, co-gasification appears to be an alternative with a high potential of application.

Meat meal has an appreciable content of volatile matter of around 60%, similar to that found in highly biodegradable biomasses. This appreciable content guarantees a good ignition point and lower requirements of oxygen in excess for a complete combustion. The presence of a high fraction of inert compounds (ca. 27%) may induce the occurrence of problems like occlusion of equipment, high production of ashes and, as a consequence, generation of chars with lower quantities and quality. The energy content of the residue is concentrated in volatile matter. No traces of sulphur or chlorine were detected and the calorific value ranges from 12 to 19 MJ/kg (Marculescu 2013).

Ashes obtained from the combustion of meat meal are rich in calcium and phosphates, and may also contain appreciable amounts of sodium, potassium and magnesium. Therefore, their application in agriculture as soil correctives is very feasible due to the high potential as a source of phosphates without heavy metals, which is the contrary situation when compared with mineral phosphates (Deydier 2005).

Because of the higher values of the lower heating value (LHV) that vary among 13–30 MJ/kg, the majority of research have focused on the valorisation through combustion, namely using fluidized beds, while the number of works about gasification and pyrolysis are minor. Due to the chemical composition of meat meal, other applications have been proposed such as the use as soil fertilizers or as a substrate in anaerobic digestion processes destined for biogas production (Cascarosa 2011).

Biomass analysis

To evaluate the potential of a biomass for energy purposes, it is necessary to know its calorific value, proximate analysis, ultimate analysis, and composition of inorganic elements (e.g., transition metals and alkaline earth metals). The physicochemical composition of forest biomass varies among species. Generally, plants consist of around 25% lignin and 75% carbohydrates or sugars. The carbohydrate fraction consists of many sugar molecules linked together in long chains or polymers. The forest biomass can be divided into two categories: cellulosic and hemicellulosic.

With regard to the conversion of biomass into energy considering thermochemical processes, other factors become significant among which the most important is moisture content, which directly influences the energy potential of the fuel.

The biomass moisture content is defined as the amount of water present in the material, expressed in percentage by weight. This value can also be determined on a wet basis (wb), dry basis (db) or dry and ash-free basis (daf). The lower the moisture content, the higher the calorific value of the material, and this has impact in the formation of tars in the combustion/gasification of biomass. Moisture content varies according to the type of biomass, being, for example, 20% for cereal grains and pellets and 50–70% for forest residues.

The quantity of volatile matter in biomass is the amount of volatile gases and solid carbonaceous particles released at temperatures around 400–500°C. It is responsible for the formation of tars in the gasifier, depending on the operating conditions. The amount of volatile substances can range from 10 to 80%, depending on the biomass concerned.

The organic elemental composition of the biomass on a daf basis is relatively uniform. The main components are carbon, hydrogen, oxygen, nitrogen and sulphur, the latter two in a lesser degree. The general formula for elemental composition of biomass is given by $CH_{1.4}O_{0.6}$. The knowledge of this parameter provides information on the calorific value of the biomass and emission levels of sulphur and nitrogen compounds.

The density of biomass refers to the mass per unit volume in an amount free of water. Values can range from 200 kg/m³ for agricultural waste to 900 kg/m³ for solid wood. The morphology is characterized qualitatively and quantitatively by the cellulose, hemicellulose, lignin, starch, lipids, proteins and sugars present in the biomass. The combination of both density and calorific value allows the calculation of the potential energy available in the gasifier.

Regarding the amount and composition of ash (i.e., inorganic components) and contaminants on a db, the former can vary by up to 20% depending on the type of biomass. The composition affects the selection of the gasifier and the behaviour of combustion at high temperatures. Composition has relevance in the efficiency of the gaseous effluent cleaning system and the prevention of problems of clogging of gasifiers due to the melting and deposition of materials that comprise the ash. It is desirable to keep levels of contaminants low, or even completely eliminate these compounds, which may include alkali metals, heavy metals, sulphur, nitrogen and chlorine.

Table 2 shows some compositions of industrial and forest biomass from the region of Alentejo, Portugal, and which were experimentally determined. Biomass with lower densities can have the drawback of requiring a larger storage area and there may be fluidity problems in thermal conversion plants. Low density can be related to low moisture content, which may be advantageous if the intended use

Table 2. Properties of some biomasses and residues that were tested.

Parameter	Unit	Biomass											
		Pine	Acacia	Hemp	Rice husk	Peach pit	Dried timber	Pine bark	Olive stone	Olive press-cake	Meat and bone	Tires	Plastics
Moisture content	wt% ar*	8.6	7.8	7.9	9.8	13.3	0	16.7	9.4	8.9	8.8	0.8	0.2
Volatile matter	wt% ar	80.6	75.7	71.9	59.9	66.4	80	57.6	57.8	70.97	50	64.5	78.3
Fixed carbon	wt% ar	8.2	12.5	4.2	14.7	19.3	19.4	24.5	19.7	19.48	8.3	29.6	20.9
Ash	wt% ar	2.6	4.0	16.0	15.6	1	0.6	1.2	13.1	0.65	32.9	5.1	0.6
Lower heating value (LHV)	MJ/kg	17.1	17.6	23.1	13.88	16.18	16.31	16.42	16.36	20.3	15.2	38.6	31.6
Carbon	wt% ar	51.6	44.2	45.7	38.8	45.49	52.0	46.24	43.22	53.87	30.6	75.5	69.2
Hydrogen	wt% ar	6.0	5.4	7.1	4.6	6.26	6.3	5.92	5.56	8.8	5.2	7.1	7.4
Nitrogen	wt% ar	2.8	1.4	3.5	1.3	0.73	0.4	0.19	1.86	2.03	8.1	0	0
Oxygen	wt% ar	22.8	33.1	15.2	29.6	33.22	40.5	29.75	26.86	26.51	52.2	11.8	23.4

*ar—as received basis.

of the biomass is for the production of pellets and briquettes. Moisture values of under 30% afford higher recovery of energy in thermal processes like combustion and gasification. Moreover, it has the advantage of generating lower ash content compared to other biomasses. The presence of inorganic elements alerts to the appearance of possible problems of scale deposits in equipment, accessories and piping in energy conversion facilities. The high content of organic elements (C, N, H) shows potential for recovery in terms of energy and organic substances.

Table 2 shows results for proximate and ultimate analyses for the characterization of biomasses and residues along with determination of calorific value, and which were performed inside the facilities of Polytechnic Institute of Portalegre, Portugal (PIP). Evaluation of these materials was done by using a Thermo Fisher Scientific Flash 2000 CHNS-O analyzer (to determine the levels of C, S, N, O and H present in the chemical structure), a calorimeter IKA C200 (to determine HHV), and a Perkin Elmer STA 6000 thermogravimetric analyser (to estimate the proximate analysis of the samples—moisture, volatile matter, fixed carbon and ash).

Thermal Gasification

Technology

The various types of biomass and residues can be treated by chemical, biochemical or thermochemical processes with the aim of producing energy in various forms including electrical, mechanical or thermal energy. Essentially, these processes may be divided into two classes: thermochemical processes (e.g., combustion, gasification, pyrolysis, liquefaction and transesterification) or biochemical processes (e.g., anaerobic digestion and fermentation). Each one may be preceded by various pre-treatment operations with the aim of preparing the materials, ensuring that the process itself occurs in an efficient manner. Physical drying, sorting, compressing and cutting/breaking operations may be cited as some examples of these pre-treatments (Basu 2013).

Combustion is the transformation of the chemical energy of fuel into heat through reactions with oxygen. For energy purposes, direct combustion occurs mainly in stoves, furnaces and boilers, with formation of steam that can be used for heating applications or for electricity production via a turbine coupled to a generator. This process is inefficient and generates appreciable amounts of gaseous contaminants with harmful effects for health and environment, forcing the implementation of expensive equipment for decontamination.

Pyrolysis is the process of converting a fuel (usually firewood) into another fuel with a better energy density, such as charcoal. The process consists of heating the material in a non-oxidative atmosphere until the volatile material is removed. The main end product (charcoal) has an energy density twice that of the source material and burns at much higher temperatures. Apart from charcoal and fuel gas, pyrolysis produces coal tar and pyroligneous acid that, when sometimes submitted to further treatments like distillation, may create a new liquid fuel with good calorific properties. Pyrolysis may be an interesting pathway for the valorization of lipid residues like used vegetable oils but not for those with high moisture content like sewage sludges (Manara et al. 2012).

Gasification is a process of converting solid fuels into a syngas through thermochemical reactions involving lower stoichiometric quantities of oxygen when compared with combustion. This atmosphere poor in oxygen may be created by adding lower amounts of air or other gases such as steam and carbon dioxide, and which are named the oxidizing agent. There are several types of gasifiers and the most commonly used are fixed or fluidized bed reactors. The syngas is usually a mixture of carbon monoxide, hydrogen, methane, carbon dioxide and nitrogen in different proportions, which can be used to generate electricity, for example, in an internal combustion engine. The environmental impact of the flue gases after the combustion of syngas is generally lower than that which results from direct combustion of feedstock. Unlike pyrolysis, gasification may tolerate feedstocks with greater moisture content.

A gasification plant is generally composed by four basic modules: a pre-treatment stage of feedstock, conversion of materials through gasification reactions, cooling and clean-up of product gas and a final stage for energy generation. Figure 1 summarizes the connection between these modules and also the main inputs and outputs of the process.

Pre-treatments usually include a grinding and a drying operation in which feedstock particle size is homogenized and a maximum moisture content of 25–30% is achieved. A torrefaction process may also be adopted, in which the feedstock is submitted to temperatures and residence times among 200–300ºC and 30–60 min, respectively, without the presence of oxygen. The aim consists in reducing moisture content and hardness of the material, giving to it a greater energy per unit of mass, hydrophobicity and brittleness, properties that will be advantageous in subsequent stages (Molino 2016, Acharya 2015).

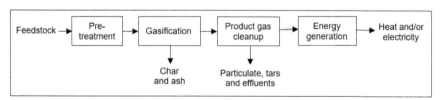

Fig. 1. Process steps in a gasification plant (adapted from Molino 2016).

Mechanism

The gasification stage inside the reactor occurs with a set of chemical reactions that may be grouped in four classes:

- *Drying*, where the material loses water until the temperature rises up till 150–160ºC;

- *Pyrolysis*, where chemical bonds among molecules in the feedstock are broken and originate gaseous light compounds (viz. CO, CO_2, H_2 and CH_4), a liquid fraction (designated by tar and composed mostly by aromatic substances) and a carbonaceous solid matrix (char and ash), that comprises the inorganic matter of the feedstock. This step is performed under temperatures between 250 and 700°C;
- *Oxidation*, in which heat is released that provides the required energy for all the other endothermic reactions, and where carbon and H_2 are oxidized with the sub-stoichiometric oxygen amounts forming more CO, CO_2 and H_2O under temperatures in the range 500–1600°C;
- *Reduction*, where the char and gaseous products originated meanwhile react among themselves to form the final syngas in the absence of oxygen. This step requires a temperature of 800–1100°C, and the greater it is, the lower are the contents of char, tar and the calorific value of the syngas (Molino 2016).

Other sub-products obtained from gasification include ash and char that may be valorized in a variety of ways instead of being discarded to landfills. Some works have proposed the use of char as catalysts in the decomposition of tar, remediation of effluents, soil fertilization and construction of roads (Di Gianfilippo, Shen 2015, Rajapaksha 2016).

Several reactor configurations are available for gasification which are described below:

- *Fixed Bed*: The feedstock is deposited over a grate and is submitted to a flow of an oxidizing agent (air, pure oxygen or steam) that trespasses the bed of material and reacts with it. The flow may be directed upwards or downwards and depending on it, the reactor is called 'updraft' or 'downdraft', respectively. Meanwhile, the feedstock is converted and falls through the grate in the form of ash, char and tar, and the syngas is captured on the top or at the bottom of the reactor according to the flow of the gaseous agent. Both gasifiers are simple and have lower costs of equipment and maintenance, but produce a syngas of poor quality;
- *Fluidized Bed*: Inert material like sand is put over a grid located at the bottom of the reactor to form a turbulent bed with the feedstock. The turbulence is induced by the injection of the oxidizing agent through the grid, enabling a good contact and heat transfer between the solid and gas phases and promoting a more efficient conversion. This configuration allows the introduction of catalysts such as dolomite or alumina to improve the production of H_2 and to attenuate the formation of tars, but needs to operate at temperatures lower than 900°C in order to prevent sintering of the bed;
- *Entrained Flow*: The feedstock in fine grains (< 1 mm) and well dried (< 15% moisture) is introduced under pressure on the top of the reactor along with the oxidizing agent, in an ambient with temperatures around 1300–1500°C and pressures among 25–30 bar; solid residues are captured at the bottom and

the syngas exits through lateral exits located on the reactor. The syngas has a good quality with lower contents of tars, but the demanding requirements for the feedstock may increase the costs of the technology;

- *Rotary Kiln*: A cylindrical reactor slightly tilted rotates around its axis in order to stir the feedstock particles with the gaseous oxidizing agent that moves in a counter-current direction. The stirring effect promotes the contact between the solid and gas phases and enhances the conversion at temperatures around 450°C. The technology has a simple operation and is less sensible to variations in the properties of the materials, but the syngas generated has a lower quality compared with other configurations, needs higher residence times and the production of dust and tar is significant;

- *Moving Grate*: In this design, feedstock particles get into a chamber with a grate placed at the base that moves the material in constant contact with the oxidizing agent. At the end of the route, ashes that are produced fall into a refuse deposit, and the syngas goes out through an exit at the top of the chamber for a possible combustion in a next stage. If the oxidation agent is air, the proportion of oxygen is about 0.5 times the required content for complete combustion (UAr1) (Molino 2016, Roche 2014, De Andrés 2011).

Figure 2 depicts the principle of operation of each configuration that was previously described.

Gasification reactions can be controlled by the adjustment of temperature, bed material used inside the reactor (e.g., sand or catalyst materials such as dolomite, alumina or olivine), type of oxidant agent (e.g., steam, air or pure O_2) and equivalence ratio (ER), i.e., the ratio between the amount of O_2 introduced and the amount required for the complete combustion of the feedstock (if air or oxygen are used as the oxidant agents). If steam is used as this agent, then the parameter to be considered is steam to biomass ratio (S/B) instead of ER (Roche et al. 2014). The change in all of these parameters promotes the production of different syngas compositions. The clean-up step of the syngas is destined to remove pollutants until their contents achieve acceptable levels before the introduction in the unit for energy production or to avoid damages to the environment, after the burning operation. These technologies can be grouped in hot, warm and cold gas clean-up, according to the temperature of the syngas in which they operate (> 300°C, 100–300°C and < 100°C, respectively). Examples of such technologies include inertial separation of particulate matter, electrostatic separation, catalytic cracking of tars (which sometimes may happen inside the reactor), filtration and purification through a cloud of water (wet scrubbers) (Woolcock 2013).

Clean-up stage produces some residues such as solid particulates and effluents (tars and wastewater) that can be remediated through adsorption in activated carbon, oxidation (with O_3 and H_2O_2) and biological treatments (Milne 1998).

Finally, the syngas decontaminated can be burned in a boiler to feed a steam cycle, gas turbine, internal combustion engine, or even injected in a fuel cell to produce electricity and heat for other purposes. In the situation of gas turbines

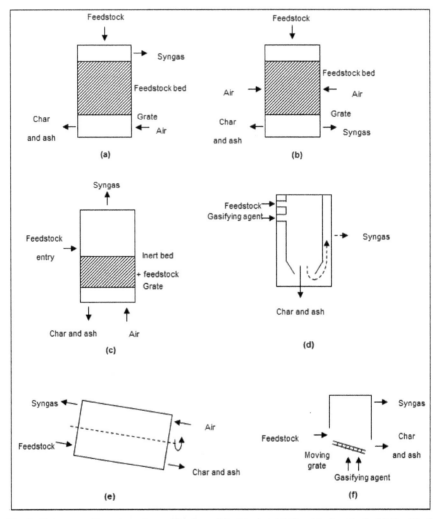

Fig. 2. Reactor configurations for gasification of feedstocks: (a) Fixed bed updraft; (b) fixed bed downdraft; (c) fluidized bed, (d) entrained flow; (e) rotary kiln; (f) moving grate (adapted from Arena 2012, Molino 2016).

and internal combustion engines, it is necessary that the clean-up process reduces, the levels of pollutants in the syngas to values around 10–100 mg/Nm³ for tars, 2.4–50 mg/Nm³ for particulates, 0.025–0.1 ppmw for metals and 20 ppmv for H₂S, just to give an idea about the degree of purity that the syngas must have. However, instead of the direct application of syngas for energy production, it is also possible to convert it into new biofuels such as bioethanol and biomethane, using biological conversion processes (e.g., fermentation) or catalysts under specific atmosphere conditions (e.g., Cu and Ni-alumina-based compounds) (Molino 2016).

Demonstration units

Brief description of the units

Inside the facilities of PIP there are two thermal gasification units at pilot scale that allow the development of experiments using different types of biomass, with a view to demonstrating this technology for energy valorization purposes. These pilot scale units are of downdraft and fluidized bed types with maximum processing capacities of around 22 and 100 kg/h. In the context of this work, some results obtained in both units will be presented.

Figure 3 displays a diagram of the gasification unit of fluidized bed type used in the experiments. The main components of the unit are:

a) a biomass feeding system with two in-line storage tanks that allow the discharge of biomass into the reactor using an Archimedes screw at a variable and controllable speed; these two storage tanks act as buffers to avoid the entrance of air through the feeding system;

b) a fluidized bed reactor with a tubular configuration, 0.5 m in diameter and 4 m of height; internally, this reactor is coated with ceramic refractory materials. Biomass enters the reactor at the height of 0.5 m from its base, and preheated air enters through the bottom side through a set of diffusers that ensure a flow of about 70 m³/h. Three temperature sensors are installed uniformly along the height in order to monitor and control the gasification temperature, that typically is set around 800°C. The syngas leaves the reactor from its top at about 700°C. The reactor operates at negative pressure gradient produced by a vacuum pump installed at the end of the process line, and temperature control is achieved by tuning the air flow rate. During the experiments, the bed was filled with dolomite acting as an agent for tar reduction;

c) a gas cooling system consisting of two heat exchangers: the former cools the syngas to about 300°C using an air co-current flow that enters the reactor,

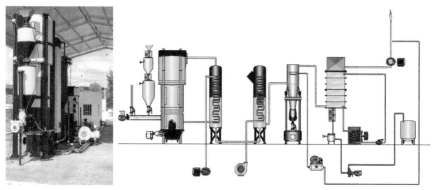

Fig. 3. Image of a fluidized bed gasification plant (left) and schematic of operation (right): (a) feeding system; (b) reactor; (c) heat exchangers; (d) bag filter and (e) condenser (Couto et al. 2017).

while this flux is simultaneously pre-heated; the second heat exchanger cools the syngas down to 150°C by a forced flow of air coming from the exterior;

d) a cellulosic bag filter that allows the removal of particulates and ashes produced during the gasification process. The cleaning of the filter is made by pressurized syngas injection in periodic intervals; particulates are collected at the bottom of the bag filter and are stored in a proper tank;

e) a condenser, where liquid condensates are removed by cooling the syngas down to room temperature on a third tube heat exchanger;

f) a control station, that allows the operation and monitoring of the several parameters along all the unit.

The application of SRF vs. RDF classification and specifications to the material flows of two mechanical-biological treatment plants of Rome: Comparison and implications, Waste Manag., 47: 195–205, 2016.

Tests with this unit were made using different biomasses at around 800°C, with mass admission rates of 40 and 63 kg/h for studying syngas composition as a function of feedstock type and operational conditions.

The downdraft unit is a PP20 Power Pallets –20 kW gasifier from AllPowerLabs as shown in Fig. 4, which is a combination of Imbert style downdraft fixed bed reactor, an electric power generator and an electronic control unit. The equipment is composed of a storage hopper of biomass, where it is simultaneously dried through the recirculation of the hot gases produced in the reactor. The biomass is supplied from the top while the air moves downwards, and which was previously heated through contact with the walls of the reactor. Ash collection is carried out in a separate tank in the lower zone of the reactor, while the syngas passes through a cyclone to remove fine particles. The syngas is conducted again to the biomass

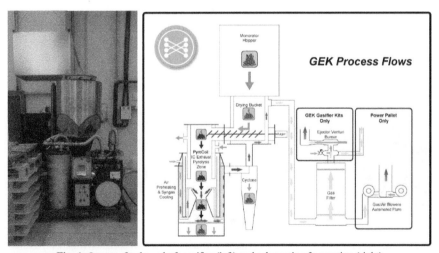

Fig. 4. Image of a downdraft gasifier (left) and schematic of operation (right).

hopper in order to dry it, as already mentioned, and then it is subjected to a new decontamination by a filter composed of biomasses of various granulometries. The syngas may be collected in this filter for analysis or directly injected into the generator. Condensates are also retained in this filter.

Analysis of the composition and properties of the syngas produced

Syngas analysis was performed in a Varian 450-GC gas chromatograph with two TCD detectors that allow the detection of H_2, CO, CO_2, CH_4, O_2, N_2, C_2H_6, C_2H_4, and which also contains a CP81069, CP81071, CP81072, CP81073 and CP81025 Varian GC columns. Helium was injected as the carrier gas. Syngas samples were collected from the units and stored in appropriate Tedlar© bags at the condenser exit every time when the gasification of a given feedstock composition reached its stationary state. These samples were then injected directly from the sampling bags in the gas chromatograph using a peristaltic pump operating at its maximum rate and equipped with a Marpren® tube. Chromatographic peaks for the different gases under analysis were identified based on their retention times and by comparing them with the reference chromatogram of the "custom solution", provided by Varian. Gas mass percentage composition was calculated on the basis of peak areas provided by the chromatographic signal, while LHV's were determined through the following expression:

$$LHV_{syngas} = \sum_{i=1}^{n}(x_i \times LHV_i)$$

with x_i expressing the volume fraction of gas component i and LHV_i the lower heating value of component i (MJ/Nm³), which may be retrieved, e.g., from (Basu 2013).

Wastes Gasification Results

Gasification

The compositions and lower heating value (LHV) of the different syngas produced, with the bubble fluidized bed reactor from different biomasses, are presented in Table 3. LHV's were calculated on basis of syngas composition determined by gas chromatography and using the standard combustion heats of the compounds obtained, at 25°C.

Generally, the results show that the produced syngas is relatively rich in carbon monoxide (mass percentage between 11 and 19%), methane (1 and 7%) and hydrogen (4 and 13%), and therefore these three gases are the main responsible for the heat content of the syngas produced. LHV's values are between 3.3 and 5.8 MJ/Nm³, as can be seen in Table 3. On the other hand, it can be seen that the syngas also contains a large amount of nitrogen (43 to 54%) and some carbon dioxide (16 to 19%) resulting from the partial combustion process that takes place simultaneously with biomass gasification.

Table 3. Experimental conditions and syngas analyses resulted from gasification of diverse biomasses.

Experimental conditions	Forest residues	*Miscanthus*	Olive bagasse	Vines pruning	Peach cob	Coffee grounds	Coffee husk
Temperature (°C)	815	745	791	815	751	736	815
Equivalence ratio, $O_2/O_{2\ Stoich.}$	0.25	0.28	0.25	0.30	0.25	0.27	0.38
Admission biomass (kg/h)	63	65	54	55	45	62	30
Air flow rate (Nm³/h)	94.0	30.0	85.0	40.0	53.0	88.0	75.0
Syngas flow rate (Nm³/h)	106.0	37.9	122.2	102.0	60.4	83.3	106.0
Condensates (l/h)	4.6	5.2	3.9	4.1	11.0	11.0	3.8
Ash (kg/h)	0.3	0.6	0.3	0.2	1.2	1.0	0.4
Syngas fraction (dry basis)							
CO_2	16.7	15.9	16.2	17.9	14.7	15.7	18.7
N_2	48.0	43.1	52.2	49.1	53.1	53.5	52.3
CH_4	4.6	6.8	4.42	2.3	3.6	1.2	1.6
CO	18.6	17.0	18.1	14.1	14.0	12,.6	11.4
H_2	8.2	10.9	4.2	12.7	7.2	12.1	12.4
Syngas LHV (MJ/Nm³)	4.9	5.8	4.3	4.0	3.8	3.3	3.4
Syngas LHV (MJ/kg biomass)	4.1	5.2	3.5	3.4	3.3	2.9	2.8

The thermal biomass gasification process involves a set of complex chemical reactions that deal with the formation of three fractions: the syngas, ashes (chars) and condensates. The most important fraction, amounting to more than 70% (wt) is made of light gases, namely, CO, H_2, CH_4, CO_2, and N_2. In fact, it is considered that in the thermal gasification process gases leave the reactor in equilibrium condition. Main equilibrium reactions to be considered in a gasification process are presented in Table 4.

In a first approach, it may be said that the gas composition, the syngas fraction and the amount of ashes and condensates, are a function of biomass nature and its elemental composition, and of the gasification operational conditions, in particular, are a function of temperature process.

Generically, the results obtained with the various biomass studied shows a relation between the biomass calorific value and the calorific value of the syngas obtained: higher biomass calorific value results in higher calorific syngas production. In fact, it demonstrates that the syngas formed with the highest calorific value was obtained by gasification of forest residues, *Miscanthus* and olive bagasse, which are the most energetic materials. This relationship between the biomass calorific content and the syngas LHV can be explained considering first, that, the biomass calorific value is related to the amount of carbon and hydrogen present in the biomass, i.e., an increased amount of carbon and hydrogen, leads to higher calorific value, and,

Table 4. Typical gasification reactions.

Type	Reaction
Carbon reactions	
1 (*Boudouard*)	$C + CO_2 \leftrightarrow 2CO + 172{,}6$ MJ/kmol
2	$C + H_2O \leftrightarrow CO + H_2 + 131.4$ MJ/kmol
3	$C + 2H_2 \leftrightarrow CH_4 - 74.8$ MJ/kmol
4	$C + 0.5O_2 \leftrightarrow CO - 111$ MJ/kmol
Oxidation	
5	$C + O_2 \leftrightarrow CO_2 - 393.8$ MJ/kmol
6	$CO + 0.5O_2 \leftrightarrow CO_2 - 284$ MJ/kmol
7	$CH_4 + 2O_2 \leftrightarrow CO_2 + 2H_2O - 803$ MJ/kmol
8	$H_2 + 0.5O_2 \leftrightarrow H_2O - 242$ MJ/kmol
Water-shift reaction	
9	$CO_2 + H_2 \leftrightarrow CO + H_2O + 41.2$ MJ/kmol
Methanization	
10	$2CO + 2H_2 \leftrightarrow CH_4 + CO_2 - 247$ MJ/kmol
11	$CO + 3H_2 \leftrightarrow CH_4 + H_2O - 206$ MJ/kmol
12	$CO_2 + 4H_2 \leftrightarrow CH_4 + 2H_2O - 165$ MJ/kmol
Steam reforming	
13	$CH_4 + 0.5O_2 \leftrightarrow CO + 2H_2 - 36$ MJ/kmol
14	$CH_4 + H_2O \leftrightarrow CO + 3H_2 + 206$ MJ/kmol

second, a larger amount of these two elements allows production of larger quantities of hydrogen and carbon monoxide, the major contributors for the calorific value of the syngas. Analysis of the dependence of producer gas properties in function of biomass admission shows, for all biomasses studied, there is a decrease in syngas calorific value with increasing quantity of biomass admitted. This is consistent with the fact of a faster acceptance of biomass by the reactor prevents the equilibrium of the reactions 1 and 2 being attained. In general terms, an increase in gasification temperature promoted the formation of a syngas with higher hydrogen and carbon monoxide contents and, consequently, a higher LHV of the syngas. This could be a result of a shifting in the gasification equilibrium, since reaction 2 is endothermic and, therefore favoured by a temperature increase. The increase of ER has a negative effect on H_2 production because the oxidation reactions are favoured when the reactions had higher contents of oxygen. On the other hand, the increase of ER has a positive effect on the reduction of tar content with increased gas yield. The use of steam as a gasifying agent in gas-phase reactions could be a way to increase the decomposition of hydrocarbons and to rise contents of H_2. The introduction of steam also leads to more tar participating in steam reforming, which led to a rapid increase of gas yield and tar reduction because of higher conversion efficiency (Silva 2017).

Co-gasification

Table 5 presents the results of co-gasification of industrial waste, namely, tires and plastics with the pilot downdraft reactor. The advantage of the co-gasification processes is that they allow reducing negative processing effects, often making feasible certain resources. The gasification of tires and plastics, in an isolated way, proved impossible to perform in this type of reactors studied, because of problems resulting from the decomposition process itself. Thus, in this case, the tests were carried out through co-gasification with acacia, a material that presents a good gasification easiness and produces relatively low content of ashes, presenting in addition great abundance in the region.

A first analysis of the results shows that the gas produced in greater quantity is carbon monoxide followed by hydrogen, this situation having occurred in any of the mixtures tested. The third one is methane. The high percentage of nitrogen is due to the fact that gasification has been carried out with atmospheric air.

An interesting aspect of the presented results is that the addition of meat meal to the acacia leads to a significant increase in the calorific value of the gas, as a consequence of the significant increase of carbon monoxide and hydrogen. If such effect is positive at lower addition percentages in the co-gasification mixtures, when this percentage increases it is not at all possible to carry out the gasification process because of difficulties in the physical consistency of the material. An explanation for an increase in calorific power with small additions of meat meal may be due to the high concentration of oxygen in the pellets of meat meal (elemental analysis

Table 5. Experimental conditions and syngas analyses of co-gasification of acacia with different residues in the downdraft reactor.

Experimental conditions	Acacia	Meat and bone		Plastics		Tires	
		20%	40%	20%	40%	20%	40%
Temperature (°C)	805	823	761	814	891	803	790
Admission biomass (Kg/h)	3.5	2	1.3	1.875	2.625	4.974	1.755
Condensed (l/h)	0.361	0.259	0.259	0.361	0.361	0.038	0.038
Ash (Kg/h)	0.04	0.033	0.033	0.04	0.04	0.231	0.231
Syngas fraction (dry basis)							
CO_2	10.4	14.4	12	11.1	7.9	10	12.2
N_2	53.9	44.3	55.9	52.5	56.7	54.3	56.1
CH_4	3.6	3.3	3	3.7	3.7	3.6	2.3
CO	5.8	15.9	5.6	8.3	5.6	10	10.7
H_2	4.0	10.7	4	5.9	4.3	6.8	6.9
Syngas LHV (MJ/Nm³)	3.34	4.56	2.3	3.09	2.53	3.64	3.14

about 52%) which, when reacting with acacia, does not require as much atmospheric oxygen (air inlet to the reactor) to perform the combustion and gasification reactions inside the reactor, so the presence of nitrogen is not as high and the concentration of CO increases substantially (see reaction 1 in Table 4). Due to the difficulties of feeding with larger percentages of meat meal, especially inside the reactor, where the pellets of meat meal completely clogged, an increase in calorific power was not observed. Other authors also reported the same phenomenon, where it is unreasonable to increase the percentage of meat meal due to the problems of sealing inside the reactor and blocking of air intake (Gulyurtlu 2005).

Considering the high carbon content of used tires (see Table 2), this may lead to the creation of conditions for the production of high quantities of ash rich in chars, during the gasification process. Several studies point to the need for increasing the co-gasification temperature, as the percentage of tires in the mixture increases, or to reduce the size of the tire particles to reduce the production of char (Straka et al. 2008). The results show that the optimum conditions places precisely at low tire percentages (below 40%). This may be one of the justifications presented so that, with the increase in mixing, the calorific value of the gas will decrease, and there will also be a lot of material that does not combust inside the reactor at the end of the test (Straka et al. 2009).

In the case of the results with plastics, a rather abnormal behavior was observed in the increase of calorific value of the gas for the mixture of 40% when the temperature of the reactor is increased. This fact is reported by several authors (Xiao 2008, Straka et al. 2008), where it is also observed that an increase of gasification temperature for biomass with these characteristics, results in an increase in the calorific value of the gas (Lee 2014).

Economic Considerations—Case Study

In this section the achieved experimental results are analysed in an economic perspective of a demand for alternative energy sources to fossil fuels, thus minimizing negative environmental impacts and making full use of the potential of gasification. The analysis will focus on a small Portuguese municipality called Vila de Rei in the district of Castelo Branco, with a population of around 3,500 habitants and an area of 192 km². It is a forest region based on wild pine and eucalyptus. Heavily forest fires here during the year of 2017, burned down about 50% of the total municipality area. The main economic activities are forestry and agro-industry with a special focus on olive groves. Due to the problem of forest fires, several measures are being taken to reduce their frequency. In addition to the need to implement a good forest management, one of the main measures is the regular cleaning of the forest. Such an approach requires high investments and therefore it makes sense to think about ways of valuing these forest residues. Energy recovery of this waste is undoubtedly a possible path with high potential and here thermal gasification can play a significant role. On the other hand, there are other agro-industrial wastes that can contribute as fuels in a thermal gasification unit coupled with energy production. Thus, in this work it was evaluated that the economic feasibility for the implementation of a thermal gasification unit admitting forest residues of the lignocellulosic type, in order to meet the thermal and electric energy requirements of all the municipality, including public buildings, effluent treatment equipment and public swimming pools.

For analysis purposes, it was considered that the energy produced by the unit will replace the electric energy bought from the national company of electricity and the natural gas utilized. The analysis was based on the results obtained in this work on gasification of the different biomasses, particularly on the composition and calorific value of the syngas obtained, and also on current prices of natural gas and electricity at market place. Table 6 shows the energy consumption of the municipality during 2016. It is observed that a variation of the consumption throughout the year, where the winter period has the greatest values. Based on these values, a 0.2 MW unit for electric energy production is required to allow the municipality to be 75% self-sustainable. In terms of thermal energy, the unit should give around 0.16 MW of thermal power.

With regard to waste fuels to be used, it was decided to consider only those that have some expression in the municipality, namely, olive and forest residues. The municipality operates an olive mill with a significant amount of waste generated that can be used. It is a two-phase winery operating at low temperature, producing about 1500 tons per season. The mill is selling the bagasse at about 7.50 €/ton. With respect to forest residues and based on the need to clean the forest for fire prevention, another option in this work was to hire external services for this purpose with a price of around 50 €/ton. This price is relatively high but incorporates a significant environmental value. Considering the amount of bagasse available and the overall energy requirement, it was estimated that 2500 ton/year of forest residues must be necessary for the gasification plant.

Table 6. Energy consumption per month by the municipality during 2016.

Month	Electrical energy (kWh)	Thermal energy (kWh)
January	325	104
February	298	102
March	271	61
April	244	103
May	217	30
June	217	28
July	190	3
August	163	5
September	217	22
October	217	39
November	244	71
December	271	120

The gasifier was based on a bubbling fluidized bed type, mounted with a gas engine and a module for thermal energy recovery. It was considered an electric energy yield of 25% and a thermal energy yield also of 25%, relative to the total energy input coming from the biomass admitted.

The economic evaluation was done by the determination of two economic indicators: the net present value (*NPV*) and the payback period (*PP*). The formula that was adopted for the calculation of *NPV* (€) is defined in Eqn. (1) (Luz 2015):

$$NPV = \sum_{n=1}^{t} \frac{CF_n}{(1+i)^n} - C_{IC} \qquad (1)$$

CF_n is the net incremental cash flow per year expressed in € (i.e., the difference between energy profits obtained from the unit and the corresponding operating costs), *n* is the year under focus, *t* is the total lifetime presumed for the unit (15 years), *i* is the discount rate (10%) and C_{IC} is the initial investment applied in the unit (€). In fact, CF_n is constant for all the life period because both energy profits and operating costs are assumed to be the same every year.

Since the first term of Eqn. (1) is a geometric progression with a ratio of $(1+i)^{-1}$, it can be rewritten in the form of Eqn. (2) for a faster calculation of *NPV*:

$$NPV = \frac{CF_1}{(1+i)^1} \times \frac{1-(1+i)^{-t}}{1-(1+i)^{-1}} - C_{IC} \qquad (2)$$

A positive result for *NPV* indicates that the unit is economically feasible during the life period of the equipment (Luz 2015).

The *PP* was also determined for each case based on the study of the accumulated cash flows that were foreseen over time. The accumulated cash flow for the year *n* (ACF_n, in €) was sequentially estimated by Eqn. (3):

$$ACF_n = \begin{cases} CF_n - C_{IC}, \text{ if } n = 1 \\ ACF_{n-1} + CF_n, \text{ if } n \geq 2 \end{cases} \qquad (3)$$

where all the variables have the same meanings as described before. The first year presenting a positive value for ACF_n corresponds to the wanted *PP* when the initial investment and succeeding costs are completely returned through the energy profits.

Taking into consideration the type of gasification unit to be installed and the cost of equipment, it is estimated that the initial investment (C_{IC}) in the unit will be around 5000 €/kW (Luz 2015, Biomass for Power Generation 2012, Nicholas Lumleya 2014). The operating costs required for the unit were estimated to be 5% of the initial investment that was made in the acquisition of the equipment (Luz 2015). Economic returns lie essentially in savings on the costs of electric energy acquired from the external grid. This analysis took as a reference the price charged for electric energy in Portugal, which amounted to be 0.1 €/kWh in off-load periods (i.e., overnight) and 0.2 €/kWh in on-load periods (i.e., during the day), and also the price of propane gas that was around 0.06 €/kWh (www.galpenergia.com 2018).

Table 7 gives a summary of the parameters considered in the economic evaluation and the results calculated for annual cash flows, *NPV* and *PP* for the unit proposed.

Results of the economic analysis indicate that there is viability in the construction of a gasification plant which uses forest and industrial residues generated by the municipality, with a positive *NPV* of 1.24 M€ and a *PP* of 4 years. Economic viability of a plant of this nature depends on the ratio among the

Table 7. Parameters for economic analysis and corresponding results for the construction of a gasification plant.

System parameters	
Type of gasifier	*Bubbling fluidized bed*
Thermal power (MW)	0.16
Electrical power (MW)	0.20
Thermal energy yield (%)	15
Electrical energy yield (%)	15
Feedstock	
Olive bagasse	
Quantity (ton/year)	1500
Biomass HHV (MJ/kg)	20
Price (€/ton)	7.5
Forest residues	
Quantity (ton/year)	2500
Biomass HHV (MJ/kg)	17
Price (€/ton)	50
Economic parameters	
Investment (€/kW)	5000
Price of thermal energy (€/kWh)	0.06
Price of electricity (€/kWh)	0.2
Operative costs (% investment)	5
Lifetime (years)	15
Discount rate (%)	10
Economic evaluation results	
Profits (annual, k€)	391
Electric energy (k€)	350
Thermal energy (k€)	41
Costs (annual, k€)	79
Feedstock (k€)	29
Operating cost (k€)	50

Table 7 contd. ...

...Table 7 contd.

Economic evaluation results	
Annual cash flow (k€)	313
Cost of installation (M€)	1
NPV (M€)	1.24
PP (years)	4

availability of residues and size of the unit. In fact, units with greater power require more amounts of feedstock, that may not be available in the region, implying a transportation of residues from longer distances.

In Table 8 gives a sensitivity analysis of the economic evaluation performed through the variation of investment cost and costs with the operation and feedstock; this variation was done for two different values relative to the base case: + 10% and + 25%. It can be concluded that the investment cost has some impact in the results of *PP*, increasing it from 4 to 5 years. However, operating and feedstock costs have a minor impact in the sensitivity analysis since the *PP* was the same for both variations (4 years). In the simulated situations, the construction of the plant was always feasible since *NPV* assumed positive values.

These results show that the investment applied in the unit must be estimated with a greater care and accuracy, because a variation on this parameter may led to an appreciable deviation in the time required for the return of the benefits.

Table 8. Sensibility analysis for the economic evaluation of a gasification plant.

Parameter	Variations			
	10%		25%	
	PP (years)	*NPV* (M€)	*PP* (years)	*NPV* (M€)
Investment	4	1.10	5	0.90
Operative costs and feedstock	4	1.17	4	1.06

Conclusions

The present study aimed to test the use of the thermal gasification process as a way to valorize waste residues into a useful gas for energy production purposes, as well as to evaluate the economic feasibility of the construction of a plant able to process residues produced in a municipality.

Based on the study and on the presented results, it was demonstrated the feasibility of transforming forest residues and highly polluting waste into valuable, hydrogen-rich gas and other products that are highly relevant in terms of heating power and are of interest to the chemical industry. It also showed that the thermal gasification unit performed very well during the tests.

In addition, a strategy for using biomass should involve investment in small production units near the biomass resource to be processed, allowing transport costs to be minimized.

It is therefore expected that this work contributes for a better management of residues giving them a valuable use for energy production, at the same time reducing the harmful effects to the environment.

Nomenclature

ar	:	as received
BSE	:	bovine spongiform encephalopathy
daf	:	dry and ash-free basis
db	:	dry basis
ER	:	equivalence ratio
HHV	:	higher heating value
LHV	:	lower heating value
MSW	:	municipal solid waste
NPV	:	net present value
PIP	:	Polytechnic Institute of Portalegre
PP	:	payback period
RDF	:	refuse-derived fuel
wb	:	wet basis
XRF	:	X-ray fluorescence

Acknowledgements

The present work received financial support from the Foundation for Science and Technology of the Portuguese Ministry of Science, Technology and Higher Education, under the grant SRFH/BD/111956/2015.

References

Acharya, B., A. Dutta and J. Minaret. 2015. Review on comparative study of dry and wet torrefaction. Sustain. Energy Technol. Assessments 12: 26–37.

Adriano Guilhermino, Gonçalo Lourinho, Paulo Brito and Nicolau Almeida. 2017. Assessment of the use of forest biomass residues for bioenergy in Alto Alentejo, Portugal: Logistics, Economic and Financial Perspectives, Waste Biomass Valor 9: 739–753.

Ahmed, I. and A.K. Gupta. 2011. Characteristic of hydrogen and syngas evolution from gasification and pyrolysis of rubber. International Journal of Hydrogen Energy 36: 4340–4347.

Alauddin, Z.A.B.Z., P. Lahijani, M. Mohammadi and A.R. Mohamed. 2010. Gasification of lignocellulosic biomass in fluidized beds for renewable energy development: a review. Renew. Sustain. Energy Rev. 14: 2852–2862.

Arena, U. 2012. Process and technological aspects of municipal solid waste gasification. A review. Waste Manag. 32: 625–639.

Basu, Prabir. 2013. Biomass Gasification, Pyrolysis and Torrefaction - Practical Design and Theory. s.l.: Academic Press, 2nd Edition.

Biomass for Power Generation, Renewable Energy Technologies: Cost Analysis Series 2012, IRENA, International Renewable Energy Agency, Volume 1: Power Sector, June 2012.

Cascarosa, E., L. Gasco, G. Gea, J.L. Sánchez and J. Arauzo. 2011. Co-gasification of meat and bone meal with coal in a fluidised bed reactor. Fuel 90: 2798–2807.

Couto, N., V.B. Silva, E. Monteiro, A. Rouboa and P. Brito. 2017. An experimental and numerical study on the Mischantus gasification by using a pilot scale gasifier. Renewable Energy 109: 248–261.

Cummins, E., K.P. McDonnel and S. Ward. 2006. Dispersion modelling and measurement of emissions from the co-combustion of meat and bone meal with peat in a fluidized bed. Bioresource Technology 97: 903–913.

De Andrés, J.M., A. Narros and M.E. Rodríguez. 2011. Air-steam gasification of sewage sludge in a bubbling bed reactor: Effect of alumina as a primary catalyst. Fuel Process. Technol. 92: 433–440.

Deydier, E., R. Guilet, S. Sarda and Patrick Sharrock. 2005. Physical and chemical characterization of crude meat and boné meal combustion residue waste or raw material? Journal of Hazardous Materials 121: 141–148.

Di Gianfilippo, M., G. Costa, S. Pantini, E. Allegrini, F. Lombardi and T.F. Astrup. 2016. LCA of management strategies for RDF incineration and gasification bottom ash based on experimental leaching data. Waste Manag. 47: 285–298.

Di Lonardo, M.C., M. Franzese, G. Costa, R. Gavasci and F. Lombardi. The application of SRF vs. RDF classification and specifications to the material flows of two mechanical-biological treatment plants of Rome: Comparison and implications. Waste Manag., vo.

Ergut, A., Y. Levendis and J. Carlson. 2017. Emissions from the combustion of polystyrene, styrene and ethylbenzene under diverse conditions. Fuel 86(12-13): 1789–1799.

Furness, D.T., L.A. Hoggett and S.J. Judd. 2000. Thermochemical treatment of sewage sludge. J. CIWEM 14: 57–65.

Fytili, D. and A. Zabaniotou. 2008. Utilization of sewage sludge in EU application of old and new methods—a review. Renew. Sustain. Energy Rev. 12: 116–140.

Gallardo, A., M. Carlos, M.D. Bovea, F.J. Colomer and F. Albarran. 2014. Analysis of refuse-derived fuel from the municipal solid waste reject fraction and its compliance with quality standards. J. Clean. Prod. 83: 118–125.

Gonçalo Lourinho, Paulo S.D. Brito, Tiago Gaio and Helia Pereira. 2015. Assessment of the potential of biomass energy from agroforestry residues in a region of Portugal, Alto Alentejo. Energy Volume 81, 1 March 2015, pp. 189–201.

Gulyurtlu, I., D. Boavida, P. Abelha, M.H. Lopes and I. Cabrita. 2005. Co-gasification of coal and meat and bone meal. Fuel 84(17): 2137–2148.

Gulyurtlu, I., D. Boavida, P. Abelha, P. Teixeira, T. Crujeira, F. Marques and I. Cabrita. 2006. Co-combustion for fossil fuel replacement and better environment. 7th European Conference on Industrial on Industrial Furnaces and Boilers, April, Porto Portugal 18–21.

Han, Z., H. Ma, G. Shi, L. He, L. Wei and Q. Shi. 2016. A review of groundwater contamination near municipal solid waste landfill sites in China. Sci. Total Environ. 569-570: 1255–1264.

Judex, J.W., M. Gaiffi and H.C. Burgbacher. 2012. Gasification of dried sewage sludge: Status of the demonstration and the pilot plant. Waste Manag. 32: 719–723.

Kaewluan, S. and S. Piparmanoma. 2011. Gasification of high moisture rubber woodship with rubber waste in a bubbling fluidized bed. Fuel Processing Technology 91: 671–677.

Kukačka, J. and R. Raschman. 2010. Possibilities of municipal plastic waste energy recovery. Odpadové fórum (Waste Management Forum) 10/2010, 14–16.

Lee, K.W., W.C. Lee, H.J. Lee and J.I. Dong. 2014. Gasification characteristics of sewage sludge combined with wood biomass. J. Mater. Cycles Waste Manag. 16: 642–649.

Lee, U., J.N. Chung and H. Ingley. 2014. High-temperature steam gasification of municipal solid waste. Rubber. Plastic and Wood. Energy Fuels 28: 4573–4587.

Lombardi, L., E. Carnevale and A. Corti. 2015. A review of technologies and performances of thermal treatment systems for energy recovery from waste. Waste Manag. 37: 26–44.

Lumley, N.P.G., D.F. Ramey, A.L. Prieto, R.J. Braun, T.Y. Cath and J.M. Porter. 2014. Techno-economic analysis of wastewater sludge gasification: A decentralized urban perspective. Biosour. Technol. 161: 385–394.

Luz, F.C., M.H. Rocha, E.E.S. Lora, O.J. Venturini, R.V. Andrade, M.M.V. Leme and O.A. Olmo. 2015. Techno-economic analysis of municipal solid waste gasification for electricity generation in Brazil. Energy Conversion and Management 103: 321–337.

Manara, P. and A. Zabaniotou. 2012. Towards sewage sludge based biofuels via thermochemical conversion—a review. Renew. Sustain. Energy Rev. 16: 2566–2582.

Marculescu, C., G. Ionescu, S. Ciuta and Constantin Stan. 2013. Energetic analysis of meat processing industry waste#. UPB Sci. Bull Series C Vol. 75(2).

McDonnell, K., J. Desmond, J.J. Leahy, R. Howard-Hildige and S. Wardm. 2001. Behaviour of meat and bone meal/peal pellets in a bench scale fluidized bed combustor. Energy 26: 81–90.

Milne, T.A. and R.J. Evans. 1998. Biomass Gasifier 'Tars': Their Nature, Formation, and Conversion. National Renewable Energy Laboratory, available at: http://www.nrel.gov/docs/fy99osti/25357.pdf.

Molino, A., S. Chianese and D. Musmarra. 2016. Biomass gasification technology: The state of the art overview. J. Energy Chem. 25: 10–25.

Nicholas, P.G. Lumleya, Dotti F. Rameyb, Ana L. Prietob, Robert J. Brauna, Tzahi Y. Cathb and Jason M. Porter. 2014. Techno-economic analysis of wastewater sludge gasification: a decentralized urban perspective. Bioresource Technology 161: 385–394.

Park, C., U. Zahid, S. Lee and C. Han. 2015. Effect of process operating conditions in the biomass torrefaction: A simulation study using one-dimensional reactor and process model. Energy 79: 127–139.

Rajapaksha, A.U., S.S. Chen, D.C.W. Tsang, M. Zhang, M. Vithanage, S. Mandal, B. Gao, N.S. Bolan and Y.S. Ok. 2016. Engineered/designer biochar for contaminant removal/immobilization from soil and water: Potential and implication of biochar modification. Chemosphere 148: 276–291.

Roche, E., J.M. De Andrés, A. Narros and M.E. Rodríguez. 2014. Air and air-steam gasification of sewage sludge. The influence of dolomite and throughput in tar production and composition. Fuel 115: 54–61.

Roche, E., J. Almeida, A. Narros and E. Rodríguez. 2014. Air and air-steam gasification of sewage sludge. The influence of dolomite and throughput in tar production and composition. Fuel 115: 54–61.

Rowhani, A. and T. Rainey. 2016. Scrap tyre management pathways and their use as fuel—A review. Energies October 2016.

Sansaniwal, S.K., K. Pal, M.A. Rosen and S.K. Tyagi. 2017. Recent advances in the development of biomass gasification technology: A comprehensive review. Renew. Sust. Energ. Rev. 72: 363–384.

Shen, Y. 2015. Chars as carbonaceous adsorbents/catalysts for tar elimination during biomass pyrolysis or gasification. Renew. Sustain. Energy Rev. 43: 281–295.

Silva, V., N. Couto, D. Eusébio, A. Rouboa, P. Brito, J. Cardoso and M. Trninic. 2017. Multi-stage optimization in a pilot scale gasification plant. International Journal of Hydrogen Energy 42: 23878.

Soni, C.G., Z. Wang, A.K. Dalai, T. Pugsley and T. Fonstad. 2009. Hydrogen production via gasification of meat and bone meal in two-stage fixed bed reactor system. Fuel 88: 920–925.

Straka, P., V. Kriz and Z. Bucko. 2008. Co-gasification of rubber with Brown coal. Acta Geodyn. Geomater 5(151): 329–334.

Straka, P., V. Kriz and Z. Bucko. 2009. Co-combustion of lignite/waste-tyre mixture in a moving bed. Fuel 90: 1202–1206.

Wang, Z., H. Richter, J.B. Howard, J. Jordan. J. Carlson and Y. Levendis. 2014. Laboratory investigation of the products of the incomplete combustion of polymers (plastics). Ind. Eng. Chem. Res. V 43: 2873–2886.

Wasilewski, R. and T. Siudyga. 2013. Energy recovery from waste plastics. CHEMIK 67(5): 435–445.

Woolcock, P.J. and R.C. Brown. 2013. Review: a review of cleaning technologies for biomass-derived syngas. Biomass and Bioenergy 52: 54–84.

Wu, M.H.H., C.L.L. Lin and W.Y.Y. Zeng. 2014. Effect of waste incineration and gasification processes on heavy metal distribution. Fuel Process Technol. 125: 67–72.

www.galpenergia.com, consulted in 01/2018. 2018.

Xiao, G., M. Ni, Y. Chi and K. Cen. 2008. Low-temperature gasification of waste tire in a fluidized bed F. Energy Convers. Manag. 49: 2078–2082.

Effect of Wood Welding Treatment on Chemical Constituents of Australian Eucalyptus Species

Benoit Belleville,[1,]* *Georges Koumba-Yoya*[2] and
Tatjana Stevanovic[2]

Introduction

The potential of wood welding as a fast and cost-effective alternative to gluing for Australian hardwood species has been demonstrated in recent studies (Belleville et al. 2016, 2017). The technique consists in assembling solid wood pieces by mechanical friction to generate heat, which induces the thermochemical changes in lignin which lead to wood welding. The results so far confirm the importance of density and occurrence of anatomical features in the weld line strength and consequently the definition of optimal welding parameter (Leban et al. 2005, Properzi et al. 2005, Belleville et al. 2016). Polymeric material and other compounds present in woods studied previously have commonly been identified to explain observed differences between Canadian wood species (Rodriguez et al. 2010, Sun et al. 2010, Belleville et al. 2013, 2016).

---

[1] School of Ecosystem and Forest Sciences, Faculty of Science, University of Melbourne, Victoria, Australia.
[2] Centre de recherche sur les matériaux renouvelables, Faculté de foresterie et de géomatique, Université Laval, Québec, Canada.
* Corresponding author: benoit.belleville@unimelb.edu.au

However, it is not clear to what extent the chemical composition of a species is related to the thermochemical reactions occurring during the welding process and ultimately the weldline strength. Such knowledge would have the potential of facilitating the identification of potential future species and provide a better understanding of past results. Consequently, the aim of the present study was to identify the thermochemical changes occurring during the welding and establish a correlation with the weldline strength of selected Eucalyptus species. The specific objectives were to (1) investigate and identify the thermochemical changes occurring at the welding interface of two Australian hardwood species; and (2) explain the influence of these chemical changes on the weldline strength.

Materials and Methods

Specimen preparation

Species selected because of their commercial value and potential for wood welding were *Eucalyptus saligna* (air dry density of 784 kg m^{-3} at 12% MC) and *Eucalyptus pilularis* (925 kg m^{-3}). The wood material was pre-conditioned at 23°C and 65% relative humidity (RH) until constant mass was reached. Clear, straight-grained wood slats 30 mm x 30 mm x 800–900 mm were selected, dressed and used to prepare welded specimens. Commercially manufactured smooth wood dowels 9.45 mm in diameter and 100 mm in length were inserted into pre-drilled holes 25 mm in depth and 7.5 mm in diameter using a manually operated fixed base drill (McMillan BD16, ¾ HP, Taiwan). Species were not mixed between slats or dowels. The welding parameters used for both species were a rotational speed of 1230 rpm and a 1.26 ratio of dowel diameter to receiving hole based on results from a previous study (Belleville et al. 2016). Following welding, the material was returned at 23°C and 65% RH for 7 days prior to testing. Each specimen consisted of one dowel inserted into a wood slat substrate. According to standard tensile strength test method ASTM D1037 (ASTM 2006) and using a universal testing machine (Instron 5569, Massachusetts, USA), the dowel excess was pulled off the wood substrate. Chemical analyses were performed on both species. Samples for pyrolysis (Py-GC/MS) and thermogravimetric analyses (TGA) analysis were collected by scraping wood from *unwelded* and *welded* samples. Wood-welded dowels were cut transversally to prepare samples for X-ray photoelectron spectroscopy (XPS) and attenuated total reflection Fourier transform infrared spectroscopy (ATR-FTIR).

Analytical analysis before and after wood-welding treatment

- ATR-FT-IR
  Normalized FT-IR spectra were obtained for each sample (welded and surrounding wood) using a Fourier transform infrared spectrometer (ATR-FT-IR/FT-NIR PerkinElmer Spectrum 400, USA). The FTIR spectra were recovered for 64 scans and collected for wave numbers ranging from 4000 to 650 cm^{-1}.

- Pyrolysis GC/MS

Pyrolysis GC/MS was performed using a filament pulse pyrolyzer (Pyro-prob 2000 CDS Analytical Inc, CDS Analytical, PA, USA) coupled to a GC-MS system. The GC-MS consists of a gas chromatograph from Varian (CP 3800) coupled with a mass spectrometer from Varian Saturn 2200 (MS/MS, 30–650 u.m.a.). An amount of 0.4 mg of sample was dried during 30 s at 100°C. The temperature of the pyrolyzer transfer line and the GC injector were both set at 250°C. The sample was pyrolyzed according to the following program: the transfer line temperature was maintained during 10 s and then increased to 550°C at a rate 20°C/s and held for 10 s. Helium was used as the vector gas and the flow rate of the carrier gas was 1.0 ml/min. A VF-5 ms capillary column was used. The oven temperature program was as follows: the temperature was kept at 45°C for 1 min and then it was increased to the final temperature of 250°C at a rate of 5°C/min and then held for 5 min. The mass spectrometer was operated in electron impact mode (EI, 70 eV, m/z = 35–400) at 1 s per scan. There were 3 repetitions for each sample examined. Each chromatogram peak was identified with the NIST Mass Spectral Library (NIST 2017).

- XPS

XPS analysis was performed with an Axis-Ultra instrument (Kratos Analytical, UK) consisting of three communicating chambers: the analysis chamber comprising the ESCA analyzer, the preparation chamber, and the introduction chamber. Base pressure in the analysis chamber was 5×10^{-10} Torr. The X-ray source was a monochromatic Al source operated at 300 W. The analyzer was run in the constant pass energy mode, with the lens system in the "hybrid" configuration and the electrostatic lens aperture in the "slot" position. This assured the highest sensitivity with an analyzed spot approximately 800 μm x 400 μm in size, equivalent to the size of the monochromatic X-ray beam. Electron counting was performed with an 8 channel detector. The electrostatic charge appearing on electrically insulating samples under X-ray irradiation was neutralized with an integrated, very low energy electron flood gun, whose parameters were calibrated against standard reference samples:

Au4f7/2: 83.95 eV   Ag3d5/2: 368.2 eV   Cu2p3/2: 932.6 eV.

These parameters were set to optimize energy resolution and counting rate. Survey scans were recorded with a pass energy of 160 eV and a step size of 1eV. Detailed high resolution spectra were recorded at 10 eV, 20 eV or 40 eV pass energy with a step size of 0.025 eV, 0.05 eV or 0.1 eV, depending on the amount of the element.

- TGA

Thermogravimetric analyses of samples were performed on TGA/SDTA851 from Mettler Toledo instrument following the procedure described by Koumba et al. (2017). Typically, thermogravimetric analyses were conducted under nitrogen from 25 to 800°C at a rate of 20°C/min and ultimately maintained at 800°C for 30 min.

Results and Discussion

ATR-FTIR before and after wood-welding treatment

The chemical changes of studied eucalyptus wood constituents during wood welding treatment have been investigated by FTIR analysis. The thermochemical changes in lignin structure were followed using the data from FT-IR spectrum for calculation of condensation index using the equation proposed by Faix et al. (1991) [Eqn. (1)]:

$$Condensation\ index\ (CI) = \frac{\Sigma\ minima\ between\ 1500\ and\ 1050\ cm^{-1}}{\Sigma\ maxima\ between\ 1600\ and\ 1030\ cm^{-1}} \qquad (1)$$

The calculated condensation indices allowed comparing the polymerization reaction of lignin during the wood-welding experiment. Thus, an increase of condensation index could be attributed to lignin condensation reactions which translate its adhesive properties and ultimately the efficiency of wood-welding.

According to the results presented in Table 1, *Eucalyptus pilularis* seems to contain less condensed substructures than *Eucalyptus saligna*, already in its original state. As can be expected, *Eucalyptus saligna* also provided significantly stronger welded joint strength when compared with *Eucalyptus pilularis* (Belleville et al. 2016). The same trend is found when the welding line of the welded samples is studied as that determined for *Eucalyptus saligna* was determined to be higher than the condensation index determined for weldline of *E. pilularis* wood weldline. Indeed, as the condensation index is proposed primarily for lignin structure considerations, the obtained results could indicate that adhesive properties of *Eucalyptus saligna* lignin are more favorable for welding treatment than those of *Eucalyptus pilularis* lignin.

On the other hand, FT-IR spectra of welded *Eucalyptus pilularis* shows more carbonyl function (C = O acid around 1740 cm^{-1}) than original *Eucalyptus pilularis*. These results could be attributed to carbohydrate solicitation during welding process which become more reactive during welding process because of lignin was less accessible in regard to index condensation.

Also, Fig. 1 shows marked differences mainly at 1230 cm^{-1} which indicate more conjugated ester linkages (C = O) that are more important for welded *Eucalyptus pilularis*.

Table 1. Condensation index calculation before and after the welding treatment.

Source	Eucalyptus pilularis		Eucalyptus saligna	
Treatment	Original wood	Welding line	Original wood	Welding line
Condensation index	0.79	0.86	0.86	0.98

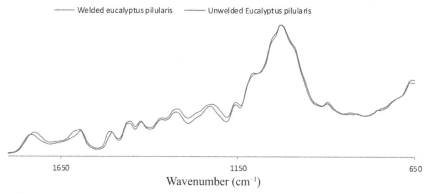

———— Welded eucalyptus pilularis   ———— Unwelded Eucalyptus pilularis

1650   1150   650

Wavenumber (cm⁻¹)

Fig. 1. FT-IR spectra of original (red line) and welded (blue line) wood of *Eucalyptus pilularis*.

The FTIR spectra of *Eucalytus saligna* (Fig. 2) shows that the relative intensity of the broad peak at 3424 cm⁻¹ increased after welding could tentatively be attributed to aldolisation reaction between aliphatic OH of lignin and carbonyl of carbohydrates.

According to Hergert (1971), the aromatic skeletal vibrations around 1510 cm⁻¹ are pure bands of lignin, whereas the 1600 cm⁻¹ aromatic skeletal is a superimposed band that has broadened by the C = O stretching mode. Moreover, the greatly weakened band of carboxylic acid C = O stretching vibrations at 1740 cm⁻¹ was lower in *Eucalyptus saligna* after welding. The thermochemical modifications following the welding process in eucalyptus species are in accordance with those observed for two Canadian hardwood species, namely *Acer saccharum* and *Betula alleghaniensis* (Belleville et al. 2013).

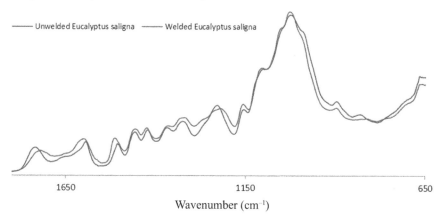

———— Unwelded Eucalyptus saligna   ———— Welded Eucalyptus saligna

1650   1150   650

Wavenumber (cm⁻¹)

Fig. 2. FT-IR spectra of original (red line) and welded (blue line) of *Eucalyptus saligna*.

Pyrolysis GC/MS

The compounds identified by Py-GC/MS have been identified as CH for carbohydrate-derivatives, G-S-H for lignin (Guayacyl, Syringyl and p-Hydroxyl)

and N as protein related. Each pyrogram was subdivided into three regions, the first one between 1 and 15 min, attributed to carbohydrate derivatives. Indeed, the compounds derived from the carbohydrates are easily fragmented into smaller compounds (Lourenço et al. 2015). The second region corresponds to lignin derivatives with phenolic units, which corresponds to range between 15 and 35 min in chromatogram. The third region, covering a mixture of compounds including extractives, lipids (fatty acid chains) and terpenoïd derivative corresponds to region of chromatogram after 30 minutes. According to the Py-GC/MS results, the most important differences between original wood and welded material are observed after 30 min, as can be seen from Fig. 3 and the results are presented in Tables 2 and 3.

Indeed, most difference between the unwelded and welded material appears by a contribution of the extractive compounds, fatty acids chains and terpenoids which were generally identified between 35 and 50 min according to theses pyrolysis GC-MS conditions.

Thus, when the material is welded, these groups of molecules, including lipids as fatty acid chains, are degraded as shown by the disappearance of the signal to 43.7 min which was present in the original material and absent after welding (Fig. 3). Such as observed in the FT-IR analysis about esterification reaction, this can be confirmed by pyrolysis GC-MS where fatty acids which were initially present in the original wood, reacted with hydroxyl groups of wood contributing to adhesive properties by formation of new covalent bonds. On the other hand, oxidation transformation of fatty acids led to formation of a variety of carboxylic acids, including citric acid (Randle et al. 1963). Interestingly, the application of citric acid during welding process has been demonstrated to contribute to the improvement of mechanical properties of the welded samples (Amirou et al. 2017).

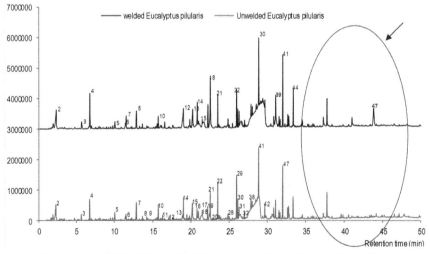

Fig. 3. Pyrolysis GC-MS analysis of *Eucalyptus pilularis* before and after welding.

Table 2. Compounds released after Py-GC/MS of unwelded and welded *Eucalyptus pilularis*.

Label	*Unwelded Eucalyptus pilularis*			*Welded Eucalyptus pilularis*		
	Compound name	Origin	rt	*Compound name*	Origin	rt
1	Propanoic acid	CH	1.90	Propanoic acid, 2-oxo	CH	1.94
2	Acetic acid	CH	2.31	Acetic acid	CH	2.35
3	2(5H)-furanone	CH	5.67	2(5H)-Furanone	CH	5.71
4	Furfural	CH	6.78	Furfural	CH	6.82
5	Cyclopentanedione	CH	10.12	1,3-Cyclopentanedione	CH	10.13
6	Butanoic acid, cyclic hydrazide	N	11.56	Furfural,5-methyl	CH	11.49
7	Oxazolidine, 2,2-diethyl-3-methyl	N	12.89	Propenamide, N-(aminocarbonyl)	N	11.58
8	1,2-cyclopentandione, 3-methyl	CH	13.67	N-butyl-tert-butylamine	N	12.91
9	2-pentanoic acid, 2-methyl	CH	14.38	1,2-cyclopentandione, 3-methyl	CH	13.68
10	Gaiacol	G	15.74	Gaiacol	G	15.75
11	2,4(3H,5H)-furanone, 3-methyl	CH	17.14	Levoglucosenone	CH	16.57
12	1H-imidiazole-2-carboxaldehyde	N	17.64	Phenol, 2-methoxy-4-methyl	G	19.00
13	4H-Pyran-4-one, 3,5-dihydroxy-2-methyl	CH	18.88	1,4:3,6-Dianhydro-alpha-D-glucopyransoe	CH	19.83
14	Crésol	H	18.98	2-furancarboxaldehyde, 5-(hydroxymethyl)	CH	20.20
15	Furancarboxaldehyde, 5-(hydroxymethyl)	CH	20.16	1,3-Di-O-acetyl-alpha-beta-D-ribopyranose	CH	20.80
16	3,4-anhydro-D-galactosan	CH	20.76	1,2-benzendiol, 3-methoxy	G	20.97
17	1,2-benzendiol, 3-methoxy	G	20.95	1-methyl-3-piperidinemethanol	CH	21.10
18	phenol, 4-ethyl-2-methoxy	G	21.45	Phenol, 4-ethyl-2-methoxy	G	21.45
19	1,3-Di-O-acetyl-alpha-beta-D-ribopyranose	CH	22.17	2-Vinyl-9-[beta-D-ribofuranosyl]hypoxanthine	CH	21.57
20	alpha-D-glucopyranoside, O-alpha-D-glucopyranosyl-(1.fwdarw.3)-beta-D-fructofuranosyl	CH	22.40	D-glycero-D-ido-heptose	CH	21.67
21	2-methoxy-4-vinylphenol	G	22.50	3,4-anhydro-D-galactosan	CH	22.21
22	Phenol, 2,6-dimethoxy	S	23.49	2-methoxy-4-vinylphenol	G	22.51
23	Eugenol	G	23.61	Phenol, 2,6-dimethoxy	S	23.49
24	Phenol, 3,4-dimethoxy	S	23.70	Phenol, 3,4-dimethoxy	S	23.72
25	Phenol, 2-methoxy-4-propyl	G	23.88	Benzaldehyde, 3-hydroxy-4-methoxy	G	24.83

Table 2 contd. ...

... Table 2 contd.

Label	Unwelded Eucalyptus pilularis			Welded Eucalyptus pilularis		
	Compound name	Origin	rt	Compound name	Origin	rt
26	D-allose	CH	24.55	3-hydroxy-4-methoxybenzoic acid	G	25.95
27	Vanillin	G	24.82	Isoeugenol	G	26.12
28	Phenol, 2-methoxy-4-(1-propenyl)	G	24.98	Phenol, 2-methoxy-4-propyl	G	26.29
29	Benzoic acid, 4-hydroxy-3-methoxy	G	25.95	1,6-anhydro-beta-D-glucopyranose (levoglucosan)	CH	26.55
30	Isoeugenol	G	26.12	Ethanone, 1-(4-hydroxy-3-methoxyphenyl	G	27.71
31	Euganol	G	26.28	3,4-altrosan	CH	27.85
32	Levoglucosan	CH	26.43	5-tert-butylpyrogallol	H	28.02
33	1-(4-hydroxybenzylidene) acetone	G	26.87	2-propanone, 1-(4-hydroxy-3-methoxyphenyl)	G	28.19
34	Acetovanillone	G	27.00	3,4-anhydro-D-galactosan	CH	28.72
35	3,4-dimethoxy-DL-phenylalanine	N	27.46	2H-1-benzopyran-3,4-diol, 2-(3,4-dimethoxyphenyl)-3,4-dihydro-6-methyl	S	28.94
36	D-glucose, 4-O-alpha-D-glucopyranosyl	CH	27.62	alpha-D-glucopyranose,4-O-beta-D-galactopyranosyl	CH	29.49
37	5-tert-butylpyrogallol	H	27.85	2-propenoic acid, 3-(4-hydroxy-3-methoxyphenyl)	G	29.55
38	2-propannoe, 1-(4-hydroxy-3-methoxyphenyl)	G	28.00	Benzen, 2-acetate-1,3-dimethoxy-5-(1-propenyl)	S	30.83
39	1-nitro-beta-D-arabinofuranose, tetraacetate	CH	28.17	Benzaldehyde, 4-hydroxy-3,5-dimethoxy	S	31.10
40	3,4-altrosan	CH	28.72	Dihydrocoumarin, 4,4-dimethyl-6-hydroxy	S	31.53
41	3,5-dimethoxyacephenone	S	28.87	Phenol, 2,6-dimethoxy-4-(2-propenyl)	S	32.02
42	Phenol, 2,6-dimethoxy-4-(2-propenyl)	S	29.67	Ethanone, 1-(2-hydroxy-4,6-dimethoxyphenyl)	S	32.67
43	1,6-anhydro-alpha-D-galactofuranose	CH	30.63	Cinnamaldehyde	G	32.81
44	2-propenic acid, 3-(4-hydroxy-3-methoxyphenyl)	G	30.82	Desaspidinol	G	33.39
45	Benzaldehyde, 4-hydroxy-3,5-dimethoxy	S	31.07	3,5-dimethoxy-4-hydroxycinnamaldehyde	S	37.80
46	3-buten-2-one, 4-(4-hydroxy-3-methoxyphenyl)	G	31.52	Octadecanoic acid	Fatty acid	41.03
47	3-hydroxy-4-methoxycinnamic acid	G	32.00	2,6,7,10-hexadecatetraenoic acid	Fatty acid	43.84

Table 3. Compounds released after Py-GC/MS of unwelded and welded Eucalyptus saligna.

Label	Unwelded Eucalyptus saligna			Welded Eucalyptus saligna		
	Compound name	Origin	rt	*Compound name*	Origin	rt
1	Propanoic acid	CH	1.90	Propanoic acid, 2-oxo	CH	1.94
2	acetic acid	CH	2.31	acetic acid	CH	2.35
3	2(5H)-furanone	CH	5.67	2(5H)-Furanone	CH	5.71
4	Furfural	CH	6.78	Furfural	CH	6.82
5	Cyclopentanedione	CH	10.12	1,3-Cyclopentanedione	CH	10.13
6	Butanoic acid, cyclic hydrazide	N	11.56	Furfural,5-methyl	CH	11.49
7	oxazolidine, 2,2-diethyl-3-methyl	N	12.89	Propenamide, N-(aminocarbonyl)	N	11.58
8	1,2-cyclopentandione, 3-methyl	CH	13.67	N-butyl-tert-butylamine	Pro	12.91
9	2-pentanoic acid, 2-methyl	CH	14.38	1,2-cyclopentandione, 3-methyl	CH	13.68
10	Gaiacol	G	15.74	Gaiacol	G	15.75
11	2,4(3H,5H)-furanone, 3-methyl	CH	17.14	Levoglucosenone	CH	16.57
12	1H-imidiazole-2-carboxaldehyde	N	17.64	Phenol, 2-methoxy-4-methyl	G	19.00
13	4H-Pyran-4-one, 3,5-dihydroxy-2-methyl	CH	18.88	1,4:3,6-Dianhydro-alpha-D-glucopyransoe	CH	19.83
14	Crésol	H	18.98	2-furancarboxaldehyde, 5-(hydroxymethyl)	CH	20.20
15	Furancarboxaldehyde, 5-(hydroxymethyl)	CH	20.16	1,3-Di-O-acetyl-alpha-beta-D-ribopyranose	CH	20.80
16	3,4-anhydro-D-galactosan	CH	20.76	1,2-benzendiol, 3-methoxy	G	20.97
17	1,2-benzendiol, 3-methoxy	G	20.95	1-methyl-3-piperidinemethanol	CH	21.10
18	phenol, 4-ethyl-2-methoxy	G	21.45	phenol, 4-ethyl-2-methoxy	G	21.45
19	1,3-Di-O-acetyl-alpha-beta-D-ribopyranose	CH	22.17	2-Vinyl-9-[beta-D-ribofuranosyl]hypoxanthine	CH	21.57
20	alpha-D-glucopyranoside, O-alpha-D-glucopyranosyl-(1.fwdarw.3)-beta-D-fructofuranosyl	CH	22.40	D-glycero-D-ido-heptose	CH	21.67
21	2-methoxy-4-vinylphenol	G	22.50	3,4-anhydro-D-galactosan	CH	22.21
22	phenol, 2,6-dimethoxy	S	23.49	2-methoxy-4-vinylphenol	G	22.51
23	Eugenol	G	23.61	phenol, 2,6-dimethoxy	S	23.49
24	phenol, 3,4-dimethoxy	S	23.70	phenol, 3,4-dimethoxy	S	23.72
25	Phenol, 2-methoxy-4-propyl	G	23.88	Benzaldehyde, 3-hydroxy-4-methoxy	G	24.83
26	D-allose	CH	24.55	3-hydroxy-4-methoxybenzoic acid	G	25.95

Table 3 contd. ...

... Table 3 contd.

Label	Unwelded Eucalyptus saligna			Welded Eucalyptus saligna		
	Compound name	Origin	rt	*Compound name*	Origin	rt
27	Vanillin	G	24.82	Isoeugenol	G	26.12
28	Phenol, 2-methoxy-4-(1-propenyl)	G	24.98	Phenol, 2-methoxy-4-propyl	G	26.29
29	Benzoic acid, 4-hydroxy-3-methoxy	G	25.95	1,6-anhydro-beta-D-glucopyranose (levoglucosan)	CH	26.55
30	Isoeugenol	G	26.12	Ethanone, 1-(4-hydroxy-3-methoxyphenyl	G	27.71
31	Euganol	G	26.28	3,4-altrosan	CH	27.85
32	Levoglucosan	CH	26.43	5-tert-butylpyrogallol	H	28.02
33	1-(4-hydroxybenzylidene) acetone	G	26.87	2-propanone, 1-(4-hydroxy-3-methoxyphenyl)	G	28.19
34	Acetovanillone	G	27.00	3,4-anhydro-D-galactosan	CH	28.72
35	3,4-dimethoxy-DL-phenylalanine	N	27.46	2H-1-benzopyran-3,4-diol, 2-(3,4-dimethoxyphenyl)-3,4-dihydro-6-methyl	S	28.94
36	D-glucose, 4-O-alpha-D-glucopyranosyl	CH	27.62	alpha-D-glucopyranose,4-O-beta-D-galactopyranosyl	CH	29.49
37	5-tert-butylpyrogallol	H	27.85	2-propenoic acid, 3-(4-hydroxy-3-methoxyphenyl)	G	29.55
38	2-propannoe, 1-(4-hydroxy-3-methoxyphenyl)	G	28.00	Benzen, 2-acetate-1,3-dimethoxy-5-(1-propenyl)	S	30.83
39	1-nitro-beta-D-arabinofuranose, tetraacetate	CH	28.17	Benzaldehyde, 4-hydroxy-3,5-dimethoxy	S	31.10
40	3,4-altrosan	CH	28.72	Dihydrocoumarin, 4,4-dimethyl-6-hydroxy	S	31.53
41	3,5-dimethoxyacephenone	S	28.87	Phenol, 2,6-dimethoxy-4-(2-propenyl)	S	32.02
42	phenol, 2,6-dimethoxy-4-(2-propenyl)	S	29.67	Ethanone, 1-(2-hydroxy-4,6-dimethoxyphenyl)	S	32.67
43	1,6-anhydro-alpha-D-galactofuranose	CH	30.63	Cinnamaldehyde	G	32.81
44	2-propenic acid, 3-(4-hydroxy-3-methoxyphenyl)	G	30.82	Desaspidinol	G	33.39
45	Benzaldehyde, 4-hydroxy-3,5-dimethoxy	S	31.07	3,5-dimethoxy-4-hydroxycinnamaldehyde	S	36.60
46	3-buten-2-one, 4-(4-hydroxy-3-methoxyphenyl)	G	31.52	Lauric acid, 2-methylbutyl ester	Fatty acid	39.89
47	3-hydroxy-4-methoxycinnamic acid	G	32.00	Myristic acid isobutyl ester	Fatty acid	43.33

The contribution of the fatty acids to the adhesive properties through formation of covalent bonds appears to be more marked for *Eucalyptus saligna* with important content of these compounds as indicated by the peaks appearing in the pyrograms of original wood, between 35 and 50 min (Fig. 4).

Most likely, the extractive components which are free molecules found in wood porous structure, are easily released upon temperature increase during welding. Consequently, the fact that *Eucalyptus saligna* has more important fatty acid content could be interpreted as their contribution to more important adhesive properties as compared to *Eucalypts pilularis* (Figs. 3 and 4). In addition, although the role of lignin as natural wood adhesive is well known, the structural changes of this polymer are difficult to follow by pyrolysis GC/MS analyses. Indeed, compounds

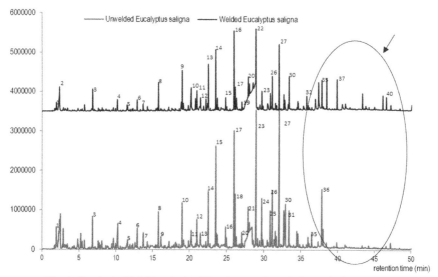

Fig. 4. Pyrolysis GC-MS analysis of *Eucalyptus saligna* before and after welding.

derived from lignin break down appearing between 15 and 35 min, Figure 3 shows a similar profile before and after the welded process in this region.

XPS

To validate the formation of new covalent bonds, XPS analysis of weldline has been performed for samples from the two studied species. The obtained results are presented in Figs. 5 and 6. Indeed, structural changes could be observed in XPS for processed samples for which the presence or absence of chemical features will be an indication of the effect of the treatment on the composition of the structural components. Changes in the intensity of the peaks to component C1s induce a more-or-less pronounced welding effect on the chemical features of wood components. Indeed, the increase in the proportions of the C1 component

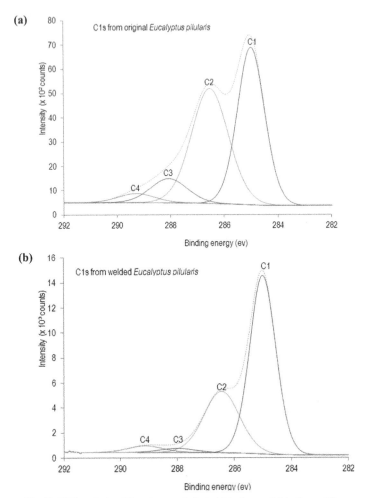

Fig. 5. XPS analysis of *Eucalyptus pilularis* (a) before and (b) after welding.

(Fig. 5) at 285 eV, corresponding to the C-C and C-H bonds confirms the formation of condensed structure such as identified by FT-IR analysis for *Eucalyptus p(luraris* welding.

Indeed, the formation of C-C bonds corresponds to the contribution of the lignin in the welding process where guayacyl units react together by radical coupling to form the units of 5–5 type. Conversely, the intensity corresponding to components C2 (287 eV) and C3 (288 eV) decreased after welding, reflecting the formation of compounds from aldolization by condensation between hydroxyl and aldehyde of polysaccharides in what concerns the decrease in intensity of the C-O bonds. As for the C-O-C bonds, their cleavage would give rise to a production of phenolic hydroxyls from lignin. The aldolization reactions are also confirmed by the decrease

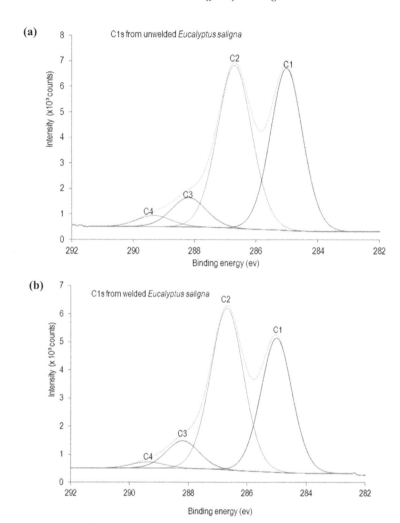

Fig. 6. XPS analysis of *Eucalyptus saligna* (a) before and (b) after welding.

of intensity in C3 component after welding since aldehyde functions are involved in these reactions (Table 4).

As for the C4 corresponding to the ester link, it is interesting to note that an almost identical intensity before and after welding is observed, reflecting possible reaction of depolymerization and repolymerization by formation of new esters link as discussed previously. On the other hand, these type O-C=O bonds could be related to derivatives of furfural, generated by hemicellulose degradation during processing.

Regarding *Eucalyptus saligna*, the XPS analyses indicate less marked variations (Fig. 6), including the intensity level of the C1 component while the intensity of C-C bonds increased slightly from 39.27% to 42.34% (Table 5).

Table 4. XPS results for original and welded *Eucalyptus piluraris* wood.

Entry	Name	Bond type	Position	Atomic %	
				Unwelded	Welded
1	C_1	C–C, C–H	285	48.85	64.53
2	C_2	C–O, C–O–C	287	43.50	30.27
3	C_3	O–C–O, C = O	288	9.20	1.98
4	C_4	O–C = O	289	3.45	3.21

Table 5. XPS of unwelded and welded *Eucalyptus saligna*.

Entry	Name	Bond type	Position	Atomic %	
				Unwelded	Welded
1	C_1	C–C, C–H	285	39.27	42.34
2	C_2	C–O, C–O–C	287	50.19	46.21
3	C_3	O–C–O, C = O	288	8.59	8.30
4	C_4	O–C = O	289	1.95	3.15

Thus, this result confirms the observed tendency in the case of *Eucalyptus pilularis*.

In addition, important changes can be observed for the C4 component, where ester groups have almost doubled in intensity level (Table 5). This result confirms the pyrolysis GC–MS analysis results, which indicated to an important level in esterification reactions for *Eucalyptus saligna* based on its important lipids content which diminished after welding.

Observed trends in the intensity of the peaks to components C1 and C2 following welding for eucalyptus species were completely different from those obtained for Canadian hardwood species by Belleville et al. (2013). One explanation might be a higher proportion of lignin in eucalyptus species compared to Canadian species favouring more condensation reactions during the welding process. The literature reports similar data in terms of eucalyptus's chemical composition. Barrichelo and Foelkel (1976) and de Oliveira Araújo (2017) reported 26.3% and 26.0% of Klason lignin content for *Eucalyptus saligna*, respectively; Timell (1967) and Koumba-Yoya and Stevanovic (2017) reported 23.1% and 22.2% of Klason lignin content for *Acer saccharum*. Timell (1967) also reported 23.8% of lignin content for *Betula alleghaniensis*.

TGA

A thermal analysis of the studied Eucalyptus wood samples was performed in order to determine the effect of treatment in welding zone. The thermogravimetric treatment of Eucalyptus samples between 25 and 800°C was followed under nitrogen (Fig. 7).

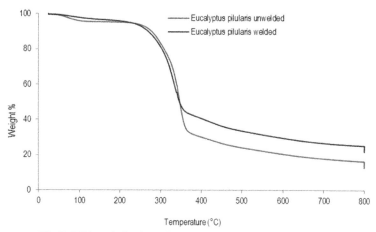

Fig. 7. TGA analysis of *Eucalyptus pilularis* before and after welding.

The thermal degradation of original *Eucalyptus pilularis* wood started at 249°C, whereas the thermal degradation of its welded sample started at 225°C. Also, the temperature corresponding to 50% weight loss of original and welded samples during thermal analysis were the same: 345°C whereas, the thermal stability is more important for welded material after 345°C. This result of higher thermal stability of welded material could be attributed to higher complexity of polymers which have increased in size following the polymerization reactions during welding. These reactions could have occurred between hydroxyl group from lignin or carbohydrate and lipids by esterification, or by giving adhesive properties through reticulation.

Conclusion

The present study was conducted with the aim of identifying the thermochemical changes occurring during the welding process in order to establish a correlation with the weldline strength of the studied Eucalyptus wood species. The thermochemical changes occurring at the welding interface of two investigated Australian hardwood species have been identified through the analysis of the results of FT-IR, Py-GC/MS, XPS and TGA studies.

Condensation indices calculated from the FT-IR data showed the presence of more condensed substructures in *Eucalyptus saligna* and a higher range of condensation following the welding treatment when compared with *Eucalyptus pilularis*. Such finding corroborates mechanical testing results where *Eucalyptus saligna* delivered stronger joints possibly as a result of better adhesive properties provided by proposed esterification reaction between lignin and fatty acids. In contrast, more carbonyl function could be attributed to carbohydrate solicitation, as expressed by more reactive during welding process due to the lignin being less accessible in regard to index condensation.

The main differences between original and welded material appeared following a degradation of extractive compounds, fatty acids chains, and terpenoids, reacting with hydroxyl groups and leading to adhesive properties by chemical linkage with new covalent bonds formation, particularly important for *Eucalyptus saligna*. On the other hand, the oxidative degradation of fatty acids was proposed as a potential source of carboxylic acid, such as citric acid, detected by Py-GC-MS, which had been previously shown to improve the mechanical properties when added during welding process.

Interestingly, a higher proportion of lignin in eucalyptus species than in previously studied Canadian hardwoods, seems to be favorable to condensation reactions during the welding process which are important for the successful wood welding.

The present study confirmed the importance of knowledge of wood chemistry for wood welding which enabled the understanding to what extent the chemical composition of a species is related to the thermochemical reactions occurring during the welding process leading to adhesion and ultimately the weldline strength. Such knowledge will ultimately facilitate the choice of wood species convenient for assembling by welding.

References

American Society for Testing and Materials. 2006. Standard test methods for evaluating properties of wood-base fiber and particle panel materials. American Society for Testing and Materials. ASTM D1037. West Conshohocken, PA, USA.

Amirou, S., A. Pizzi and L. Delmotte. 2017. Citric acid as waterproofing additive in butt joints linear wood welding. Eur. J. Wood Prod. DOI: 10.1007/s00107-017-1167-x.

Barrichelo, L.E. and C.E. Foelkel. 1976. Estudos para produção de celulose sulfato de seis espécies de eucalipto. Instituto de pesquisas e estudio florestais ESALQ/USP, Piracicaba 12: 77–95.

Belleville, B., T. Stevanovic, A. Cloutier, A. Pizzi, M. Prado, S. Erakovic, P. Diouf and M. Royer. 2013. An investigation of thermochemical changes in Canadian hardwood species during wood welding. Eur. J. Wood Prod. 71(2): 245–257.

Belleville, B., B. Ozarska and A. Pizzi. 2016. Assessing the potential of wood welding for Australian eucalypts and tropical species. Eur. J. Wood Prod. 74(5): 753–757. DOI: 10.1007/s00107-016-1067-5.

Belleville, B., S. Amirou, A. Pizzi and B. Ozarska. 2017. Optimization of wood welding parameters for Australian hardwood species. BioResources 12(1): 1007–1014. DOI:10.15376/biores.12.1.1007–1014.

de Oliveira Araújo, S., D.M. Neiva, J. Gominho, B. Esteves and H. Pereira. 2017. Chemical effects of a mild torrefaction on the wood of eight *Eucalyptus* species. Holzforschung 71(4): 291–298. DOI: 10.1515/hf-2016-0079.

Faix, O. 1991. Condensation indices of lignins determined by FTIR-spectroscopy. Holz als Roh-und Werkstoff 49: 356.

Koumba-Yoya, G. and T. Stevanovic. 2017. Study of organosolv lignins as adhesives in wood panel production. Polymers 9(2): 46. DOI:10.3390/polym9020046.

Leban, J.-M., A. Pizzi, M. Properzi, F. Pichelin, P. Gelhaye and C. Rose. 2005. Wood welding: A challenging alternative to conventional wood gluing. Scan J. For. Res. 20(6): 534–538.

Lourenço, A., D.M. Neiva, J. Gominho, A.V. Marques and H. Pereira. 2015. Characterization of lignin in heartwood, sapwood and bark from Tectona grandis using Py–GC–MS/FID. Wood Sci. Technol. 49: 159–175. DOI 10.1007/s00226-014-0684-6.

NIST. 2017. Mass Spectral Library. National Institute of Standards and Technology. Gaithersburg, MD, USA.

Hergert, H.L. 1971. Infrared spectra. pp. 267–293. *In*: Sarkanen, K.V. and C.H. Ludwig (eds.). Lignins. Occurrence, Formation, Structure and Reactions. Wiley interscience, New York, USA.

Properzi, M., J.-M. Leban, A. Pizzi, S. Wieland, F. Pichelin and M. Lehmann. 2005. Influence of grain direction in vibrational wood welding. Holzforschung 59(1): 23–27.

Randle, P.J., P.B. Garland, C.N. Hales and E.A. Newsholme. 1963. The glucose fatty acid cycle its role in insulin sensitivity and the metabolic disturbances of diabetes mellitus. The Lancet 281(7285): 785–789.

Rodriguez, G., P. Diouf, P. Blanchet and T. Stevanovic. 2010. Wood dowel bonding by high-speed rotation welding—application to two Canadian hardwood species. J. Adhes. Sci. Technol. 24: 1423–1436.

Sun, Y., M. Royer, P.N. Diouf and T. Stevanovic. 2010. Chemical changes induced by high speed rotation welding of wood: application to two Canadian hardwood species. pp. 397–414. *In*: Pizzi, A. and K.L. Mittal (eds.). Wood Adhesives, Brill, The Netherlands.

Timell, T.E. 1967. Recent progress in the chemistry of wood hemicelluloses. Wood Sci. Technol. 12: 89–103.

Impregnation of Wood Products

Past, Present and Future Work

Diane Schorr,[1] *Stéphanie Sabrina Vanslambrouck*[2] and *Véronic Landry*[3,*]

Introduction

Wood is a material appreciated for its warmth, visual appearance and versatility. However, as for all other organic materials, it is prone to aesthetic and structural changes when exposed to UV light, heat, microorganisms, pollution, water and humidity, acid rains, etc. (Evans 2008). Moreover, wood is a soft material compared to other building materials. As such, indentations are difficult to avoid and lead to decrease in aesthetics. In order to limit wood degradation or alteration, it must be protected to ensure long-term performance.

A most common way to protect wood products is the application of paint and coating on wood surfaces. The two main functions of coating are to protect and decorate. Lignin, a polymer representing between 25 and 35% of the softwood composition and 18 to 25% of the hardwood one (Aloui 2006), is highly sensitive to UV light and undergoes chemical degradation when exposed to it. Coatings, especially those with high pigment volume concentration and UV absorbers, were

---

[1] Postdoctoral fellow, FP Innovations, 319 rue Franquet, Québec, G1P 4R4.
  Email: diane.schorr@fpinnovations.ca
[2] Postdoctoral fellow, Laval University, 2425 rue de la terrasse, G1V 0A6, Québec city, Canada.
  Email: stephanie-sabrina.vanslambrouck.1@ulaval.ca
[3] Associate Professor, Laval University, 2425 rue de la terrasse, G1V 0A6, Québec city, Canada.
* Corresponding author: veronic.landry@sbf.ulaval.ca

shown to be effective to protect wood against photodegradation (Evans 2008, Aloui 2006, Teaca and Bodirlau 2016). Likewise, thermoset coating application also provides a good protection against chemicals (detergents, acid rain, etc.), heat, water and mechanical damages (ex.: low depth scratches). Coatings can also be used to alter the color of wood surfaces, accentuate wood grain patterns, change gloss, etc.

Coatings have gained mechanical, chemical and thermal resistance over the last decades (Bongiovanni et al. 2002, Cristea et al. 2011, Graziola et al. 2012, Kaboorani et al. 2017). They are also becoming more and more environmentally friendly as they need to fulfill new regulations on volatile organic compounds (VOC) (Wolkoff 1995, de Meijer 2001). However, coatings' benefits are limited to the wood products surface. They cannot fulfill functions where the entire wood thickness is solicited (ex.: improving wood dimensional stability, increasing surface hardness against mar and in depth indentation, limiting wood biodeterioration, etc.). As a result, in-depth treatments must be performed to modify wood behavior. One of the solutions to reach a higher level of performance is to impregnate wood substrates with chemicals of different types, from inorganic fillers to monomers and polymers (Schneider 1994, Ellis and O'Dell 1999, Mahmoud et al. 2000, Gindl et al. 2003, Devi et al. 2006). Chemicals can be either simply deposited in the wood structure, they can react with functional groups of wood polymers (hydroxyl, carboxyl, etc.) (Mallon and Hill 2002, Bach et al. 2005, Khalil et al. 2014, Hubbe et al. 2015) or they can polymerize within the wood structure (*in situ* polymerization) (Soplan and Guven 1999, Pfriem et al. 2012, Dong et al. 2014, Mattos et al. 2015, He et al. 2017).

Wood impregnation was the subject of numerous papers over the last few decades. The type of processes used, the process parameters (time, pressure level, etc.), the selected impregnants (reactive, non-reactive) and the penetration pathway within the wood structure (cell wall penetration and/or lumen penetration) are all factors controlling the final performance of the wood products. The energy source (energy level, penetration depth of the energy source, etc.), selected to dry or cure the chemicals impregnated into the wood structure also plays an important role in the final properties and the level of performance obtained (Erickson and Balatinecz 1964, Islam et al. 2008, Malkov et al. 2003, Tondi et al. 2013, Ding et al. 2008).

As scientific studies highlighted that the performance reached following chemical impregnation is closely related to the level (depth and retention) of impregnation, a short description of the impregnation processes reported in the literature will first be presented in this chapter. Then, a review of the most studied chemical impregnation systems will be presented. As the industry and the academic community are clearly moving to greener and safer technologies, the remaining and the main focus of this chapter will be based on up-coming technologies, such as bio-based and low VOC ones. These approaches aim to decrease the chemicals' impact on the environment and ensure the health and safety of the workers and the end users.

Impregnation Processes

The process selected for the impregnation controls totally or partially the impregnation depth, the percentage of retention and the resulting properties. The processes can involve vacuum and/or pressure steps to ensure good penetration (Trey et al. 2010, Wang et al. 2012, Bao et al. 2016, Stolf et al. 2017). Dipping, spraying, and roller pressing are also impregnation methods reported, although less frequently, in the literature (Chao et al. 2003, Kučerová 2012, Neyses et al. 2016, Mahltig et al. 2008). Vacuum-pressure processes are by far the most common processes reported in the literature.

Full-cell process, also called Bethell process, was first proposed as a method of impregnating wood with preservatives (Freeman et al. 2003, Hill 2007). The main intent of this process is to impart resistance to decay, fire, insects, and wood-boring marine animals to exterior wood products. This process permits to impregnate the full wood cells, meaning that both cell walls and lumens can be filled with the impregnants. This type of process generally leads to a high chemical retention as chemicals penetrate in lumens deeply and in some cases in the cell walls.

To perform the Bethell process, the autoclave containing the wood pieces is first placed under vacuum. Then, the liquid to be impregnated is injected and pressure is applied. Other processes, all including at least one vacuum step, are also reported in the literature (Schneider et al. 2003, Reinprecht 2016). The most popular ones are: the VAV process (vacuum-atmospheric pressure on liquid-vacuum), the Lowry process (also known as PV, increased pressure on liquid-vacuum), the Ruping process (also known as PPV, increased pressure on wood-more increased pressure on liquid vacuum) and several combined technologies. Figure 1 summarizes the different steps involved in each of these processes.

These processes and their variants are used to impregnate a wide range of chemicals in wood substrates, as reported in the following sections. Even if they enable a good chemical retention, most of them are time-consuming (batch process) and they require large volumes of chemicals. Although these techniques are well accepted to confer durability to exterior wood products, the two drawbacks mentioned above result in low industrial acceptance for interior applications (ex.: appearance products). As the improvement desired from impregnation treatments is mainly aesthetic (ex.: indentations' decrease) for these products as opposed to exterior applications where life-threatening situations can be encountered if protection is not efficient enough, simpler processes are generally preferred. Thus, other options are suggested in the literature: dipping in hot and cold baths, spraying, brushing and roller pressing are some of the options (Kučerová 2012, Neyses 2016, Avila et al. 2012).

The penetration of impregnants is the function of several factors related to the process, the impregnants and the wood species. Process related factors include the presence or absence of vacuum and pressure steps, the length of these steps and the level of pressure or vacuum applied (Ewald 1966, Brich et al. 1999, Badillo et al. 2011). Impregnants-related factors include viscosity, molecular weight and

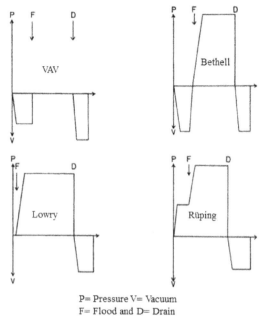

P= Pressure V= Vacuum
F= Flood and D= Drain

Fig. 1. (a) VAV process (vacuum-atmospheric pressure on liquid-vacuum), (b) the Bethell process, (c) the Lowry process (increased pressure on liquid-vacuum), the Ruping (increased pressure on wood-more increased pressure on liquid vacuum).

compatibility (polar/unipolar) with wood constituents (Larnøy et al. 2005, Furuno et al. 2004). Finally, wood species-related factors include cell length, punctuations' number, punctuation aperture size and the thickness of the punctuation membrane (Tondi et al. 2013, Ahmed et al. 2011). Ahmed et al. (2011) also demonstrated that low wood humidity promotes fluid penetration. Wood surface and impregnation temperature also influence the penetration depth of the impregnant (Acda et al. 2007, McIntyre and Eakin 1984, Fernandes et al. 2012).

Most Common Chemical Impregnation Systems

Several chemical impregnation systems have shown great promises to improve wood properties. Literature reports the improvement of hardness, elastic modulus, dimensional stability, durability, wettability, etc. (Gabrielli and Kamke 2010, Stamm 1959, Donath et al. 2004, Zhang et al. 2006, Gindl et al. 2004, Devi et al. 2003, Gindl et al. 2003). Chemical impregnation treatments can be divided in two main categories. The first one relates to the impregnation of monomers, oligomers or polymers followed by *in situ* polymerization/cross-linking. In this case, the treated wood is densified. The second type of treatment is the impregnation of chemicals reacting with cell wall functional groups, mainly hydroxyl groups. This is generally performed to enhance the wood's dimensional stability as the presence of hydroxyl groups is related to wood swelling and shrinkage. This chapter will mostly focus on the first category of treatment, while still considering impregnation systems leading to a higher density and penetrating or reacting with cell walls.

Most of the studies published on wood densification employ either condensation reaction resins or vinyl monomers polymerizing through a free radical mechanism. These two types of systems, their main characteristics and the results obtained using each of these systems are presented below.

Condensation reaction resins impregnation studies

Wood impregnation by condensation reaction resins is divided into three steps. First, the pre-polymer is prepared by a two-step process. An addition reaction is first performed between the selected nucleophile and the electrophile. This is followed by a condensation reaction of the addition products to form low molecular weight oligomers, still soluble in water. The resulting resins are impregnated in wood products. The impregnated wood is then placed under heat to ensure proper curing (cross-linking) (Gindl et al. 2003, Devi et al. 2013). Pressure is often used to ensure good penetration in the wood structure.

The most studied systems in this category are those based of formaldehyde. These systems are well known as they have been used extensively over the last decades for wood composites panels. Several nucleophiles can be used for the addition reaction: phenol, melamine, urea, triethylamine, carbamide, etc. (Gindl et al. 2003, Wang et al. 2012, Cox 2009). The first three are by far the most studied for impregnation systems. While phenol-formaldehyde (PF) resins are highly resistant to harsh environments (high humidity, immersion, heat, etc.), and lead to low formaldehyde emissions while in service, they confer a reddish color to the densified wood. On the other hand, melamine-formaldehyde (MF) and urea-formaldehyde (UF) resins are not colored but they present lower resistance to water and humidity than their phenol counterparts (Pizzi and Mittal 2011). MF resins are also expensive. UF resins, while being less expensive, are sensitive to hydrolysis, thus release formaldehyde emissions in service, which limits their use. MF and UF resins are often used together to ensure good performance, low cost and acceptable formaldehyde emissions. One drawback of the condensation reaction resins is the pH of the catalysts, which is either alkaline or acid. That leads to wood degradation and strength loss over time. This said, several studies worth of interest using condensation reaction resins are reported in the literature and presented below.

Numerous studies report the impregnation of PF resins in wood substrates. They reveal that PF resins, due to their polarity and thus compatibility with wood cell walls, penetrate and swell the cell walls (Stamm and Seborg 1936, Stamm and Elwin 1953, Farmer 1967). That leads to reduce hygroscopicity (Hill 2006) and to dimensionally stable wood composites. To impart such properties to wood products, PF resins have to penetrate in wood cell walls sufficiently. To do so, low molecular weight resins have to prioritize. A study from Furuno et al. (2004) revealed that PF resin, with number average molecular weight (Mn) of 290–480 g/mol, penetrated the cell walls and increased dimensional stability. On the other hand, PF resin with Mn of 820 g/mol were found mostly in the cell lumens and imparted little to no stability.

PF resin impregnation was performed on several wood species using different penetration processes and Mn resins. Nichida et al. (2007) impregnated solutions containing 10 and 20% of PF in Japanese cedar (*cryptomeria japonica*). To do so, a pressurized injection process was employed. Different nuclear magnetic resonance (NMR) spectroscopy analyses were performed to study the impact of PF impregnation on wood polymers mobility. ^{13}C cross-polarization/magic angle spinning (CP-MAS) NMR spectra revealed that PF resin permeated near carbohydrate polymers as well as lignin regions. On the other hand, ^{13}C pulse saturation transfer/magic angle spinning (PST-MAS) NMR results revealed that the molecular mobility of cellulose endocyclic groups was suppressed by the PF resin impregnation. Finally, the ^{1}H MAS NMR spectra exposed that the PF resin impregnation to less water to be bound inside of the wood structure. These results explained the higher stability of the PF treated wood when exposed to humidity changes. Figure 2 compares the nanostructures of PF resin impregnated in Japanese cedar.

Gabrielli and Kamke (2010) reported high anti-swelling efficiency values resulting from PF resin impregnation. They impregnated low (172 g/mol) and high (780 g/mol) molecular weight PF resins in hybrid poplar wood, a cross between Eastern and Black Cottonwood (*Populus deltoides* and *trichocarpa*). They impregnated aqueous solutions of 5, 10 and 20% PF solids content for each of the two resins. They used a viscoelastic thermal compression (VTC) process to promote PF resin penetration, which involves dynamic heat, steam and mechanical compression schedule. They found that PF resins were successfully impregnated using the VTC process. All PF resins impregnated wood present reduce volumetric swelling compared with the unmodified control. This study also revealed that the stabilizing effect is correlated to the resin retention and that the higher Mn resin was retained to a greater extent in the VTC process. Therefore, the wood samples impregnated with the higher Mn resin provided a greater stabilizing effect. These results highlighted that while Mn plays a role in the stabilizing effect, the process and thus the retention, is also of significant importance.

Formaldehyde emissions is a constant worry when using formaldehyde-based resins, especially when low molecular weight resins are used, such as those favored for wood impregnation. Urea was added in a low molecular weight PF resin treating solution used for the impregnation of jelutong (*Dyera costulata*) and sesenduk (*Endospermum diadenum*) strips. The presence of urea in the PF solution is efficient to reduce formaldehyde emissions (Adawiah et al. 2012). All solutions tested in this study, with and without urea, led to higher modulus of rupture, modulus of elasticity and dimensional stability than the untreated wood samples, while rendering high resistance against white rot fungus.

MF resins solutions were also impregnated in Norway spruce (*Picea abies*) by a short-term vacuum treatment and long-term immersion using the solvent exchange treatment (Gindl et al. 2003). The latter specimens were immersed in distilled water until they had taken up so much water that they sunk. They were subsequently placed

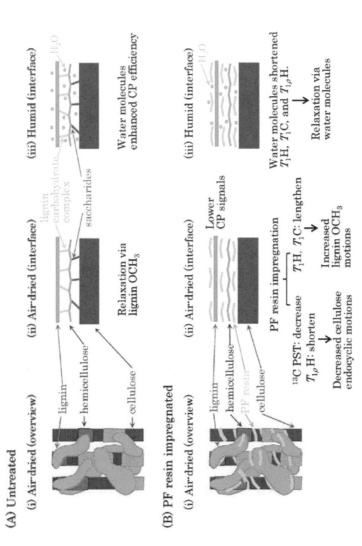

Fig. 2. Nanostructures of PF resin impregnated Japanese cedar (Nichida et al. 2007).

in MF resin for 70 h to enable the resin diffusion into the wood. FTIR spectroscopy and UV microspectrophotometry analyses revealed that long-term immersion is necessary to reach sufficient penetration of MF resin into the wood structure. A two-days immersion resulted in a 4 mm deep penetration. The low penetration depth obtained for the vacuum treatment was explained by the presence of aspirated pits, limiting the liquid flow in Norway spruce wood. Following the three days immersion (solvent exchange treatment), which enables the diffusion through the cell walls, the hardness was improved significantly to reach hardwood beech hardness values. Lumina of cells remain largely free of MF, which was found mainly in cell walls. Indentation experiments revealed that 2 mm penetration depth was necessary to reach the optimum increase in hardness. An impregnation larger than 2 mm only resulted in a small addition of hardness improvement.

Hazarika and Maji (2013) prepared wood polymer nanocomposites using a low Mn MF-furfuryl alcohol (MFFA) co-polymer, n-methylol acrylamide (NMA, a cross-linking agent), and montmorillonite (MMT) into fig tree wood (*Ficus hispida*). They selected MF resin for its high water resistance and its capacity of forming hydrogen bonds. Furfuryl alcohol was chosen as several studies report that its impregnation, also known as furfurylation, results in wood with high durability, improved dimensional stability, and increased hardness. Furfuryl alcohol, produced from the hemicellulosic part of agricultural wastes, is well-known to polymerize *in situ* and permanently swells the wood cell walls. MMT was added as it was demonstrated that MMT improves several properties, such as flame retardancy, dimensional stability, surface hardness, modulus of elasticity and water repellency. The clay selected in this study was modified using 2-acryloyloxy ethyl trimethyl ammonium chloride (ATAC) or cetyl trimethyl ammonium bromide (CTAB) surfactants. The former was selected for its capacity to cp-polymerize with the NMA double bond and the latter for its long alkyl chain that can expand the distance between the clay platelets. The measurements of the flexural and tensile properties revealed that the MFFA polymer alone already increases the mechanical performance, although the addition of clay imparts much more improvement. Clays treated with ATAC and CTAB at the same time led to the most significant improvements.

In a similar way, Cai et al. (2007a, 2007b, 2008, 2010) impregnated aspen wood with a MUF resin. Different hydrophilic and hydrophobic MTT clays were added in the MUF resin aqueous solutions. They found a remarkable improvement of surface hardness, modulus of elasticity, dimensional stability, and water repellency.

Vinyl polymers impregnation systems

Several studies report the impregnation of vinyl monomers in wood substrates. Among the monomers impregnated in different wood species, there are: ethylene, propylene, vinyl chloride, vinyl acetate, vinylidene chloride, acrylonitrile, acrylate, methacrylate, styrene, etc. (Kumar 2007). Vinyl monomers are non-polar chemicals, and, as such, they show limited compatibility with cell walls. Therefore, they are found mainly in lumens. As they do not penetrate cell walls, they present only limited improvement in dimensional stability with water uptake and limited decay resistance. Their impregnation was however found to improve several mechanical properties. Acrylates and methacrylates are, by far, the most studied vinyl monomers. As for other vinyl monomers, they polymerize by free radical polymerization, initiate either by radiation (ultraviolet or electron beam, but also cobalt 60 and X-rays) (Cai and Blanchet 2015, Trey et al. 2010, Berejka et al. 2005, Cleland et al. 2009) and/or thermal initiation (thermal initiator needed such as peroxides or azo compounds) (Ding et al. 2013) or chemical initiation (Wang and Matyjaszewski 1995).

The most studied monomer within this group is methyl methacrylate (MMA), alone or pair with other monomers. The aim of impregnating MMA in wood species is mainly to increase the hardness of wood products. Wan and Wang (2009) reported the impregnation of six Canadian wood species with MMA. The highest hardness (11.31 MPa) was recorded for red maple (*Acer Rubrum*) impregnated with MMA. In this study, the MMA polymerization was performed by hot pressing using a thermal initiator. In a similar way, Ding et al. (2011a, 2013b) impregnated the hybrid poplar (*Populus deltoids* x *Populus nigra*) with MMA. They obtained Janka hardness values 2.5 to four times higher for impregnated samples.

Stolf and Lahr (2004) carried out a study on the impregnation of Caribbean pine (*Pinus caribaea*) and eucalyptus with styrene and MMA monomers using a vacuum pressure process. They noted a significant increase in hardness, parallel and perpendicular to the wood grain. Zhang et al. (2005) impregnated sugar maple (*Acer saccharum*) with MMA as well as two other methacrylates: 2-hydroxyethylmethacrylate (HEMA) and ethylene glycol dimethacrylate (EGDMA). Significant increases in Brinell hardness and hardness modulus were recorded. The use of HEMA, acting as a cross-linking agent, enables an increase in the mechanical performance of the impregnated samples more than MMA alone.

In a similar way, Cai and Blanchet (2015) reported in their study on the polymerization of 1,6-hexanediol diacrylate (HDDA) and trimethylolpropane triacrylate (TMPTA), that using electron beam could lead to well polymerized acrylate wood samples. Several passes had to be performed at 4.5 MeV and 100 KGy with between each of them a waiting time of 7 min. This treatment led to a significant increase in the surface hardness of sugar maple (*Acer saccharum*) by nearly 200%, reaching 16 MPa. Trey et al. (2010) reported that the use of electron beam technology resulted in a conversion rate of polyethylene glycol diacrylate significantly higher than that achieved using conventional heat-convective cooking technologies (convection).

Bio-based Wood Impregnation Systems

As environmental concern keep growing over the years, new impregnation products are constantly proposed and studied. Indeed, bio-based monomers, oligomers and polymers are becoming preferred to the classical options described above.

Densification with a bio-based oil or wax

Oils and waxes are commercially used in water repellent finishes and wood preservatives. Literature also reports their use to impregnate wood substrates by using thermal treatment (oil and waxes at high temperatures to decrease their viscosity) and vacuum-pressure processes. Although these processes are not new, a renewed interest is to be expected as they used bio-based products, which are mostly safe and easy to use. The resulted densified wood has improved properties such as dimensional stability (Wang et al. 2015), photodegradation (Lesar et al. 2011a) and even improved mechanical properties (Kakavas et al. 2014). Due to environmental concerns, natural oils such as vegetable oil (soybean oil, linseed oil) or even waxes, such as beeswax, Carnauba wax, Montan wax (obtained by solvent extraction of certain types of lignite or brown coal), are preferred. Fadl and Basta (2005) modified the wood by vapor phase acetylation or by swelling phase acetylation before impregnating it with novolac resin and linseed oil. They observed that using the vapor phase acetylation before impregnation led to higher dimensional stability compared to the liquid phase acetylation. However, liquid phase acetylation improved the impregnation process of resins and oils compared to the vapor phase acetylation. They observed that increasing the curing time and the impregnation temperature with linseed oil improved water repellency and dimensional stability of the wood contrary to the impregnation with novolac resin where these properties decreased. Neither the novolac resin nor the linseed oil led to covalent bonds with the wood cell walls, however both formed hygroscopic insoluble polymer. Awoyemi et al. (2009) studied the curing treatment of the wood in oil with and without cooling time after oil treatment. They showed that the wettability, the water uptake and the swelling properties decreased with a better retention caused by the addition of the process in oil cooling. The authors explained these results by a better oil impregnation during the cooling process. To improve other properties such as decay resistance, some additional products can be mixed with the wax such as boron compounds to limit the moisture content and the wood sorption wood. As a result, mold and fungi growth was decreased (Lesar et al. 2011b).

Densification with glycerol

Glycerol is a polyol ($C_3H_8O_3$), which is available in important quantities as a byproduct of the biodiesel production (Maminski et al., 2011). It could be used on its own to impregnate and densify wood thus enhancing its mechanical or physical properties. Thuvander et al. (2001) impregnated green wood (*Acacia auriculiformis*) with a solution of glycerol as a bulking agent before the drying step. They studied the tensile strength of the dried wood with and without such treatment. An improvement in the tensile strength parallel to the grain of the impregnated wood was noticed. They confirmed that the impregnation with glycerol helped to prevent and to limit fractures of the cell walls induced by the drying stresses. These damages were mainly responsible for the decrease of mechanical properties, such as tensile strength. They observed an infiltration of glycerol in the cell walls and in the lumens, but without filling it completely. Wallström and Lindberg (1999) also showed that the impregnation of glycerol in dried wood leads to cell wall penetration. They observed that the glycerol concentration was low in the middle lamella, higher in the S1 layer and the most important concentration of glycerol was found in the S2 layer of the cell walls (Fig. 3).

Ghazali and Yusoff (2011) showed that impregnating acacia (*Acacia auriculiformis*) dried wood with pure glycerol improved their parallel compressive strength up to 30% compared to the raw wood. They used a vacuum-pressure process for the impregnation and the impregnated wood was dried at 102°C. Scanning electron microscopy (SEM) revealed that after the compression test, raw wood showed more cell wall damages compared to the impregnated wood. It was also reported that impregnation of glycerol could lead to drawbacks. For example, Yan et al. (2011) impregnated dry poplar wood (*Populus cathayana* Rehd.) with a glycerol solution (99%) during 48 h. The impregnation was preceded by a vacuum step of 30 min. They studied the impact of the glycerol on the stress relaxation of the wood at different temperatures. By raising the temperature, the stress relaxation increased. The results were aggravated with the glycerol. They explained this by

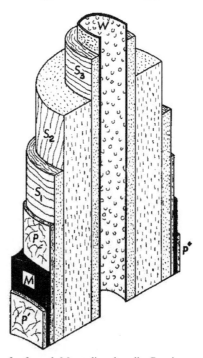

Fig. 3. Cell wall diagram of softwood; M: medium lamella; P: primary wall; S1: outer layer of the secondary wall; S2: median layer of the secondary wall; S3: internal layer of the secondary wall; W: warty membrane (Siau 1995).

the breaking of the molecular chains of hemicelluloses and lignins caused by the strong hydrogen bonds created between the glycerol and the wood.

In summary, benefits from wood densification with glycerol were closely related to the wood state (green or dry). The densification could improve mechanical properties, such as tensile strength and compressive strength. On the other hand, it could decrease other properties, such as the stress relaxation of the wood.

Densification with glycerol and carboxyl groups bearing chemicals (anhydrides or carboxylic acids)

Use of maleic anhydride

Several studies reported the impregnation of carboxylic acids and anhydrides to improve wood properties, such as dimensional stability and mechanical properties. A well-known example is acetylated wood, such as the one prepared from the Accoya© process (Alexander and Bongers 2013). Essoua Essoua et al. (2015) impregnated lodgepole pine (*Pinus contorta*) and white pine (*Pinus strobus*) woods with maleic anhydride. They optimized the process depending on the curing temperature and time. They observed characteristic peaks of elongation of ester bonds C = O by Fourier-transform infrared spectroscopy (FT-IR). Dimensional stability was found to be related to the mono- and di-ester concentrations, which are in turn related to the curing temperature and time. The authors showed that a higher curing temperature leads to more diester bonds (Table 1), a more important aging resistance and a better dimensional stability.

Several authors focused on wood or cellulose impregnation with a mixture of glycerol and anhydride or acid. Maleic anhydride could react either with the hydroxyl groups of glycerol to form a cross-linked polymer or with the hydroxyl groups of the wood polymers. A highly cross-linked polymer could be formed in the wood. Roussel et al. (2001) studied the esterification reaction of polyglycerol (pre-polymerized glycerol) and maleic anhydride prior to their impregnation in a

Table 1. Percentage of mono- and di-ester bonds for pine wood impregnation with maleic anhydride (Essoua Essoua et al. 2015).

Wood Species	Treatment		WPG (%)	Monoester bonds (%)	Diester bonds (%)
	Drying time (h)	T(°C)			
Lodgepole pine	24	180	21.41	6.55	6.59
		160	32.31	15.77	3.24
		140	37.24	18.09	2.33
	18	180	29.08	16.52	2.56
	12	180	30.91	10.99	4.41
White pine	24	180	37.25	12.11	3.26
		160	39.91	15.40	2.28
		140	42.49	21.64	1.79
	18	180	46.58	16.29	1.15
	12	180	61.58	20.14	1.14

pinewood species (*pinus*) by a vacuum-atmospheric pressure on liquid vacuum (VAV) process. Titration experiments revealed a decrease of the acid and hydroxyl contents when the esterification occurred. ^{1}H NMR analyses showed that the peak corresponding to the proton attached to the double bond of the maleic anhydride disappeared with the increase of the reaction temperature or by adding catalysts (2-butanone peroxide 2% w and cobalt naphthenate 2% w). This disappearance confirmed that the polymerization occurred in the impregnated wood. In the second part of their study, Pinewood was impregnated with polyglycerol/maleic anhydride in the presence of a mix of catalysts (2-butanone peroxide 2%w and cobalt naphthenate 2% w). A weight gain of 13–44% depending on the concentration of the polyglycerol/maleic anhydride in the solution was observed after treatment. The impregnated wood samples showed better properties, such as better resistance to fungal degradation and better dimensional stability. An extraction of pinewood prior to a vacuum pressure impregnation was found to be favorable. The optimized solution used for this impregnation was an aqueous solution of polyglycerol/ maleic anhydride at 30% w content in presence of the two catalysts (2-butanone peroxide 2% w and cobalt naphthenate 2% w). By impregnating cellulose with maleic anhydride and glycerol, Uraki et al. (1994) proved by gel permeation chromatography (GPC) that these two compounds react together to form a polyester with a molecular weight higher than 1000 g/mol. Moreover, ^{13}C solid state NMR confirmed the formation of covalent bonds between the polyester and the cellulose.

Use of citric acid

Essoua Essoua et al. (2016) investigated the impregnation of white pine (*Pinus strobus*) and lodgepole pine (*Pinus contorta*) woods with a solution of citric acid and glycerol to improve the technical performance of outdoor wood products. They observed a density increase of 30 to 50%. Treated wood showed a dimensional stability increase of 52.4% and 53.5% respectively for the lodgepole pine and for the white pine, as well as better decay resistance and hardness. However, a reduction of coating adhesion was observed. This could be due to the polymerization and condensation of polymer inside the cell walls and the lumen of the wood, which limits the mechanical anchorage of the coating in the lumens. Schorr et al. (2018) studied a similar system. They replaced the liquid catalyst (HCl) by nanoclay and investigated the impact of the curing temperature. They observed that with the nanoclay, the impregnated wood still showed good dimensional stability and hardness, while improving the coating adhesion in comparison to the samples prepared by using the liquid catalyst without surpassing coating adhesion values for the raw wood. They explained this by the roughness analyses performed by optical profilometry of the treated wood surfaces (Fig. 4). Wood surfaces were rougher when using nanoclay instead of HCl catalyst, leading to a more important mechanical anchorage. However, it was not enough to surpass the coating adhesion onto the raw wood. All impregnated wood showed higher density than raw wood. However, no significant differences of density were observed between those impregnated wood.

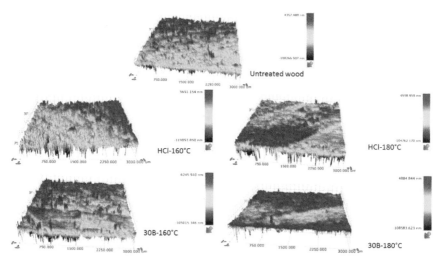

Fig. 4. 3D optical profilometer images of the surface roughness of untreated and treated specimens (with HCl and nanoclay 30B catalysts) at 160°C and 180°C (Schorr et al. 2018).

Berube et al. (2017) focused on citric acid and glycerol mixture impregnation. They optimized the reaction by varying the citric acid/glycerol molar ratio, the liquid catalyst concentration and nature as well as the impregnation depth. When investigating the polymer alone (not within the wood structure), they confirmed the esterification reaction by FT-IR and the polymerization/cross-linking by differential scanning calorimetry (DSC) with an increase of the glass transition temperature.

Impregnation of white pine (*Pinus strobus*) and lodgepole pine (*Pinus contorta*) woods with a glycerol/citric acid solution leads to an improvement of the dimensional stability up to 50–70%. The polymer was fixed in the cell wall and filled completely part of the wood lumens. Berube et al. (2017) observed that the results depended on the wood species and its anatomical structure, the impregnation depth and especially the catalysts used. In this study, the best performance was obtained by using p-toluenesulfonic acid (p-TSA) or HCl as catalysts (3% w content) with a molar ratio of citric acid/glycerol of 1.2:1.

All the studies reported above lead to densified wood with a bio-based polymer, formed with glycerol and different compounds bearing carboxylic groups (anhydride or carboxylic acid). These polymers were found in the cell walls and in the lumens. Improvements from this type of impregnants depend on the impregnation depth (related to the impregnation process) but also on the nature of the catalysts (controlling the type of esterification, mono or multi esterification), the wood species and the curing time and temperature. As reported, these systems succeeded to enhance mechanical properties, dimensional stability and even the biological resistance and the resistance to aging.

Impregnation of biopolyesters

Environmental concerns lead to the development of new polymers. The most produced bio-based polymer is the polylactic acid (PLA) obtained from the polymerization of lactic acid or lactide monomers. It was used as a polymer matrix with different fibers (kenaf, bamboo, corn, wood, etc.), wood flour or mixed with plastics to form composites (Bajpai et al. 2014). Several studies also reported the use of PLA for wood densification. Noël et al. (2009a) synthesized pre-polymerized PLA from lactic acids. They impregnated dry beech wood (*Fagus*) with this prepolymer by a vacuum-pressure process for 2 hours. The densified wood was placed at 120°C for different periods of time in order to ensure the curing of the PLA. Three different mixtures were used to impregnate the wood samples: the pre-polymerized polymer alone (PrP) or with catalysts such as sulfuric acid (PPSA) or tin II octoate (PPTO) (Fig. 5).

They observed that PLA polymerized into beech cell walls but did not react with the wood hydroxyl groups to form covalent bonds. These results showed that the time and temperature conditions are important parameters to control the final performance of wood samples. Indeed, with a short curing time (1 hr) at 120°C, the products obtained were softened. With a long curing time (until weight stabilization), a more reticulated polymer was obtained in the impregnated wood. This was proved by the molecular mass results obtained by gel permeation chromatography (GPC) of the extracted polymer. Long curing time led to mechanical properties improvement. By comparing the physical and mechanical properties of the wood samples densified with the three different solutions (PrP, PPSA and PPTO), they observed that with the pre-polymerized PLA alone, the densified wood showed high density and a good hardness compared to the raw wood. These results showed that with this PrP solution, the densified wood could be of interest for wood flooring applications. Adding the catalysts PPSA and PPTO provoked damages leading to a softened wood material because of the catalyst acidity. With an additional curing step, densified wood became stable and resistant to biological attacks. However, some properties such as shearing, bending, hardness strength were decreased because of

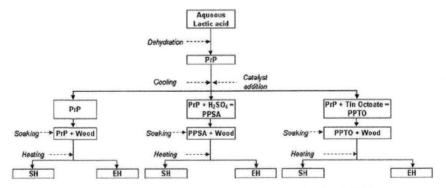

Fig. 5. Steps involved in the wood treatment with PLA (Noel et al. 2009).

the middle lamella deterioration caused by the acidity (Noël et al. 2009b). Grosse et al. (2016) studied the impregnation of PLA in wood with some other chemicals. They used maleic anhydride to improve the polymer bonding on the wood constituents. They also investigated tannins as additives to improve the biological resistance of the treated wood and observed that the higher the curing temperature, the more the leaching rate decreased for all the specimens. Dimensional stability was improved for all the treated wood samples. Treated wood also demonstrated a good biological resistance, especially for the one treated with maleic anhydride. Finally, they confirmed that tannins were interesting additives for two main reasons: they are hydrophobic and do not damage the wood during the treatment. Maleic anhydride, with its high acidity, could lead to cell wall damages. Noël et al. (2015a) studied European beech (*Fagus sylvatica*) and Sycamore maple wood (*Acer pseudoplatanus*) densification with other polyacids, such as poly(butylene succinate acid) (PBS), polybutylene adipate (PBA), poly(glycolic acid) (PGA). They impregnated oven-dried beech wood. They investigated the polymerization *in situ* by two steps: first the impregnation and then a curing treatment at low temperature (103°C during 24 hr) or at high temperature (120°C during 6 hr). Oligomers PGA and PLA were impregnated in the cell walls due to a high wood swelling (30%) after treatment (Table 2). The use of moderate temperature during impregnation leads to a decrease of the oligomer viscosity and thus to a better penetration in the cell walls. Wood impregnated by these oligomers softened the wood material. Glass transition temperature (T_g) of the wood constituents decreased and an increase of storage modulus was observed.

They made the assumption that this softening and plasticizing state was due to the weak cohesive forces and the mobility of lignocellulosic constituents in the matrix. They hypothesized that these wood treatments imply a degradation of the hemicelluloses leading to a lower thermal stability and lower mechanical properties of the treated wood. In comparison with a swelling of less than 2% of the samples after impregnation, PBS and PBA led to a low cell penetration, filling only the lumens (Table 2). The wood treatment by these oligomers led to a similar or better

Table 2. Weight uptake, weight loss and volume swelling of treated wood with different oligomers (Noel et al. 2015).

Ref.	Weight uptake (Wu)		Weight loss (WL)		Volume swelling (S)	
	Wu after impregnation (%)	Wu after curing time (%)	WL water leaching (%)	WL solvent extraction (%)	S after impregnation (%)	S after curing time (%)
OLA1LT	76	54	31	37	3	23
OLA1HT	81	56	29	39	0.9	24
OGA1LT	118	84	15	11	24	28
OGA1HT	150	92	19	14	9	26
OBS2LT	68	65	7	37	2	1.0
OBS2HT	64	64	13	51	0.6	1.7
OBA2LT	64	61	5	51	1.2	0.3
OBA2HT	68	67	10	56	0.1	1.3

thermal stability than those of the raw material and the mechanical properties remain unchanged. With the impregnation of PBS in wood, the dimensional stability of the treated wood was studied. By impregnating at moderate temperature and with an *in situ* polymerization, they observed a dimensional stability around 30–40%. Meanwhile, by adding a wet step between the impregnation and the heating or by using wood with a high moisture content before the impregnation, the dimensional stability increased up to 60–70% (Noël et al. 2016). These results confirmed that the temperature and the humidity during the treatment have a great influence on the specimen properties, once impregnated and cured. They hypothesized that high moisture content of wood could favor a partial dissolution of the oligomer, which appears to enable a more important diffusion in the wood structure. For all oligomers studied, they confirmed by thermogravimetric analysis (TGA) that the degree of oligomer polymerization was similar for the compounds only and *in situ* polymerization in solid wood. The higher the temperature was and the longer the curing time was, the higher was the average molecular mass observed by GPC (Noël et al. 2015b). These treatments using different oligomers led to different mechanical, thermal and physical properties. These differences could be related to the monomers used which consequently led to specific properties, such as thermal stability or to the different zone of penetration in wood (wall cells and/or lumens).

These studies show that wood densification with bio-based oligomers depends on several parameters, such as the oligomer nature, the presence of catalyst, the humidity content of wood and the curing temperature. Oligomers could penetrate in the cell walls or only in the lumens. Depending on the parameters selected, wood properties could be modified and the final use of the densified wood has to be reviewed in consequence.

The paper presents a new method of wood surface impregnation with biopolymers and resins, such as chitosan, zein, gelatine and guaiac resin, by using 1-ethyl-3-methylimidazolium chloride ionic liquid as solvent-carrier. The compounds mentioned diffuse into the wood, and by ionic liquid removal and drying, are able to form a uniform coating. It confers higher moisture resistance, dimensional stability and greater hardness to the treated wood (references). The main advantages of the proposed method of impregnation are the usage of a new solvent (1-ethyl-3-methylimidazolium chloride), and considerably lower temperatures (40°C) than conventional processes of impregnation (100–200°C). Another advantage of the proposed method is the higher solubility of the impregnants in the ionic liquid, as opposed to water or other molecular solvents (references).

Densification with a sol-gel process

Silanes are used as modification agents in plastics, textiles, building and paper industry. They can be used to increase the hydrophobicity of the matrix and to improve the scratch resistance property. Moreover, bridges with covalent bonds between inorganic and organic can be created in the composites. Sol-gel process is obtained by the hydrolysis of the alkoxysilanes [Eqn. (1)]. By hydrolysis, the silanol created condense in a three dimensional unit leading to the formation of colloid oligomeric particles, called sols. These particles continue to crosslink in order to form a highly condensed gel. This sol-gel process depends on multiple parameters, such as pH, temperature, H_2O:Si ratio, etc. (Mai and Militz 2004). The alkoxysilanes have been known and studied for wood coating applications. The modified coating showed improved properties such as a lower rate of moisture sorption, a better thermal stability and also a better UV resistance for the coated products (Tshabalala et al. 2003, Sano et al. 2016, Donath et al. 2006). The silanols obtained by hydrolysis of alkoxysilanes could react with the hydroxyls of the wood constituents, forming Si-O-cellulose (Equation 2).

$$R - Si - O - R' + H_2O \longrightarrow R - Si - OH + R' - OH \qquad (1)$$

$$R - Si - OH + OH - cellulose \longrightarrow R - Si - O - cellulose + H_2O \quad (2)$$

These Si-O-C bonds are highly hydrolysable. To decrease their tendency to hydrolyze, alkyl groups could be grafted to the alkoxysilane (or alkylalkoxysilane). Wood densification by a sol-gel process of alkoxysilane can be performed under vacuum pressure process according to two strategies: one being in a monomeric silane state with fiber saturated wood samples and the other in a sol state with an already prehydrolysed oligomer before impregnation (Donath et al. 2004). The level of impregnation depends on the process, which will have an impact on the wood properties. The silane nature and their concentration in the wood also have an impact on the final properties. Indeed, Donath et al. (2004) confirmed by SEM and by weight gain measurements on the impregnated pine sapwood (*Pinus sylvestris*) and beech sapwood (*Fagus sylvatica*) that the prehydrolysed silane (sol state) favored a filling of the lumen with a weight gain of 20–30% whereas the monomeric silane (silane state) infiltrated most important in the cell walls with a weight gain of 10–20% (Fig. 6).

The dimensional stability was improved up to 30% for all the silane state impregnated samples. For the sol state impregnated samples, it depended on the type of silane used (tetraethyl orthosilicate silane-TEOS, tetramethoxy silane-MEOS, Phenyltriethoxysilane-PTEOS). Indeed, the larger the alkyl group of the silane was, the better the dimensional stability and moisture sorption were. After ageing, the dimensional stability was also studied. For all the treatment, the dimensional stability from silane state or sol state showed a slightly decreased value compared to the non-aged impregnated wood. Panov and Terziev (2009) evidenced that the temperature of polymerization and the catalyst content had to be optimized to pursue a reliable *in situ* polymerization and then to obtain a low leaching of the silane compounds from the treated wood. Impregnation with arylalkoxysilane showed good stabilization effect. The impregnation with ethylphenyldiethoxysilane (EtPhSi(OEt)$_2$) improved wood durability and even more with boric acid as an additive. De Vetter et al. (2010) studied the impregnation of conditioned scots pine sapwood (*Pinus sylvestris* L.) with different formulation of organosilicon. They suggested that treatments with organosilicon had an effect as

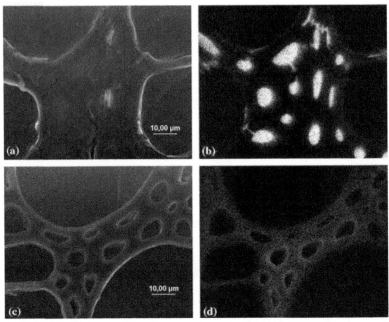

Fig. 6. (a, c) SEM micrographs of beech wood specimens treated with MTES (1 500·, cross sections); (b, d) the corresponding SEM-EDX mappings of silicon in beech. (a, b) Treatment with pre-hydrolyzed MTES (sol state); (c, d) treatment with monomeric MTES (silane state) (Donath et al. 2004).

a water repellent. The moisture uptake rate could be influenced by their presence in the wood. However, the authors suggested that in order to improve the decay resistance, the organosilicon had to have a functional group such as amino group. Indeed, Giudice et al. (2013a) evidenced that using an aminoalkylalkoxysilane for the impregnation of the *Araucaria angustifolia* wood improved its dimensional stability (up to 72%) and its decay resistance. The best performance was found with the aminopropyl triethoxysilane. They also studied the thermal stability of treated wood with aminopropyl methyldiethoxysilane, aminopropyl triethoxysilane and a mix of both as fireproof impregnants. They confirmed that the higher the weight gain and the cross-linking of the polymer in the cell walls, the more efficient the fireproof impregnants were. The authors explained the higher thermal stability of the treated wood by the new condensed Si-O-C bonds formed between the inorganic polymer and the wood constituents. They showed that the best performance for fireproof impregnants was obtained for the aminopropyl triethoxysilane followed by the mixture of the aminopropyl alkylalkoxysilane (Giudice et al. 2013b). Ying Lin and Fu (2012) impregnated conditioned poplar wood (*Populus euramericaha*) with a silicone sol under vacuum pressure process. They confirmed the presence of silicon in the treated wood by FT-IR and they observed their presence in the cell walls and in the lumens by SEM and energy dispersive X-ray analysis (SEM-EDXA). TGA and goniometry experiments evidenced improvements of thermal stability and water repellency of the treated wood. The authors observed that the elastic modulus, the bending strength and the hardness were all enhanced for the treated wood, especially when the weight gain increases (Table 3).

Broda et al. (2017) studied the impregnation of waterloged wood elm wood (*Ulmus americana*) with a silane treatment with methyltrimethoxysilane (MTMOS) in order to prevent dramatical dimensional changes, like shrinkage colapse and loss of shape during drying step. They confirmed the process effectiveness. However, they pointed out the fact that classic process with poly(ethylene glycol) (PEG) compared to the use of MTMOS showed a better dimensional stability.

Wood densification with alkylalkoxysilane, arylalkoxylsilane and other silanes with different functional groups could be performed by sol-gel process under a vacuum-pressure method. Dry or wet wood could be impregnated with silane monomers as silane state or with prehydrolysed oligomers. Wood impregnation varied depending on these states. With silane monomers, the cell wall penetration was more important than with a sol state (prehydrolysed). The silane nature and its functional groups, the impregnation condition and the polymer condensation are important parameters which could influence the properties of the treated wood, such as the dimensional stability, the mechanical properties, the thermal stability and even the decay resistance.

Table 3. Mechanical properties of untreated wood and silicone treated wood (Ying Lin and Fu 2012).

Materials	WPG (%)	Modulus of elasticity (Gpa)	Bending strength (Mpa)	Hardness (kN)		
				Transversal	Tangential	Radial
Untreated wood	-	10.20	82.09	4.13	2.34	2.32
Composites	16.5	13.22	87.38	5.37	3.43	3.81
	22.8	15.02	94.5	5.86	3.53	3.67
	30.3	15.36	93.27	6.77	3.75	4.47

New method of wood densification by a two-step process—delignification and impregnation

Recently, a new method was investigated to obtain a translucide densified wood. Several studies focused on making a densified and translucide wood by a two-step process. Lignin is one of the main wood constituents with cellulose and hemicelluloses. Moreover it is the most light absorbing compound in the wood. The first step was to delignify wood. The lignin was removed from dry wood with a solution containing sodium chlorite in an acetate buffer at 80–100°C for 12 hrs. The second step was to infiltrate a polymer into it in order to increase the wood density and to improve the optical transmittance of the material. A pre-polymerized methyl methacrylate solution was prepared and the wood was infiltrated by the polymer solution using a vacuum infiltration process. The treatment was completed by the final reticulation of the polymer in the delignified wood by heating the samples. Different wood species were used, such as beech wood (*Fagus*) and balsa wood

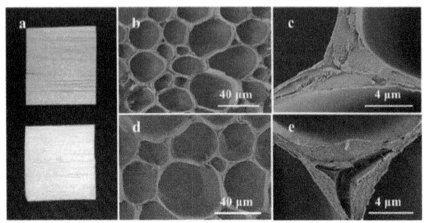

Fig. 7. Delignification of wood: (a) an optical image of wood before (up) and after (down) delignification. Low magnification (b, c) images of original wood (OW) cross section showing the micro structure of wood. Low magnification (d, e) images of delignified wood (DLW) cross section supporting the presence of a well-preserved wood structure (Li et al. 2016).

(*Ochroma pyramidale*) (Li et al. 2016, Yaddanapudi et al. 2017). Lignin content was studied before and after the delignification. The Klason lignin decreased from 24.9% for the raw wood to 2.9% for the delignified wood. From SEM observations, the anatomical structure of the wood and the nanoscale cell wall organization were preserved while large voids in the lignin rich cell walls was observed (Fig. 6) (Li et al. 2016). By FT-IR, Yaddanapudi et al. (2017) confirmed the delignification with the reduction of the absorption intensity at 3429 cm^{-1} caused by the disappearance of the hydroxyls groups from the lignin. The delignified wood became white but not transparent while the shape was conserved. The specific surface area increased from 1.2 m^2/g to 19.8 m^2/g for the Balsa wood (Li et al. 2016). The enhancement of the specific surface area facilitates the impregnation of polymethyl methacrylate (PMMA) in the wood.

In both studies, the authors confirmed the infiltration of PMMA in the wood. Indeed, by SEM, they observed PMMA in the hollow spaces of the cell walls. The hollow lumen spaces were also filled with PMMA. The optical transmission reached 85%. This property decreased with the increase of the sample thickness and so with the volumetric fraction of cellulose. The densified translucide wood demonstrated better mechanical properties than the delignified or even the raw wood (Fig. 8).

The authors explained the obtained results for the delignified wood by the lack of strong hydrogen bonds binding across the cell walls. The enhancement for the densified wood was explained by the synergy created between PMMA and the cellulose. Indeed, strong hydrogen bonds between PMMA carbonyls and the cellulose hydroxyls were created at the interface between cellulose/PMMA in the cell walls. Moreover, the PMMA was also well integrated in the lumen. Other authors presented the same results with Cathaypoplar wood (*Populus cathayana* Rehd). However, the polymer infiltrated in the delignified wood was an epoxy

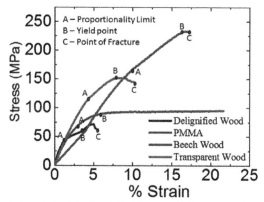

Fig. 8. Stress vs. strain relationship for delignified wood, beech wood and transparent wood (Yaddanapudi et al. 2017).

resin (Zhu et al. 2016). They also evidenced a high optical transmittance (up to 90%) and better mechanical properties (strength, modulus and hardness) improved by 500% in the radial plan and slightly improved in the longitudinal plan. Some authors pursued this investigation by adding quantum dots such as $\gamma\text{-Fe}_2\text{O}_3@$ $\text{YVO}_4\text{:Eu}^{3+}$ nanoparticles dispersed in homogeneous silica with MMA before pre-polymerization (Gan et al. 2017, Li et al. 2017b). It resulted in a luminescent transparent wood (Fig. 9). This material still showed high optical transmittance, excellent mechanical properties but also a better dimensional stability than raw wood. Indeed, after 60 days of soaking the luminescent, wood gain volume was 15.8%, compared to 40.4% for the raw wood. The authors suggested that this new material could be interesting as luminescent buildings/furniture or even as green LED lighting equipment, luminescent magnetic switches, and anticounterfeiting facilities.

Finally, Li et al. (2017a) demonstrated that instead of a delignification, it could be possible to modify the lignin and still obtain a lignin retained transparent wood. They only removed a small fraction of the chromophore responsible of the absorbance of the light in an environmentally friendly process. 80% of the lignin was still present in the wood before the infiltration of PMMA in the lumens. They still observed a high optical transmittance in the lignin retained transparent wood. The removal of some chromophoric structures of lignin is interesting for the species where the delignification step is difficult, like for the softwood species. Moreover, in keeping lignin in the transparent wood, the mechanical properties remained good, better than for the transparent delignified wood. These studies on the densified transparent/translucide wood showed a new material, which could be interesting for applications in the building sector to decrease the energy consumption by creating a new material which could lead to new light transmitted building. A French company, namely Woodoo, founded in 2016, developed a patented treatment that delignify the wood and densify it with a biopolymer. They claimed that this new material

Fig. 9. (a) raw wood; (b) delignified wood; (c) transparent wood; (d) luminescent wood with a γ-Fe2O3@YVO4:Eu3+ NP concentration of 0.1 wt%, and (e)-luminescent wood with the γ-Fe2O3@ YVO4:Eu3+ NP concentration of 0.1 wt% under UV light excitation at 254 nm (Gan et al. 2017).

showed several properties improvements such as fire resistance, decay resistance and mechanical properties while being a translucent material (Boitouzet 2017).

Conclusions

The chemical impregnation of wood substrates is an efficient way to improve several properties such as the hardness, flexural and tensile properties, dimensional stability, decay resistance, etc. The extent of improvement is the function of several factors: process parameters, impregnants used, wood anatomy, etc. The most studied systems are the condensation and the vinyl polymerization reactions. The former reside on diffusion into cell walls, while the latter are dependent only on penetration into lumens. As a result, they lead to improvement of different properties. Condensation reaction system imparts good dimensional stability and decay resistance. Vinyl polymerization systems were mostly developed to improve mechanical performance. As the industry and the academia is clearly moving to low VOC and bio-based systems, new systems are developed and reported in the literature recently. Impregnation with bio-based oil or glycerol have already been applied in wood drying or to improve some mechanical properties. However, adding new steps such as acetylation or new products such as carboxylic acids into those impregnations introduced new pathways for the densification of wood with an additional reticulation step inside wood. Depending on the treatment, the impregnated wood showed better dimensional stability, UV resistance and even better hardness properties. Since several years, the densification with polyesters such as bio-sourced polylactic acid or with organosilicon compounds also showed great potential to obtain a densified wood. Finally, a two-step bio-based system is also interesting for wood densification. The densified wood showed similar properties to those of processes implicating delignification before the impregnation, the densified wood becoming also transparent and even luminescent. However, none of these technics are yet commercially available. These systems need to be further investigated in order to be commercially viable.

References

Acda, M., J.J. Morrell and K.L. Levien. 2007. Effect of process variables on supercritical fluid impregnation of composites with tebuconazole. Wood and Fiber Science 29(3): 282–290.
Adawiah, M.R.A., A. Zaidon, F.N. Izreen, E.S. Bakar, S.M. Hamami and M.T. Paridah. 2012. Addition of urea as formaldehyde scavenger for low molecular weight phenol formaldehyde-treated compreg wood. Journal of Tropical Forest Science 24(3): 348–357.

Alexander, J.P. and H.P.M. Bongers. 2013. Acetylated wood. Patent WO2013113850 A1. https://www. google.com/patents/WO2013113850A1.

Aloui, F. 2006. Rôle des absorbeurs UV inorganiques sur la photostabilisation des systemes bois-finition transparente. Ph.D. thesis, Université Henri Poincaré, France.

Avila, C.B., W.G. Escobar, A. Cloutier, C.-H. Fang and P.V. Carrasco. 2012. Densification of wood venners combined with oil-heat treatment. Part III: Cell wall mechanical properties determined by nanoindentation. Bioresources 7: 1525–1532.

Awoyemi, L., P.A. Cooper and T.Y. Ung. 2009. In-treatment cooling during thermal modification of wood in soy oil medium: soy oil uptake, wettability, water uptake and swelling properties. European Journal of Wood and Wood Products 67(4): 465–470.

Bach, S., M.N. Belgacem and A. Gandini. 2005. Hydrophobisation and densification of wood by different chemical treatments. Holzforschung 59(4): 389–396.

Badillo, G.M., L.A. Segura and J.B. Laurindo. 2011. Theoretical and experimental aspects of vacuum impregnation of porous media using transparent etched networks. International Journal of Multiphase Flow 37(9): 1219–1226.

Bajpai, P.K., I. Singh and J. Madaan. 2014. Development and characterization of PLA-based green composites: A review. Journal of Thermoplastic Composite Materials 27(1): 52–81.

Bao, M., X. Huang, Y. Zhang, W. Yu and Y. Yu. 2016. Effect of density on the hygroscopicity and surface characteristics of hybrid poplar compreg. J. Wood Sci. 62(5): 441–451.

Berube, M.-A., D. Schorr, R.J. Ball, V. Landry and P. Blanchet. 2017. Determination of *in situ* esterification parameters of citric acid-glycerol based polymers for wood impregnation. Journal of Polymers and the Environment.

Boitouzet, T. 2017. Woodoo. Retrieved 2017-11-13, from https://woodoo.fr/.

Bongiovanni, R., F. Montefusco, A. Priola, N. Macchioni, S. Lazzeri, L. Sozzi and B. Ameduri. 2002. High performance UV-cured coatings for wood protection. Prog. Org. Coatings 45(4): 359–363.

Brich, M.A., V.P. Kozhin and V.K. Shchitnikov. 1999. Kinetics of the wood impregnation process. Modeling and experiment. Journal of Engineering Physics and Thermophysics 72(4): 590–598.

Broda, M., B. Mazela, I. Dąbek, A. Dutkiewicz, H. Maciejewski, R. Markiewicz, M. Grzeszkowiak and S. Jurga. 2017. Influence of organosilicon compounds and nanocellulose treatment on the dimensional stability of waterlogged elm wood.

Cai, X., B. Riedl, S.Y. Zhang and H. Wan. 2007. Effects of nanofillers on water resistance and dimensional stability of solid wood modified by melamine-urea-formaldehyde resin. Wood and Fiber Science 39(2): 307–318.

Cai, X. and P. Blanchet. 2010. Acrylate wood densification: Effect of vacuum time and nanoparticles on chemical retention, penetration, and resin distribution. Wood Fiber Sci. 42(3): 318–327.

Cai, X. and P. Blanchet. 2011. Effect of vacuum time, formulation, and nanoparticles on properties of surface-densified wood products. Wood Fiber Sci. 43(3): 326–335.

Cox, P.E. 2009. Wood hardening procedures. Patent. US 20090142613 A1. https://www.google.com/ patents/US20090142613.

Cristea, M.V., B. Riedl and P. Blanchet. 2011. Effect of addition of nanosized UV absorbers on the physico-mechanical and thermal properties of an exterior waterborne stain for wood. Prog. Org. Coatings 72(4): 755–762.

de Meijer, M. 2001. Review on the durability of exterior wood coatings with reduced VOC-content. Prog. Org. Coatings 43(4): 217–225.

De Vetter, L., J. Van den Bulcke and J. Van Acker. 2010. Impact of organosilicon treatments on the wood-water relationship of solid wood. Holzforschung 64: 463–468.

Devi, R.R., I. Ali and T.K. Maji. 2003. Chemical modification of rubber wood with styrene in combination with a crosslinker: effect on dimensional stability and strength property. Bioresource Technology 88(3): 185–188.

Devi, R.R. and T.K. Maji. 2006. Effect of chemical modification with styrene and glycidyl methacrylate on the properties of pinewood. Indian J. Eng. Mater. Sci. 13(2): 149–154.

Devi, R.R. and T.K. Maji. 2013. Effect of nanofillers on flame retardancy, chemical resistance, antibacterial properties and biodegradation of wood/styrene acrylonitrile co-polymer composites. Wood Sci. Technol. 47(6): 1135–1152.

Ding, W.D., A. Koubaa, A. Chaala, T. Belem and C. Krause. 2008. Relationship between wood porosity, wood density and methyl methacrylate impregnation rate. Wood Material Science and Engineering 3(1-2): 62–70.

Ding, W.D., A. Koubaa and A. Chaala. 2013. Mechanical properties of MMA-hardened hybrid poplar wood. Industrial Crops and Products 46: 304–310.

Donath, S., H. Militz and C. Mai. 2004. Wood modification with alkoxysilanes. Wood Science and Technology 38(7): 555–566.

Donath, S., H. Militz and C. Mai. 2006. Weathering of silane treated wood. Holz als Roh-und Werkstoff. 65(1): 35.

Dong, Y., Y. Yan, S. Zhang and J. Li. 2014. Wood/polymer nanocomposites prepared by impregnation with furfuryl alcohol and nano-SiO$_2$. Bioresources 9(4): 6028–6040.

Ellis, W.D. and J.L. O'Dell. 1999. Wood-polymer composites made with acrylic monomers, isocyanate, and maleic anhydride. J. Appl. Polym. Sci. 73(12): 2493–2505.

Erickson, H.D. and J.J. Balatinecz. 1964. Liquid flow paths into wood using polymerization techniques-Douglas-fir and styrene. Forest Products Journal 14: 293–299.

Essoua Essoua, G.G., P. Blanchet, V. Landry and R. Beauregard. 2015. Maleic anhydride treated wood: effects of drying time and esterification temperature on properties. Bioresources 10(4): 6830–6860.

Essoua Essoua, G.G., P. Blanchet, V. Landry and R. Beauregard. 2016. Pine wood treated with a citric acid and glycerol mixture: biomaterial performance improved by a bio-byproduct. Bioresources 11(2): 3049–3072.

Evans, P.D. 2008. Weathering and photoprotection of wood. pp. 69–117. *In*: Schultz, T.P., H. Militz, M.H. Freeman, B. Goodell and D.D. Nicholas (eds.).

Ewald, V.A.K. 1966. U.S. Patent No. 3,233,579. Washington, DC: U.S. Patent and Trademark Office.

Fadl, N.A. and A.H. Basta. 2005. Enhancement of the dimensional stability of natural wood by impregnates. Pigment & Resin Technology 34(2): 72–87.

Fernandes, J., A.W. Kjellow and O. Henriksen. 2012. Modeling and optimization of the supercritical wood impregnation process—Focus on pressure and temperature. The Journal of Supercritical Fluids 66: 307–314.

Freeman, M.H., T.F. Shupe, R.P. Vlosky and H.M. Barnes. 2003. Past, present, and future of the wood preservation industry: wood is a renewable natural resource that typically is preservative treated to ensure structural integrity in many exterior applications. Forest Products Journal 53(10): 8–16.

Furuno, T., Y. Imamura and H. Kajita. 2004. The modification of wood by treatment with low molecular weight phenol-formaldehyde resin: a properties enhancement with neutralized phenolic-resin and resin penetration into wood cell walls. Wood Science and Technology 37(5): 349–361.

Gabrielli, C.P. and F.A. Kamke. 2010. Phenol–formaldehyde impregnation of densified wood for improved dimensional stability. Wood Science and Technology 44(1): 95–104.

Gan, W., S. Xiao, L. Gao, R. Gao, J. Li and X. Zhan. 2017. Luminescent and transparent wood composites fabricated by poly(methyl methacrylate) and γ-Fe2O3@YVO4:Eu3+ nanoparticle impregnation. ACS Sustainable Chemistry & Engineering 5(5): 3855–3862.

Ghazali, H. and R. Yusoff. 2011. Effets of glycerol impregnation on compressive strength parallel to grain of acacia auriculiformis upon drying. Borneo Science 29: 1–9.

Gindl, W., F. Zargar-Yaghubi and R. Wimmer. 2003. Impregnation of softwood cell walls with melamine-formaldehyde resin. Bioresour. Technol. 87(3): 325–330.

Gindl, W., C. Hansmann, N. Gierlinger, M. Schwanninger, B. Hinterstoisser and G. Jeronimidis. 2004. Using a water-soluble melamine-formaldehyde resin to improve the hardness of Norway spruce wood. Journal of Applied Polymer Science 93(4): 1900–1907.

Giudice, C.A., P.V. Alfieri and G. Canosa. 2013a. Decay resistance and dimensional stability of Araucaria angustifolia using siloxanes synthesized by sol–gel process. International Biodeterioration & Biodegradation 83(Supplement C): 166–170.

Giudice, C.A., P.V. Alfieri and G. Canosa. 2013b. Siloxanes synthesized "*in situ*" by sol–gel process for fire control in wood of Araucaria angustifolia. Fire Safety Journal 61(Supplement C): 348–354.

Graziola, F., F. Girardi, R. Di Maggio, E. Callone, E. Miorin, M. Negri, K. Mueller and S. Gross. 2012. Three-components organic-inorganic hybrid materials as protective coatings for wood: Optimisation, synthesis, and characterisation. Prog. Org. Coatings 74(3): 479–490.

Grosse, C., M. Noel, M.F. Thévenon and P. Gérardin. 2016. Optimising wood chemical modification with lactic acid oligomers by screening of processing conditions and chemical additives. 47th IRG Annual Meeting.

Hadi, Y.S., I.S. Rahayu and S. Danu. 2013. Physical and mechanical properties of methyl methacrylate impregnated jabon wood. Journal of the Indian Academy of Wood Science 10(2): 77–80.

Hazarika, A. and T.K. Maji. 2013. Effect of different crosslinkers on properties of melamine formaldehyde-furfuryl alcohol copolymer/montmorillonite impregnated softwood (*Ficus hispida*). Polymer Engineering & Science 53(7): 1394–1404.

He, S., Y. Guo, T. Stone, N. Davis, D. Kim, T. Kim and M. Rafailovich. 2017. Biodegradable, flame retardant wood-plastic combination via *in situ* ring-opening polymerization of lactide monomers. J. Wood Sci. 63(2): 154–160.

Hill, C.A. 2007. Wood Modification: Chemical, Thermal and Other Processes (Vol. 5). John Wiley & Sons.

Hubbe, M.A., O.J. Rojas and L.A. Lucia. 2015. Green modification of surface characteristics of cellulosic materials at the molecular or nano scale: a review. Bioresources 10(3): 6095–6206.

Islam, M.N., K. Ando, H. Yamauchi, Y. Kobayashi and N. Hattori. 2008. Comparative study between full cell and passive impregnation method of wood preservation for laser incised Douglas fir lumber. Wood Science and Technology 42(4): 343–350.

Kaboorani, A., N. Auclair, B. Riedl and V. Landry. 2017. Mechanical properties of UV-cured cellulose nanocrystal (CNC) nanocomposite coating for wood furniture. Prog. Org. Coatings 104: 91–96.

Kakavas, K., D. Birbilis and T. Tsioukas. 2014. Determining the tensile strength and other properties of European Beech (Fagus Sylvatica) wood impregnated with Rapeseed oil. Journal of International Scientific Publications: Materials, Methods and Technologies 8: 819–823.

Khalil, H.P.S.A., R. Dungani, I.A. Mohammed, M.S. Hossain, N.A.S. Aprilia, E. Budiarso and E. Rosamah. 2014. Determination of the combined effect of chemical modification and compression of Agatis wood on the dimensional stability, termite resistance, and morphological structure. Bioresources 9(4): 6614–6626.

Kučerová, I. 2012. Methods to measure the penetration of consolidant solutions into dry wood. Journal of Cultural Heritage 13(3): S191–S195.

Larnøy, E., H. Militz and M. Eikenes. 2005. Uptake of chitosan based impregnation solutions with varying viscosities in four different European wood species. European Journal of Wood and Wood Products 63(6): 456–462.

Lesar, B. and M. Humar. 2011. Use of wax emulsions for improvement of wood durability and sorption properties. European Journal of Wood and Wood Products 69(2): 231–238.

Lesar, B., M. Pavlič, M. Petrič, A.S. Škapin and M. Humar. 2011a. Wax treatment of wood slows photodegradation. Polymer Degradation and Stability 96(7): 1271–1278.

Lesar, B., A. Straže and M. Humar. 2011b. Sorption properties of wood impregnated with aqueous solution of boric acid and montan wax emulsion. Journal of Applied Polymer Science 120(3): 1337–1345.

Li, Y., Q. Fu, S. Yu, M. Yan and L. Berglund. 2016. Optically transparent wood from a nanoporous cellulosic template: combining functional and structural performance. Biomacromolecules 17(4): 1358–1364.

Li, Y., Q. Fu, R. Rojas, M. Yan, M. Lawoko and L. Berglund. 2017a. Lignin-retaining transparent wood. ChemSusChem. 10(17): 3445–3451.

Li, Y., S. Yu, J.G.C. Veinot, J. Linnros, L. Berglund and I. Sychugov. 2017b. Luminescent transparent wood. Advanced Optical Materials 5(1): 1600834.

Mahltig, B., C. Swaboda, A. Roessler and H. Böttcher. 2008. Functionalising wood by nanosol application. Journal of Materials Chemistry 18(27): 3180–3192.

Mahmoud, A.A., A.M.F. Eissa, M.S. Omar, A.A. El-Sawy and A.F. Shaaban. 2000. Improvement of white pinewood properties by impregnation with thiourea-formaldehyde resin and orthophosphoric acid. J. Appl. Polym. Sci. 77(2): 390–397.

Mai, C. and H. Militz. 2004. Modification of wood with silicon compounds. Treatment systems based on organic silicon compounds—a review. Wood Science and Technology 37(6): 453–61.

Malkov, S., P. Tikka and J. Gullichsen. 2003. Towards complete impregnation of wood chips with aqueous solutions. Part I. A retrospective and critical evaluation of the penetration process. Paperi ja puu. 85(8): 460–466.

Mallon, S. and C.A.S. Hill. 2002. Covalent bonding of wood through chemical activation. Int. J. Adhes. 22(6): 465–469.

Maminski, M., P. Parzuchowski, P. Borysiuk and P. Boruszewski. 2011. Glycerol as a renewable resource for wood adhesives. Proceedings of 2nd International Conference on Environmental Engineering and Applications, Singapore 13–16.

McIntyre, C.R. and D.A. Eakin. 1984. U.S. Patent No. 4,466,998. Washington, DC: U.S. Patent and Trademark Office.

Mattos, B.D., P.H. G. de Cademartori, A.L. Missio, D.A. Gatto and W.L.E. Magalhaes. 2015. Wood-polymer composites prepared by free radical *in situ* polymerization of methacrylate monomers into fast-growing pinewood. Wood Sci. Technol. 49(6): 1281–1294.

Matsumura, J., R.E. Booker, L.A. Donaldson and B.G. Ridoutt. 1998. Impregnation of radiata pine wood by vacuum treatment: identification of flow paths using fluorescent dye and confocal microscopy. Iawa Journal 19(1): 25–33.

Neyses, B. and D. Sandberg. 2016. Application of a new method to select the most suitable wood species for surface densification. *In*: Forest Products Society International Convention. Forest Products Society.

Neyses, B., O. Hagman, D. Sandberg and A. Nilsson. 2016. Development of a continuous wood surface densification process: the roller pressing technique. pp. 17–24. *In*: Proceedings of International Convention of Society of Wood Science and Technology: Forest Resource and Products: Moving Toward a Sustainable Future. Society of Wood Science and Technology.

Nishida, M., T. Tanaka, T. Miki, Y. Hayakawa and K. Kanayama. 2017. Integrated analysis of solid-state NMR spectra and nuclear magnetic relaxation times for the phenol formaldehyde (PF) resin impregnation process into soft wood. RSC Advances 7(86): 54532–54541.

Noël, M., E. Fredon, E. Mougel, D. Masson, E. Masson and L. Delmotte. 2009a. Lactic acid/wood-based composite material. Part 1: Synthesis and characterization. Bioresource Technology 100(20): 4711–4716.

Noël, M., E. Mougel, E. Fredon, D. Masson and E. Masson. 2009b. Lactic acid/wood-based composite material. Part 2: Physical and mechanical performance. Bioresource Technology 100(20): 4717–4722.

Noël, M., W. Grigsby, I. Vitkeviciute and T. Volkmer. 2015a. Modifying wood with bio-polyesters: analysis and performance. International Wood Products Journal 6(1): 14–20.

Noël, M., W.J. Grigsby and T. Volkmer. 2015b. Evaluating the extent of bio-polyester polymerization in solid wood by thermogravimetric analysis. Journal of Wood Chemistry and Technology 35(5): 325–336.

Noël, M., W.J. Grigsby, G.A. Ormondroyd and M.J. Spear. 2016. Influence of water and humidity on chemically modifying wood with polybutylene succinate bio-polyester. International Wood Products Journal 7(2): 80–88.

Panov, D. and N. Terziev. 2009. Study on some alkoxysilanes used for hydrophobation and protection of wood against decay. International Biodeterioration & Biodegradation 63(4): 456–461.

Pfriem, A., T. Dietrich and B. Buchelt. 2012. Furfuryl alcohol impregnation for improved plasticization and fixation during the densification of wood. Holzforschung 66(2): 215–218.

Reinprecht, L. 2016. Chemical protection of wood. pp. 145–217. *In*: Wood Deterioration, Protection and Maintenance. John Wiley & Sons, Ltd.

Pizzi, A. and K.L. Mittal. 2011. Wood Adhesives. CRC Press.

Roussel, C., V. Marchetti, A. Lemor, E. Wozniak, B. Loubinoux and P. Gérardin. 2001. Chemical modification of wood by polyglycerol/maleic anhydride treatment. Holzforschung 55(1): 57–62.

Sano, K., H. Kanematsu and T. Tanaka. 2016. Overview of silane-based polymer coatings and their applications. pp. 493–509. *In*: M. Hosseini and A.S.H. Makhlouf (eds.). Industrial Applications for Intelligent Polymers and Coatings. Cham, Springer International Publishing.

Siau, J.F. 1995. Wood: influence of moisture on physical properties. [Blacksburg, Va.], Virginia Polytechnic Institute and State University, Department of Wood Science and Forest Products.

Schneider, M.H. 1994. Wood polymer composites. Wood Fiber Sci. 26(1): 142–151.

Schneider, P.F., K.L. Levien and J.J. Morrell. 2003. Internal pressure measurement techniques and pressure response in wood during treating processes. Wood Fiber Sci. 35(2): 282–292.

Schorr, D., P. Blanchet and G.G. Essoua Essoua. 2018. Glycerol and citric acid treatment of Lodgepole Pine. Journal of Wood Chemistry and Technology. In press.

Solpan, D. and O. Guven. 1999. Improvement of mechanical stability of beechwood by radiation-induced *in situ* copolymerization of allyl glycidyl ether with acrylonitrile and methyl methacrylate. J. Appl. Polym. Sci. 71(9): 1515–1523.

Stamm, A.J. 1959. The dimensional stability of wood. For Prod. J. 9(10): 375–381.

Stolf, D.O. and F.A.R. Lahr. 2004. Wood-polymer composite: physical and mechanical properties of some wood species impregnated with styrene and methyl methacrylate. Materials Research 7(4): 611–617.

Stolf, D.O., M. da S. Bertolini, A.L. Christoforo, T.H. Panzera, S.L. Moni Ribeiro Filho and F.A.R. Lahr. 2017. *Pinus caribaea* var. hondurensis wood impregnated with methyl methacrylate. J. Mater. Civ. Eng. 29(6): 05016004.

Teaca, C.-A. and R. Bodirlau. 2016. Photochemical behavior of wood based materials. pp. 91–107. *In*: Rosu, D., M., V.P. (eds.). Photochemical Behavior of Multicomponent Polymeric-based Materials, Springer International Publishing.

Thuvander, F., L. Wallström, L.A. Berglund and K.A.H. Lindberg. 2001. Effects of an impregnation procedure for prevention of wood cell wall damage due to drying. Wood Science and Technology 34(6): 473–480.

Tondi, G., M.F. Thévenon, B. Mies, G. Standfest, A. Petutschnigg and S. Wieland. 2013. Impregnation of Scots pine and beech with tannin solutions: effect of viscosity and wood anatomy in wood infiltration. Wood Science and Technology 47(3): 615–626.

Trey, S.M., J. Netrval, L. Berglund and M. Johansson. 2010. Electron-beam-initiated polymerization of poly(ethylene glycol)-based wood impregnants. ACS Appl. Mater. Interfaces 2(11): 3352–3362.

Tshabalala, M.A., P. Kingshott, M.R. VanLandingham and D. Plackett. 2003. Surface chemistry and moisture sorption properties of wood coated with multifunctional alkoxysilanes by sol-gel process. Journal of Applied Polymer Science 88(12): 2828–2841.

Uraki, Y., K. Hashida, N. Watanabe, Y. Sano, T. Sasaya and H. Fujimoto. 1994. Novel wood processing by maleic acid-glycerol mixture system: Improvement of water resistance and mechanical property of cellulose by the processing. Journal of Wood Chemistry and Technology 14(3): 429–449.

Wallström, L. and K.A.H. Lindberg. 1999. Measurement of cell wall penetration in wood of water-based chemicals using SEM/EDS and STEM/EDS technique. Wood Science and Technology 33(2): 111–122.

Wang, B.J. and Y.H. Chui. 2012. Performance evaluation of phenol formaldehyde resin-impregnated veneers and laminated veneer lumber. Wood Fiber and Science 44(1): 5–13.

Wang, W., Y. Zhu, J. Cao and X. Guo. 2015. Thermal modification of Southern pine combined with wax emulsion preimpregnation: Effect on hydrophobicity and dimensional stability. Holzforschung 69(4): 405–413.

Wolkoff, P. 1995. Volatile organic compounds—Sources, measurements, emissions, and the impact on indoor air quality. Indoor Air. 3: 1–73.

Yaddanapudi, H.S., N. Hickerson, S. Saini and A. Tiwari. 2017. Fabrication and characterization of transparent wood for next generation smart building applications. Vacuum 146: 649–654.

Yan, L., J. Cao, W. Gao, X. Zhou and G. Zhao. 2011. Interaction between glycerin and wood at various temperatures from stress relaxation approach. Wood Science and Technology 45(2): 215–222.

Ying Lin, L. and F. Fu. 2012. The composite wood impregnated with silicon sol solution 466-467: 121–126.

Zhang, Y., S.Y. Zhang, Y.H. Chui and H. Wan. 2006. Effect of impregnation and *in-situ* polymerization of methacrylates on hardness of sugar maple wood. Journal of Applied Polymer Science 99(4): 1674–1683.

Zhu, M., J. Song, T. Li, A. Gong, Y. Wang, J. Dai, Y. Yao, W. Luo, D. Henderson and L. Hu. 2016. Highly anisotropic, highly transparent wood composites. Advanced Materials 28(26): 5181–5187.

Part III

Structural Polymers and Extractives From Lignocellulosic Sources—Properties and New Applications

Latest Applications of TEMPO-Oxidized Cellulose Nanofibres Obtained from an Ultrasound-assisted Process

*Éric Loranger** and *Claude Daneault*

Introduction

In recent years, considerable interest has been shown in application of ultrasound as an advanced oxidation process for the treatment of hazardous contaminants in water (Naddeo et al. 2014). Sonochemistry has been demonstrated as a promising method for destruction of aqueous pollutants (Mahvi 2009). The advantage is that there are no additives introduced into the ultrasonic system and no byproducts generated by ultrasonic technology. The efficiency of ultrasound in the processing of vegetal materials extraction has been already proved (Ebringerova and Hromadkova 2010). The known ultrasound benefits, such as swelling of vegetal cells and fragmentation due to the cavitational effect associated with the ultrasonic treatment, act by increasing the yield and by shortening the extraction time. The effect of ultrasound on lignocellulosic biomass has been employed in order to improve the extractability of hemicelluloses (Minjares-Fuentes et al. 2016), cellulose (Pappas et al. 2002), lignin (Sun and Tomkinson 2002) or to get clean cellulosic fibres from used paper (Scoot and Gerber 1995), but only few attempts to improve the susceptibility of

Lignocellulosic Materials Research Centre, Université du Québec à Trois-Rivières, 3351 boulevard des Forges, Trois-Rivières (QC), Canada G9A 5H7.
* Corresponding author: Eric.loranger1@uqtr.ca

lignocellulosic materials to biodegradation by using ultrasound power has been described. Even if the oxidation power of ultrasound alone cannot be disputed, the technico-economic future of this technique is clearly oriented on coupling this technology with other processes. Thus, the combination of ultrasonic waves with perfectly controlled techniques constitutes an interesting path of development. With our research, we have privileged this synergistic avenue in order to produce nanofibrillated cellulose.

Referring to the literature, different types of cellulose microfibrils have been produced by physical and chemical treatments of cellulose fibres (De Nooy et al. 1995, Saito and Isogai 2005, Elazzouzi-Hafraoui 2008): mechanical treatment, acid hydrolysis and catalyzed oxidation. In the case of mechanical treatment, the resulting products consist mainly of microfibrils bundles, and requires a large amount of energy. Thus, it has not been sufficient to individualize cellulose microfibrils using only mechanical treatment. To overcome this problem, catalyzed oxidation is an alternate promising route. This technique is based on the conversion of alcohol groups of cellulose into carboxylic acids using a 2,2,6,6-tetramethyl-1-piperidine oxoammonium radical (TEMPO) in the presence of sodium hypochlorite and sodium bromide. With such reagents, the oxidation is selective, as it oxidizes exclusively the primary hydroxyl groups, while leaving untouched the secondary ones (De Nooy et al. 1995). While this chemical modification promotes individualized fibres through electrostatic repulsion, it also provides a versatile route, which allows the introduction of other functionalities for developing new materials. In this context, the oxidized cellulose containing carboxylic groups can serve as templates to bind different molecules of interest (fluoprobes, peptides, antibodies, etc.), opening thus new horizons for various applications. In the first part of our work, efforts have been made to study the coupling of three fluorescent amino acids and evaluate the grafting of trytophan-based peptides (Trp-Ps) onto TEMPO oxidized cellulose nanofibres (TOCN). The advantage of using TOCN rather than other cellulose products is that the coupling reaction between TOCN and the amino acid is very fast in mild conditions and without alteration of the polysaccharide (Hoare and Koshland 1967). This coupling is a nucleophilic reaction between the amine group ($-NH_2$) of amino acid and carboxyl group ($-COOH$) of the TOCN, and is commonly catalyzed by the coupling agent hydrochloride 1-ethyl-3-(3-dimethylaminopropyl) carbodiimide (EDAC) in the presence of an activation agent such as N-hydroxysuccinimide (NHS) (Bulpitt and Aeschlimann 1999). The preparation of novel biocomposites based on TEMPO-oxidized cellulose nanofibres will open new areas for applications in the food industry, cosmetics, medicine, packaging, and other areas.

Another route toward the preparation of nanocelluloses relies on acid treatment. By dissolving the amorphous domains of the cellulose, this lead to the formation of cellulose nanocrystals (CNC) having dimensions, which depend on the type of acid, acid concentration, time, and temperature of the hydrolysis reaction and the origin of the cellulose (Araki et al. 1998, Azizi Samir et al. 2005). One of the main disadvantages of this hydrolytic treatment is the depolymerization of cellulose chains that promotes a dramatic decrease in the microfibril length and width, which negatively affects the structure of microfibrils. At the present time, applications of crystalline nanocellulose (NCC or CNC) are very limited which is not the case with cellulose nanofibres (CNF).

TOCN is often used as a reinforcement agent in composite materials and they are considered to be a cellulosic nanomaterial (5–20 nm wide single-sized microfibrils), composed of crystalline and amorphous domains (Mishra et al. 2011). The richness of these nanofibres (TOCN) over crystalline nanocellulose (NCC or CNC), resides in the conservation of the amorphous regions with a strong presence of carboxylic groups on its structure and a large specific surface. As showed earlier, this oxidized cellulose-containing carboxyl and hydroxyl groups can serve as templates for grafting other molecules of interest, thus opening up new horizons for many applications (biocomposites, packaging, electronics, biomedical, etc.). In our research, we have combined the oxidation of cellulose by 4-acetamido-TEMPO (Saito et al. 2007, Saito et al. 2006) under ultrasound in a pilot plant process. Although the technology has been shown to be feasible on a small scale (Mishra et al. 2011), the commercialization of sonolysis is still a challenge. Currently, we use high-intensity ultrasound (170 kHz) generated in a semi-industrial sonoreactor tube in combination with a TEMPO system to oxidize the primary hydroxyl groups on the cellulose to carboxylate groups, in order to produce TEMPO-oxidized cellulose nanofibres (TOCN).

Lately, the use of electroactive composite materials has become an important research area in polymer science and in several technological and medical applications. In that way, conducting polymers are promising because of their functional properties, allowing them to be used in many application fields such as batteries, sensors, antistatic coating among others (Ambade et al. 2013, Wang et al. 2014). Lately, polyacetylene, polyaniline, polythiophene or polypyrrole (PPy) have been massively studied but, it appears that PPy is one of the most promising conducting polymers for most of the applications, in electronics (supercapacitors) and biomedical field (biosensors, blood purification). This is related to PPy biocompatibility, environmental and thermal stability and high electrical conductivity (Vaitkuviene and Kaseta 2013). However, PPy presents poor processability and mechanical properties which is limiting its commercial applications (Sangawar and Moharil 2012). In order to overcome these physical problems, considerable attention has been focused on a possible reinforcement of the matrix with natural fibres such as bacterial and algal cellulose, dextrin, carboxymethylcellulose, chitosan (Cabuk et al. 2014, Zare et al. 2014). Among them, cellulose who is the most abundant natural polymer on earth, presents a great interest. Indeed, using cellulose-based materials is advantageous since cellulose is renewable, unlimited and biodegradable. Among the advantages of using oxidized cellulose fibres are their large specific surface area, their microporous structure and surface charges for good adherence and their high mechanical properties (Syverud and Stenius 2009). The polymerization of PPy on the surface occurs when the TEMPO oxidized cellulose nanofibres are dipped in a solution of iron chloride in the presence of pyrrole. Usually, the bonds obtained between the PPy and the reinforcement are weak electrostatic interactions (Jradi et al. 2012). In order to overcome this limitation, grafting of 1-(2-cyanoethyl) pyrrole on TEMPO oxidized cellulosic nanofibres (TOCNF) before performing the deposition of PPy, is a possible way. The 1-(2-cyanoethyl) pyrrole offers the possibility of creating stronger covalent bond (amides) between carboxylic groups and the conducting polymer (Lee and Schmidt 2010). Thus the cellulose is first grafted (PyCNFo) before the synthesis of the final composite [p(PyTOCNF)] via polymerization of polypyrrole using $FeCl_3$. We have then synthetized TEMPO-oxidized cellulose nanofibres (TOCN), and polypyrrole (PPy) composites using three other different processes not involving grafting. Depending on their final application, the choice of a fabrication technique as to be done is relies on an appropriate fabrication method, these composites could be considered in the design of high-performance electrodes for supercapacitor, batteries or sensors.

New generation of food packaging based on nanomaterials, has attracted much attention in recent years and represent a better alternative, in response to conventional food packaging (Huang et al. 2015, Cushen et al. 2012). These new materials can be the key for maintaining food quality, freshness and product security to slow down and to prevent microbial development and the associated physiological and chemical changes in food produced by microbial and enzymatic activity (Bradley et al. 2011, Lavoine et al. 2015). Microbial and enzymatic activities are not the only responsible partly for most of the degradation reactions of foodstuffs. Oxygen, either directly or indirectly, plays an important role in the loss of organoleptic and nutritional properties as well as food spoilage by aerobic microorganisms. It is therefore necessary to use materials with good oxygen barrier properties (Duncan 2011). However, residual oxygen levels that are sufficiently high also promote food spoilage. To address this problem, oxygen scavengers or antioxidants may also be incorporated (Busolo and Lagaron 2015). For all these reasons, the food industry has shown increasing interest in the development of active packaging with antioxidant properties, which can be incorporated into mono or multilayer structures, or coated on the packaging layer (Arrua et al. 2010). In this regard, the industry is seeking solutions with bio-based products for the new generation of food packaging. Among these new solutions, cellulose has been widely explored since it can confer biodegradability and a recyclable character for packaging (Lavoine et al. 2014). Cellulose nanofibres have demonstrated barrier properties to oxygen in addition to its excellent mechanical Properties due to the fibres that are mostly entangled and are forming a tight network (Nair et al. 2014, Lavoine et al. 2012). Nevertheless, cellulose has a major inconvenience for packaging, because of its affinity for water. Indeed, cellulose films have very poor barrier properties to water vapour because of its hydrophilicity. To resolve this problem, we have coupled TOCN (nanofibres obtained by TEMPO oxidation) with a semi-conductive polymer, polypyrrole and polyvinyl alcohol (PVA). The PPy coating was an interesting alternative due to its hydrophobic character. Moreover, PPy confers attractive properties for its active packaging as antibacterial and antioxidant properties. In addition, the PVA chains allow better tensile strength. The goal was to prepare a packaging that simultaneously combines biodegradability and the barrier properties to gas of cellulose with the physico-chemical properties of PPy.

Ultrasonic Technology in TEMPO-Oxidized Cellulose Nanofibres Production

During oxidation with the TEMPO system, 4-acetamido 2,2,6,6-tetramethyl-1-piperidinyloxy always acts in the presence of a co-oxidant. In an aqueous environment, the most often used system is the TEMPO/NaBr/NaOCl. The catalytic oxidation mechanism of the primary alcohols is generally the most accepted, and is presented in Fig. 1 (Paquin et al. 2013). As explained earlier, the introduction of negative charges during the oxidation reaction with the TEMPO-NaOCl-NaBr on the surface of the fibres induces repulsive forces that ease their individualization and their dispersion in the fibrous suspension.

This effect facilitates, afterwards, the nanocellulose preparation using a light mechanical treatment (refining, ultrasounds) after the oxidation step. In the fine chemistry field, from organic to environmental chemistry, the use of sonochemistry is well known. On the other hand, the transposition up to the industrial scale of promising laboratory results is not an obvious thing. The laboratory experiments of Qin et al. (Qin et al. 2011) showed that treating cotton with the TEMPO/NaOCl/NaBr system in the presence of ultrasound has allowed achieving higher levels of carboxylic groups. Therefore leading to the production of more nanocellulose and more stability in an aqueous environment. In all these studies, the sonochemical reactors were used either in ultrasonic bath (40 kHz) or ultrasound probes (20 kHz) at low power (300–600 W), mostly operated in batches. This is really a laboratory approach since these reactors must work at low concentrations to keep a homogeneous environment. The inhomogeneity of the ultrasonic field is problematic in scaling up since the geometric parameters of the reactor influence the efficiency

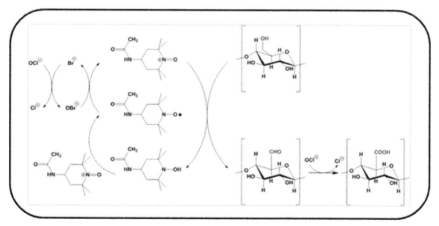

Fig. 1. Oxidation mechanism of cellulose with 4-acetamido-TEMPO in an aqueous environment at pH 10–11.

of the ultrasounds. The results of the research work of Mishra (Mishra et al. 2011) on the use of ultrasound to catalyze '*in situ*' oxidation reaction of bleached Kraft fibres with the TEMPO reagent has served a reference to develop a 40-liters pilot sonoreactor with variable frequency (40–170 kHz) and a nominal input power capacity of 2000 W shown in Fig. 2 (Loranger et al. 2011, Paquin et al. 2013).

The concept of a sonoreactor has been previously discussed in detail (Loranger et al. 2011). Briefly, a 40 L flow-through sonoreactor with a nominal input power capacity of 2000 W was used (Fig. 2) (Loranger et al. 2011). It was operated in a semi-continuous mode. The sonoreactor which was built with 24 Vibra-bar and a stainless steel pipe (316 L) was supplied by Ultrasonic Power Corporation. The pH of the reaction solution can be adjusted in a mixing tank by means of a programmable logic controller (PLC) with two separate metering pumps (acid, base). The temperature of the sonoreactor was also adjusted by PLC control. The mixing tank and the ultrasonic pipe were equipped with a cooling jacket to control the temperature during experimentation.

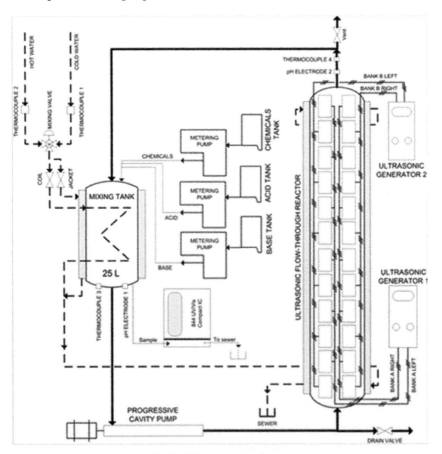

Fig. 2. Pilot sonoreactor diagram.

Before oxidation, 500 g of dried hardwood bleached Kraft pulp was pre-soaked for two days in 15 L of distilled water at room temperature. The soaked pulp was then disintegrated for 10 min in a laboratory disintegrator to obtain a uniform fibre suspension at about 1% consistency. To the cellulose suspension was added 25 g of 4-acetamido TEMPO and 62.5 g of sodium bromide. A solution of NaOCl (3.1 mmol/g) was added to the mixture at room temperature under gentle agitation during the first 30 min. NaBr was added as a co-catalyst to increase the reaction velocity but too much NaBr can decrease the ability to detect the effect of ultrasound because of its doping effect on the reaction system (Mishra et al. 2011). The other reaction conditions were pH 10.5, 25°C and 90 minutes. The reaction was stopped after 90 min by adding 1 L of H_2O_2 (1%), and the final pH of the suspension was adjusted to 7 by adding 0.5 M NaOH or HCl as required. Finally, the oxidized fibre suspension obtained was filtered and thoroughly washed with distilled water on a cotton membrane, and stored at 4°C to be used for further analysis. The degree of polymerization was evaluated at 200 and the original fibrous morphologies were unchanged even after oxidation with sufficient reagents quantities. When a bleached softwood Kraft pulp (approximately 90% of cellulose and 10% hemicelluloses) was oxidized using the TEMPO/NaBr system in water at pH 10 with 3.75 mmol/kg of NaOCl as the primary oxidant, a significant amount of sodium carboxylate groups and small amounts of aldehyde groups (50 to 100 mmol/kg of pulp) are formed on cellulose by oxidation. As presented earlier, an ultrasound-assisted TEMPO-system (US-TEMPO) will facilitate the oxidation of cellulose for the production of cellulose nanofibres (Mishra et al. 2011, Rataz et al. 2011). After 90 min of reaction at 25°C, the carboxyl group's content reached 1600 mmol per kilogram of pulp in an aqueous environment (pH 10). The acoustic frequency used was 170 kHz (125 W) which was found earlier to be optimum (Mishra et al. 2011, Loranger et al. 2011). Thus, this system gave a 10 to 15% increase in carboxyl content over reaction without ultrasounds and 10% increase in nanocellulose yield under optimized conditions. Moreover, the results obtained with the sonoreactor show a diminution of 50% in the reaction time, 50% of the chemical product consumption, an yield of 95% and a crystallinity index of 70% equivalent to the original Kraft pulp. Using acoustic cavitation gives a positive effect on the polymerization degree; the nanofibres produced are also very thermally stable. From the NMR analysis data, it can be concluded that like TEMPO-system, oxidation with US-TEMPO is also regioselective, and occurs mainly at C6 primary hydroxyl groups (Mishra et al. 2011).

As presented ealier, the maximum value of carboxylate groups (1600 mmol/kg) was achieved at 125 W (16,4 W/L) and at a frequency of 170 kHz. The carboxylate content of the TEMPO-oxidized cellulose nanofibres was determined by an electrical conductivity titration method (Saito and Isogai 2005) using a Dosimat 765 (Metrohm) titrator. In this procedure, the sodium carboxylates groups in TEMPO oxidized cellulose are converted to their free carboxyl form by treating the sample with 0.1 M HCl solution three times and, finally, thoroughly washed with de-ionized water to remove the excess acid. The oxidized pulp prepared in this way was transferred to a 600 mL beaker containing 450 mL of 0.001 N NaCl solution and well mixed. 5 mL of 0.1 N HCl was added to the fibre suspension before starting the titration of the carboxylate groups with a 0.1 N NaOH standardized solution. At the end of the titration, the fibres were filtered, washed and dried in an oven at 105°C to determine the exact weight of the sample. The carboxyl content expressed in mmol/g was then calculated by the software.

The designing of the pilot sonoreactor (Paquin et al. 2013) has finally allowed us to consider an industrial production '*in situ*' of cellulose nanofibres in an alkaline environment (TEMPO/NaBr/NaOCl) in the presence of ultrasounds (Fig. 3). In this system, the mechanical dispersion of the TEMPO-oxidized cellulose gel (TOCN gel nanofibres) was prepared by high shear dispersion of oxidized pulp in a wet colloid milling apparatus (MK 2000/4) from IKA Works, Inc. (USA). The oxidized pulp suspension (2 L) was pumped from an agitated tank between a conical rotor and a stator in the MK mill. The mill was operated in closed loop for a given amount of time (1 h) under the optimized operating conditions. The normal operating conditions of this laboratory setup were: 3% consistency, 0.073 mm gap, 200 mL/min recirculation rate, 25°C, and pH 7 (Loranger et al. 2012). As already explained, carboxylate groups created on the cellulose microfibrils by TEMPO-mediated oxidation have anionic charge in water. These charges induce electrostatic repulsion between the microfibrils which consequently help the disintegration process and contribute to the formation of individual microfibrils.

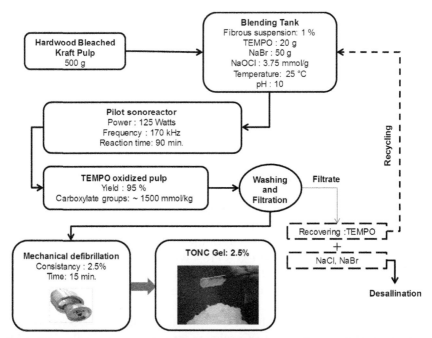

Fig. 3. Schematic flow diagram of TEMPO-oxidized cellulose nanofibres production assisted by ultrasounds

With this mechanical dispersion, the width and length of nanofibres are estimated at 3.5 and 306 nm. Nevertheless a small proportion of microfibrillated cellulose is still present.

The high aspect ratios, uniform widths and complete nanodispersion of individual TOCN are different from the others, which clearly indicates the advantages of TOCN and US-TEMPO (Table 1). The contribution of ultrasound in the process of US-TEMPO in terms of reduction of chemicals, reaction time and clear the energy consumption allows the process to be more economically viable. The richness of these nanofibres, contrary to crystalline nanocellulose, resides in the conservation of the amorphous regions, of a strong presence of carboxylic groups on its structure, and a large specific surface, that is characteristics, which have a considerable interest for polymerists.

TEMPO-Oxidized Cellulose Nanofibres Coupling with Amino Acids

The coupling of TOCN with amino acids is a modified version of the two-step attachment procedure proposed by Jiang et al. (2004). In the first step, 4 mL of the TOCN suspension (~ 2.0 mg of dried TOCN) was added to 5 mL of deionized water under moderate magnetic stirring. After which, 2.3 mL of 50 mg/mL N-hydroxysuccinimide (NHS) aqueous solution were added to the above suspension and mixed. Under fast stirring, 1.2 mL fresh N-ethyl-N'-(3-dimethylaminopropyl) carbodiimide hydrochloride (EDAC) aqueous solution (10 mg/mL) was added quickly, and the mixture was continually stirred at room temperature for 30 min while keeping the pH at 7.5–8 by adding 0.5 M NaOH and/or HCL. The suspension was then dialyzed in deionized water for 8 h to remove excess EDAC, NHS and the byproduct, urea. The deionized water used for dialysis was changed every hour. In the second step, 1 mL of 10 mg/mL amino acid in deionized water was added to 9 mL of the estered TOCN under moderate magnetic stirring overnight at room temperature. The suspension was then dialyzed thoroughly in deionized water for 72 h to remove unbound amino acid. The deionized water used for dialysis was changed every 6 or 8 h. Three amino acids, L-Tryptophan (Trp), L-Phenylanaline (Phe) and L-Tyrosine (Tyr) were attached to TOCN in this study. Hereafter, the amino acid grafted TOCN will be referred to as TOCN-amino acid (TOCN-Trp, TOCN-Phe and TOCN-Tyr). In the present work, effort has been made to probe the coupling of fluorescent amino acids and peptides onto TEMPO-oxidized cellulose nanofibres (TOCN) using a two-step coupling method at room temperature. The choice of fluorescent amino acids in this study is to further characterize their coupling with TOCN using fluorescence spectroscopy in addition to other techniques. The advantage of using TOCN rather than other cellulose products is that the coupling reaction between TOCN and the amino acid is very fast in mild conditions and without alteration of the polysaccharide (Hoare and Koshland 1967, Barazouk and Daneault 2011). This coupling is a nucleophilic reaction between the amine group ($-NH_2$) of amino acid and carboxyl group ($-COOH$) of the TONC,

Table 1. Differences between TOCN ultrasounds, TOCN, MFC and NCC.

	SONOCHEMISTRY	Isogai et al. 2011		
	TEMPO-OXIDIZED Cellulose nanofibers Ultrasound «in situ» TOCN	TEMPO-OXIDIZED Cellulose nanofibers TOCN	Microfibrillated cellulose MFC	Nanocrytalline Cellulose NCC
Preparation Method	TEMPO-oxidized with ultrasound «in situ» Mild disintegration in water	TEMPO-oxidized Mild disintegration in water	High pressure Homogenization treatment	Hydrolysis (64% H_2SO_4) Disintegration of residues in water
Yield	> 96%	> 90%	~ 100%	< 50%
Morphology	Uniform width of 3-4 nm, Length < 2 μm	Uniform width of 3-4 nm, Length < 2-3 μm	Uneven width of 10-2000 nm,	Uneven width of 5-100 nm, Length < 300 nm
Energy consumption	< 5 MJ / kg	< 7 MJ / kg	700-1400 MJ / kg	< 7 MJ / kg

and is commonly catalyzed by the coupling agent hydrochloride 1-ethyl-3-(3-dimethylaminopropyl) carbodiimide (EDAC) in the presence of an activation agent such as N-hydroxysuccinimide (NHS) (Fig. 4) (Barazouk and Daneault 2011).

We have succeeded in coupling TOCN with fluorescent amino acids using a two-step grafting procedure. We have also shown that the coupling reaction is pH dependent. At pH below 7 or higher than 11, there was no coupling reaction between TOCN and the amino acids. At pH = 8, a good coupling has been obtained. TEM images (Barazouk and Daneault 2011, Barazouk and Daneault 2012a,b) of all samples revealed that TOCN fibres were well individualized (no aggregation) even after grafting of TOCN with different amino acids. We have further confirmed the grafting of TOCN with amino acids spectroscopically. In fact, before the coupling, TOCN does not show any characteristic absorption band and does not fluoresce but after its coupling with amino acid, TOCN clearly shows the characteristic absorption and fluorescence features of the grafted amino acid. The formation of amides bonds as a result of coupling between TOCN and amino acid is evidenced by the presence of a small shoulder at 1545 cm^{-1} in FTIR experiments (Barazouk and Daneault 2011, Barazouk and Daneault 2012a,b). This absorption is characteristic of the C–N bond of the amide. Finally, the XPS studies further confirmed that N-1 peak (*400 eV) of TOCN amino acid is due to the nitrogen atom of the amide bond formed between TOCN and amino acid (Barazouk and Daneault 2011, Barazouk and Daneault 2012a,b). This coupling approach can also be applied to attach biomolecules (proteins) and other useful molecules containing amine groups, such as polymers and inorganic nanoparticles, onto TOCN. To the best of our knowledge, this is the first report dealing with detailed spectroscopic studies of oxidized nanocelluloses grafted with fluorescent amino acids (Barazouk and Daneault 2011, Barazouk and Daneault 2012a,b).

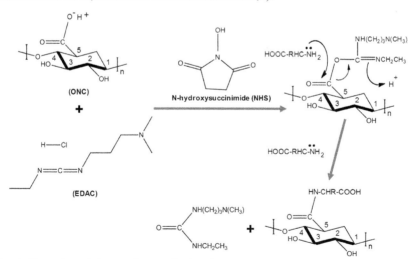

Fig. 4. Illustration of the coupling mechanism between TEMPO-oxidized cellulose nanofibres (TOCN) and an amino acid (H$_2$N-CHR-COOH, where *R* is an organic substituent).

TEMPO-Oxidized Cellulose Nanofibres Composites with Polypyrrole

In our work, we have made composite with TOCN and polypyrrole with two main way: grafing and chemical polymerization. In order to graft the 1-(2-cyanoethyl) pyrrole onto the TOCN, it needs to be reduced in N-(3-aminopropyl) pyrrole. The product was formed in agreement with the data gathered in the literature (Naji et al. 2003, Rajesh Bisht et al. 2005). A yellow oily product was obtained and stored at 4°C for further uses. The reduction was confirmed by 1H NMR and FTIR analysis. Yield about 90% (Naji et al. 2003, Abu-Rabeah et al. 2005) $\delta H(400 \text{ MHz; CDCl}_3)$: 1.90 (2H, quint, CH_2-2), 2.70 (2H, t, CH_2-3), 3.96 (2H, t, CH_2-1), 6.15 (2H, d, CH-β), 6.66 (2H, d, CH-α). FTIR spectrum (cm^{-1}) give 3364 and 3300 (NH) while Bisht et al. (2005) have reported 3370 and 3295. To synthesize PyTOCN, a part of N-(3-aminopropyl) pyrrole was introduced in a pH 7 buffer solution containing TOCN at a concentration of 0,5% and mixed for 10 min. In order to start the grafting, N-Hydroxysuccinimide (NHS) and 1-ethyl-[3-(dimethylamino) propyl]-3-ethylcarbodiimide HCl (EDC) were added (Abu-Rabeah et al. 2005). The quantities of N-(3-aminopropyl) pyrrole, NHS and EDC have been calculated to graft 10, 20 and 50% of the carboxylic groups of the TOCN (1600 mmol/kg). The solution was stirred for 3 hours before being placed in a dish and dried in a forced air oven for two days. The cast films were analyzed with FTIR and EDX to confirm grafting (Abu-Rabeah et al. 2005).

Various strategies are possible in order to produce composite films with TOCN and polypyrrole. Transparent films were prepared from PyCNFo suspension at a concentration of 0.5%. The film was dipped in a solution of $FeCl_3$ (0.2 M) for 10 min at room temperature. Then 2 mL of pyrrole was added and dispersed over the entire film. After 30 min of polymerization, the film was black and completely coated with PPy nanoparticles. The composite was thoroughly rinsed with distilled water and placed into a solution of HCl at 0.1 M, according to the literature (Hu et al. 2011) to enhance the conductivity. Finally, the film was dried between two hot plates at 80°C. When needed, pure PPy was synthesized by mixing 1 mL of pyrrole with 10 mL of $FeCl_3$ solution at 0.3 M. The reaction lasted 30 min after which, the solid residue was collected with a Buchner funnel and thoroughly washed with distilled water before it was air dried on a petri dish. Pure TOCN films were prepared with a solution of TOCN (50 mL) at 0.5% using a gel of TOCN at 2.6% and water. The mixture was thoroughly mixed until a homogenous medium was obtained. The solution was then poured in an aluminum dish of 90 mm to be first, oven dried for 1 h at 70°C and second, two days at ambient air.

In some situation, to reduce shrinkage when drying, a small quantity of polyvinyl alcohol (PVA) may be introduced in the TOCN solution before drying. For that matter, some TOCN/PVA may be required in order to isolate the effect of the PPy alone. In order to prepare TOCN/PVA films, TOCN suspensions were used in a concentration of 0.5%. A small amount of PVA solution (0.08 g in 2 mL of water) was added to 50 mL of TOCN suspension. The solutions were stirred for 1 hour at room temperature at a moderate speed to prevent bubble formation in the solution. The solution was later poured in an aluminum dish of 90 mm. The film was oven dried first, for 1 h at 70°C and afterwards, in ambient temperature conditions fortwo days. TOCN/PVA film were then immersed in a solution of FeCl$_3$ at 0.3M for 20 min. Next, 2 mL of pyrrole was added and distributed over the entire film. After 30 min of polymerization, the film was completely dark, being completely coated with PPy particles. The composite was thoroughly rinsed with distilled water before being dried between two hot plates at 80°C.

TEMPO-Oxidized Cellulose Nanofibres Composites with Polypyrrole for Organic Conductive Films

The development of biocomposites with a base of cellulose nanofibres has become a centre of interest in the biomedical domain, electronics (conductor circuits, transistors, and solar panels), packaging and thermic isolation. The main goal of this part of the study was to compare three composites resulting from three different synthesis process (Fig. 5) (Bideau et al. 2016a). The first process was *in situ* chemical polymerization which results in the composite commonly found in the literature while the second process involved the grafting of 1-(2-Cyanoethyl) pyrrole on TOCN before polypyrrole (PPy) polymerization. The 1-(2-cyanoethyl) pyrrole offers the possibility of creating covalent bonds (amides) between carboxylic groups and the PPy chains which are stronger bonds (Lee et al. 2010). Thus, to obtain a functionalized TOCN that can be conjugated with polypyrrole, we first reduced 1-(2-cyanoethyl) pyrrole into N-(3-aminopropyl pyrrole) using LiAlH$_4$ in anhydrous ether.

After reduction, PyTOCN was synthesized by coupling N-(3-aminopropyl) pyrrole to carboxylic groups on TOCN via EDC/NHS chemistry (Fig. 6) (Bideau

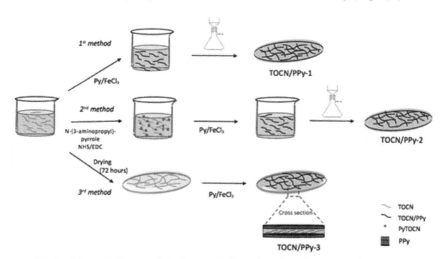

Fig. 5. Schematic diagram of the three methods used to TOCN/PPy composite synthesis.

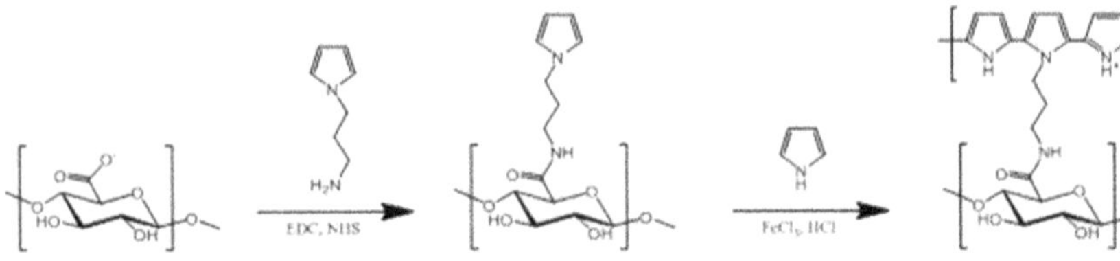

Fig. 6. Chemical structure of p(PyTOCN) composite.

et al. 2016b). In a second step, we polymerized polypyrrole onto PyTOCN from pyrrole by using $FeCl_3$ as oxidants (Fig. 6). Finally, the third method was carried out to synthesize multilayer composites films, which were obtained via polymerization of pyrrole on the surface of the TOCN film (Bideau et al. 2016a). These three methods led respectively to the synthesis of the composite TOCN/PPy-1, TOCN/PPy-2 and TOCN/PPy-3.

PPy particles distribution is detrimental for the films characteristic like electrical and mechanical properties. The chemical polymerization of pyrrole on the surface of TOCN films (third method) shows the best mechanical and thermal properties (TOCN/PPy-3). Indeed, the results revealed a significant effect of TOCN film on the improvement of the mechanical properties of TOCN/PPy, whereas the PPy layer on the top confers good thermal protection to the cellulosic fibres. Moreover, the presence of PPy layers allowed an increase in the hydrophobic character of the composite. On the contrary, TOCN/PPy-1 presented the worst mechanical properties due to the low number of interactions between cellulosic fibres. Indeed, the first method, which consists of the polymerization *in situ* of pyrrole around the cellulosic fibres, reduces cellulose hydrogen bonds potential, which confer mechanical properties. Nonetheless, this composite (TOCN/PPy-1) has demonstrated a better conductive capacity than TOCN/PPy-3 (51.6 against 21.7 S/cm) (Bideau et al. 2016a).

From our proposed methods, the grafting of N-(3-aminopropyl) pyrrole (TOCN/PPy-2) was interesting because it presented intermediate properties of all composites, and could be a good compromise between the mechanical and electrical properties. Indeed, several analysis performed on TOCN/PPy-2 at different grafting degrees has shown that the deposition of PPy nanoparticles onto the surface of PyTOCN has increased its mechanical properties (stress and elongation at break), particularly at high grafting percentage (Bideau et al. 2016b). Moreover, it is the first time in the literature that the PPy is accounted as a mechanical reinforcement for a composite. From our proposed mechanism, the grafting of 1-(2-cyanoethyl) pyrrole played a leading role in improving these properties since allowing for the connection between chains of conducting polymer and cellulose fibres (Fig. 7). The polypyrrole also allowed improving the thermal resistance and the hydrophobic character of the films by forming a protective layer. The combination of polypyrrole and 1-(2-cyanoethyl) pyrrole also allowed obtaining a high conductivity (about 47.6 S/cm).

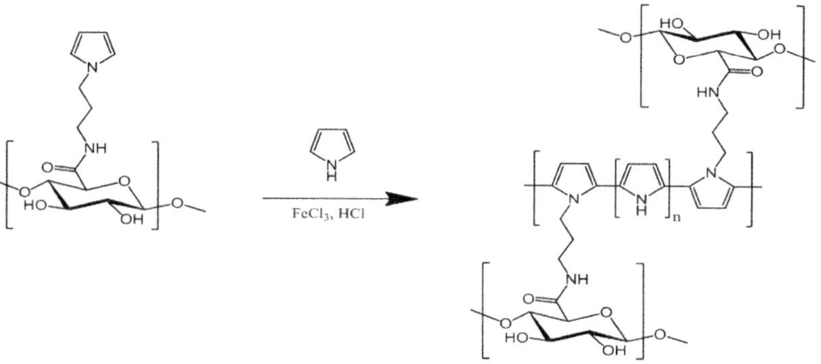

Fig. 7. Chemical structure and proposed pathway of the covalent bond formation between the PyTOCN fibres via the polymerization of polypyrrole.

Therefore, we can conclude that the method used directly influence the composite conductivity and PPy particles distribution is detrimental for the films' mechanical properties. Thus, for electrical applications such as in batteries, capacitors, sensors or solar cells for example, the use TOCN/PPy-1 and TOCN/PPy-2 seem to offer the best match but for flexible electrodes, TOCN/PPy-3 appears to be the best.

TEMPO-Oxidized Cellulose Nanofibres Composites with Polypyrrole for Food Preservation

When food cannot be consumed immediately after production, food packaging must create a protective atmosphere around the food product. The packaging must improve the shelf life of perishable foods and protect the food from dirt, dust, oxygen, light, pathogenic microorganisms, and moisture by acting as an effective barrier to moisture, gases (CO_2 and O_2) or even by possessing antimicrobial properties. Microbial and enzymatic activities are not the only responsible party for most of the degradation reactions of foodstuffs. Oxygen, either directly or indirectly, plays an important role in the losses of organoleptic and nutritional properties as well as food spoilage by aerobic microorganisms. It is therefore necessary to use materials with good oxygen barrier properties (Duncan 2011).

Cellulose nanofibres have even demonstrated barrier properties to oxygen in addition to its excellent mechanical proprieties due to the fibres that are mostly entangled and are forming a tight network (Lavoine et al. 2012, Nair et al. 2014). Nevertheless, cellulose is a major inconvenience for packaging due to its affinity for water. Indeed, cellulose films have very poor barrier properties to water vapour because of its hydrophilicity. To resolve this problem, we have coupled TOCN (nanofibres obtained by TEMPO oxidation) with a semi-conductive polymer, polypyrrole and polyvinyl alcohol (PVA). The polypyrrole (PPy) coating was an interesting alternative due to its hydrophobic character. Moreover, PPy confers attractive properties for its active packaging as antibacterial and antioxidant properties (Bideau et al. 2016c, Zare et al. 2014). In addition, the PVA chains allow better tensile strength. The goal was to prepare a packaging that simultaneously combines the biodegradability and the barrier properties to gas of cellulose with the physico-chemical proprieties of PPy.

Multilayer composite films (TOCN/PVA-PPy) synthesis was made in two steps. For the first time, TOCN/PVA film was prepared with a small quantity of PVA (6.7 wt %). The film was then immersed in a solution of iron chloride, an inexpensive oxidant, to initiate the *in situ* polymerization of pyrrole on the surface of the film, a second time (Fig. 8) (Bideau et al. 2016c).

The results of the tests of food simulation by contact for our multilayered composite are shown in Fig. 9 (Bideau et al. 2017). The TOCN/PVA-PPy film attests to an improved conservation of bananas after five days compared to the control or the TOCN film. With the control (five days in ambient air), banana browned and was dehydrated before showing rot, as evidenced in Fig. 9. In the presence of the cellulose film, bananas remained hydrated, but began to deteriorate (Fig. 9b),

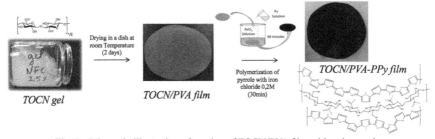

Fig. 8. Schematic illustration of coating of TOCN/PVA film with polypyrrole.

Fig. 9. Schema of food packaging test by contact (a); food simulation after 5 days, with the TOCN film (b), TOCN/PVA-PPy (c); schema of food packaging test by noncontact (d); food simulation flasks (left: TOCN/PVA-PPy and right: TOCN/PVA films) (e); piece of banana after 5 days (f).

as we can see with the brown colour of the banana from the oxygen oxidation. This oxidation is also responsible for the degradation of the banana. As seen in Fig. 9c, the banana pieces are retained even after five days. This result shows the notable barrier properties of the composite. To limit carbon and iron transfer from the composite to the food, we have considered a noncontact packaging method. The flasks were sealed so that the only possible exchanges with the outside are made through the film (Fig. 9e). In Fig. 9f, we can see that banana is better preserved with the TOCN/PVA-PPy than the TOCN or the TOCN/PVA films. No brown colour is visible, so there is no trace of oxidation, unlike the other two trials. This test demonstrates again the good barrier property against oxygen, due to the presence of PPy particles.

Nanocellulose has opened vast possibilities of utilizing cellulose-based materials to be substitutes for plastics. The use of TOCNs in composite films with a coating was found to substantially reduce the oxygen permeability of the material (Bideau et al. 2017). The improvement in the barrier properties can be attributed to the polypyrrole particles, which coat the dense network formed by nanofibrils. In addition, we have demonstrated two different applications of food packaging for this composite. The combination of barriers and antioxidant properties allows the TOCN/PVA-PPy films to improve the shelf life of perishable foods and protect the food from dirt, dust or oxygen (Bideau et al. 2017). The dark colour of the films may also protect food from light degradation. The biodegradable character could also reduce packaging waste generated by conventional plastics.

Conclusion

For many years, the production of nanocellulose was a centre of interest to many research groups. However, now the research moves towards the finding of new applications. In this chapter, applications of TEMPO-oxidized nanocellulose (TOCN) from wood has been presented. Most of the applications are possible from their unique surface carboxylates groups and their amorphous region. Combined with a large specific surface, we have shown that this kind of nanocellulose-containing carboxyl and hydroxyl groups can serve as templates for grafting molecules of interest, opening up new horizons for many applications (biocomposites, packaging, electronics, biomedical, etc.). At the present time, industrial application of nanocelluloses is very limited even if this material has many advantages.

In our research, we have done the oxidation of cellulose by 4-acetamido-TEMPO under ultrasound. The known ultrasound benefits, such as swelling of vegetal cells, fragmentation due to the cavitational effect and radical species associated with the ultrasonic treatment will help catalyze the TEMPO oxidation process. Although the technology has been shown to be feasible and advantageous on a small scale, the commercialization of sonolysis is still a challenge. In our group, we have succeeded in a first step toward industrialization by using high-intensity ultrasound (170 kHz) generated in a semi-industrial sonoreactor tube.

Successful applications of TOCN were made in amino acids grafting, organic conductive films and food packaging. Coupling TOCN with fluorescent amino acids using a two-step grafting procedure was done with well individualized (no aggregation) even after grafting of TOCN. Organic conductive films with polypyrrole have shown one of the best conductivity of such film while being well under inorganic metals. Finally, polypyrrole and TOCN combined have shown interesting antioxidants and gaz barrier properties that could be used in the food packaging industry. These small steps are towards the finding of industrial applications for nanocelluloses, which could finally propel this material to usual uses in our everyday life.

References

Abu-Rabeah, K., B. Polyak, R.E. Ionescu, S. Cosnier and R.S. Marks. 2005. Synthesis and characterization of a pyrrole-alginate conjugate and its application in a biosensor construction. Biomacromolecules 6: 3313–3318.

Ambade, R.B., S.B. Ambade, N.K. Shrestha, Y.C. Nah, S.H. Sung-Hwan Han, W. Lee and S.H. Lee. 2013. Polythiophene infiltrated TiO_2 nanotubes as high-performance supercapacitor electrodes. Chem. Commun. 49: 2308–2310.

Araki, J., M. Wada, S. Kuga and T. Okano. 1998. Flow properties of microcrystalline cellulose suspension prepared by acid treatment of native cellulose. Colloids Surf. A 142: 75–82.

Arrua, D., M.C. Strumia and M.A. Nazareno. 2010. Immobilization of caffeic acid on a polypropylene film: synthesis and antioxidant properties. J. Agric. Food Chem. 58: 9228–9234.

Azizi Samir, M.A.S., F. Alloin and A. Dufresne. 2005. Review of recent research into cellulosic whiskers, their properties and their application in nanocomposite field. Biomacromolecules 6: 612–626.

Barazouk, S. and C. Daneault. 2011. Spectroscopic characterization of oxidized nanocellulose grafted with fluorescent amino acids. Cellulose 18: 643–653.

Barazouk, S. and C. Daneault. 2012a. Tryptophan-based peptides grafted onto oxidized nanocellulose. Cellulose 19(2): 481–493.

Barazouk, S. and C. Daneault. 2012b. Amino acid and peptide immobilization on oxidized nanocellulose: Spectroscopic charaterization. Nanomaterials 2: 187–205.

Bideau, B., E. Loranger and C. Daneault. 2016a. Comparison of three polypyrrole-cellulose nanocomposites synthesis. J. Adv. Nanomaterials 1(2): 105–114.

Bideau, B., L. Cherpozat, E. Loranger and C. Daneault. 2016b. Conductive nanocomposites based on TEMPO-oxidized cellulose and poly(N-3-aminopropylpyrrole-co-pyrrole). Industrial Crops and Products 93: 136–141.

Bideau, B., J. Bras, S. Saini, C. Daneault and E. Loranger. 2016c. Mechanical and antibacterial properties of a nanocellulose-polypyrrole multilayer composite. Mat. Sci. and Engineering C. 69: 977–984.

Bideau, B., J. Bras, N. Adoui, E. Loranger and C. Daneault. 2017. Polypyrrole/nanocellulose composite for food preservation: barrier and antioxidant characterization. Food Packaging and Shelf Life 12: 1–8.

Bisht, R.V., W. Takashima and K. Kaneto. 2005. An amperometric urea biosensor based on covalent immobilization of urease onto an electrochemically prepared copolymer poly(N-3-aminopropyl pyrrole-co-pyrrole) film. Biomaterials 26: 3683–3690.

Bradley, E.L., L. Castle and Q. Chaudhry. 2011. Application of nanomaterials in food packaging with consideration of opportunites for developing countries. Trends Food Sci. Technol. 22: 604–610.

Bulpitt, P. and D. Aeschlimann. 1999. New strategy for chemical modification of hyaluronic acid: preparation of functionalized derivatives and their use in the forrmation of novel biocompatible hydrogels. J. Biomed. Mater. Res. 47: 152–169.

Busolo, M.A. and J. Lagaron. 2015. Antioxidant polyethylene films based on a resveratrol containing clay of interest in food packaging applications. Food Packaging and Shelf Life 6: 30–41.

Cabuk, M., Y. Alan, M. Yavuz and H.I. Unal. 2014. Synthesis characterization and antimicrobial activity of biodegradable conducting polypyrrole-graft-chitosan copolymer. Appl. Surf. Sci. 318: 168–175.

Cushen, M., J. Kerry, M. Morris, M. Cruz-Romero and E. Cummins. 2012. Nanotechnologies in the food industry recent developments, risks and regulation. Trends Food Sci. Technol. 24(1): 30–46.

De Nooy, A.E., A.C. Besemer and H. van Bekkum. 1995. Highly selective ntrosyl radical-mediated oxidation of primary alcohol groups in water-soluble glucans. Carbohydr. Res. 69: 89–98.

Duncan, T.V. 2011. Applications of nanotechnology in food packaging and food safety: Barrier materials, antimicrobials and sensors. J. Colloid Interf. Sci. 363: 1–24.

Ebringerova, A. and Z. Hromadkova. 2010. An overview on the application of ultrasound in extraction, separation and purification of plant polysaccharides. Cent. Eur. J. Chem. 8(2): 243–257.

Elazzouzi-Hafraoui, S., Y. Nishiyama, J.L. Putaux, L. Heux, F. Dubreuil and C. Rochas. 2008. The shape and size distribution of orystalline nanoparticles prepared by acid hydrolysis of native cellulose. Biomacromolecules 9: 57–65.

Hoare, D.G. and D.E. Koshland Jr. 1967. A method for the quantitative modification and estimation of carboxylic acid groups in proteins. J. Biol. Chem. 242: 2447–2453.

Hu, W., S. Chen, Z. Yang, L. Liu and H. Wang. 2011. Flexible electrically conductive nanocomposite membrane based on bacterial cellulose and polyaniline. J. Phys. Chem. B 115: 8453–8457.

Huang, J.Y., X. Li and W. Zhou. 2015. Safety assessment of nanocomposite for food packaging application. Trends Food Sci. Technol. 45: 187–199.

Isogai, T., T. Saito and H. Fukuzumi. 2011. TEMPO-oxidized cellulose nanofibres. Nanoscale 3: 71–85.

Jiang, K., L.S. Schadler, R.W. Siegel, X. Zhang, H. Zhang and M. Terrones. 2004. Protein immobilization on carbon nanotubes via a two-step process of diimide-activated amidation. J. Mater. Chem. 14: 37–39.

Jradi, K., B. Bideau, B. Chabot and C. Daneault. 2012. Characterization of conductive composite films based on TEMPO-oxidized cellulose nanofibres and polypyrrole. J. Mater. Sci. 47: 3752–3762.

Lavoine, N., I. Desloges, A. Dufresne and J. Bras. 2012. Microfibrillated cellulose-Its barrier properties and applications in cellulosic materials: a review. Carbohydr. Polym. 90: 735–764.

Lavoine, N., I. Desloges and J. Bras. 2014. Microfibrillated cellulose coatings as new release systems for active packaging. Carbohydr. Polym. 103: 528–537.

Lavoine, N., I. Desloges, B. Manship and J. Bras. 2015. Antibacterial paperboard packaging using microfibrillated cellulose. J. Food Sci. Technol. 52: 5590–600.

Lee, J.Y. and C.E. Schmidt. 2010. Pyrrole-hyaluronic acid conjugates for decreasing cell binding to metals and conducting polymers. Acta Biomater. 6: 4396–4404.

Loranger, E., M. Paquin, C. Daneault and B. Chabot. 2011. Comparative study of sonochemicals: effects in an ultrasonic bath and in a large-scale flow-through sonoreactor. Chem. Eng. J. 178: 359–365.

Loranger, E., A.O. Piche and C. Daneault. 2012. Influence of high shear dispersion on the production of cellulose nanofibres by ultrasound-assisted TEMPO-oxidation of kraft pulp. Nanomaterials 2(3): 286–297.

Mahvi, A.H. 2009. Application of ultrasonic technology for water and wastewater treatment. Iranian J. Publ. Health. 38(2): 1–17.

Minjares-Fuentes, R., A. Femenia, M.C. Garau, M.G. Candelas-Cadillo, S. Simal and C. Rossello. 2016. Ultrasound-assisted extraction of hemicelluloses from grape pomace using response surface methodology. Carbohydrate Polymer 138: 180–191.

Mishra, S., J. Thirree, A.S. Manent, B. Chabot and C. Daneault. 2011. Ultrasound-catalyzed TEMPO-mediated oxidation of native cellulose for production of nanocellulose: Effect of process variables. BioResources 6(1): 143–212.

Naddeo, V., A. Cesaro, D. Mantzvinos, D. Fatta-Kassinos and V. Belgiorno. 2014. Water and wastewater disinfection by ultrasound irradiation a critical review. Global Nest J. 16(3): 561–577.

Nair, S.S, J.Y. Zhu, Y. Deng and A.J. Ragauskas. 2014. High performance green barriers based on nanocellulose. Sustainable Chemical Processes 2: 23–30.

Naji, A., M. Cretin, M. Persin and J. Sarrazin. 2003. Electrical characterization of the ionic interactions in N-[3-(dimethylpyridyl-2-yl) aminopropyl) polypyrrole and N-(3-aminopropyl) polypyrrole membranes. J. Membr. Sci. 212: 1–11.

Pappas, C.P.A., I. Tarantilis, T. Daliani, M. Mavromoustakos and M. Polissiou. 2002. Comparison of classical and ultrasound-assisted isolation procedures of cellulose from kenaf and eucalyptus. Ultrason. Sonochem. 9(1): 19–23.

Paquin, M., E. Loranger, V. Hannaux, B. Chabot and C. Daneault. 2013. The use of Weissler method for scale-up a kraft pulp oxidation by TEMPO-mediated system from a batch mode to a continous flow-through sonoreactor. Ultrasonics Sonochemistry 20: 103–108.

Qin, Z.Y., G.L. Tong, Y.C.F. Chin and J.C. Zhou. 2011. Preparation of ultrasonic-assisted high carboxylate content cellulose nanocrystals by TEMPO oxidation. BioResources 6(2): 1136–1146.

Rataz, A., S.P. Mishra, B. Chabot and C. Daneault. 2011. Cellulose nanofibres by sonocatalysed-TEMPO-oxidation. Cellulose 18: 585–593.

Saito, T., S. Kimura, Y. Nishiyama and A. Isogai. 2007. Cellulose nanofibres prepared by TEMPO-mediated system. Biomacromolecules 8(8): 2485–2491.

Saito, T., Y. Nishiyama, J.L. Putaux, M. Vignon and M. Isogai. 2006. Homogenous suspensions of individualized microfibrils from TEMPO-catalyzed oxidation of native cellulose. Biomacromolecule 7(6): 1687–1691.

Saito, T. and A. Isogai. 2005. Ion-exchange behavior of carboxylate groups in fibrous cellulose oxidized by the TEMPO mediated system. Carbohydr. Polym. 61: 183–190.

Sangawar, V.S. and N.A. Moharil. 2012. Study of electrical, thermal and optical behavior of polypyrrole filled PVC: PMMA thin film thermoelectrics. Chem. Sci. Trans. 1: 447–455.

Scott, W.E. and P. Gerber. 1995. Using Ultrasound to Deink Xerographic Waste. TAPPI 78: 125–130.

Sun, R.C. and J. Tomkinson. 2002. Comparative study of lignins isolated by alkali and ultrasound-assisted alkali extraction from wheat straw. Ultrasonics Sonochem. 9(2): 85–93.

Syverud, K. and P. Stenius. 2009. Strength and barrier properties of MFC flms. Cellulose 16: 75–85.

Vaitkuviene, A. and V. Kaseta. 2013. Evaluation of cytotoxicity of polypyrrole nanoparticles synthesized by oxidative polymerization. J. Hazard. Mater. 250-251: 167–174.

Wang, Z., P. Tammela, P. Zhang, M. Strømme and I. Nyholm. 2014. Efficient high active mass paper-based energy storage devices containing free-standing additive less polypyrrole-nanocellulose electrodes. J. Mater. Chem. A2: 7711–7716.

Zare, E.N., M.M. Lakouraj and M. Mohseni. 2014. Biodegradable polypyrrole/dextrin conductive nanocomposite: synthesis, characterization, antioxidant and antibacterial activity. Synth. Met. 187: 9–16.

Processing and Characterization of Modified Nanocellulose/ Polyester Composites

Jelena Rusmirović,[1,*] *Milica Rančić*[2] and *Aleksandar Marinković*[3]

Introduction

Materials, along with energy and information, represent one of the main pillar of industries in the world economy of the 21st century. Nanotechnology is recognized as one of the most promising area of technological development. Due to nanotechnology development and recent concern about environmental issues, there is an increasing demand for products made from renewable and sustainable non-petroleum-based resources. In this regard, natural fibers have been gaining much more interest because of their promising characteristics such as biodegradability, renewability and lower price. Among these natural fibers, cellulose is the most plentiful biopolymer which exists in a wide variety of living species including plants, bacteria and some animal species like tunicates, has been the subject of extensive research in nanotechnology. Functional nanomaterials are especially

[1] Innovation Center, Faculty of Technology and Metallurgy, University of Belgrade, Karnegijeva 4, 11120 Belgrade, Serbia.
[2] Faculty of Forestry, University of Belgrade, Kneza Višeslava 1, 11030 Belgrade, Serbia.
  Email: milica.rancic@sfb.bg.ac.rs
[3] Faculty of Technology and Metallurgy, University of Belgrade, Karnegijeva 4, 11120 Belgrade, Serbia.
  Email: marinko@tmf.bg.ac.rs
* Corresponding author: jrusmirovic@tmf.bg.ac.rs

attractive because they enable the creation of materials with new or improved properties possessing potential applications in specific fields (Dufresne 2012a).

Cellulose is a naturally occurring polymer found abundantly in nature and used as structural building blocks conferring its mechanical properties to higher plant cells (Dufresne 2012a). The worldwide production of this biopolymer is estimated to be between 10^{10} and 10^{11} t each year, and most of that, about 6×10^9 t, are processed by industries such as paper, textile, material and chemical industries (Lavoine et al. 2012). Beside its impressive mechanical properties, cellulose represents renewable, biodegradable, low-weight and non-toxic biopolymer. Although cellulose represents the main building block of wood, there are other major sources such as plants fibers (cotton, hemp, flax, etc.), marine animals (tunicate), algae, fungi, invertebrates, and some bacteria. Regardless of its source, cellulose can be present in the leaf (sisal), in the fruit (cotton) or in the stalk or the rigid structure of plant as wood or flax (Dufresne 2013).

Recently, nanotechnology has gained a great attention in many industries, and opened up many new possibilities in the forest industry and cellulose-based products. Nanotechnology is defined as the understanding and control of matter with at least one dimension measuring from 1 to 100 nm. By combining mechanical, chemical or enzymatic treatment, cellulose fibers can be converted into cellulose nanofibers (CNFs) or cellulose nanocrystals (CNCs) that possess outstanding properties compared to the original cellulosic fiber, but also when compared with other materials normally used as reinforcements in composite materials such as Kevlar or steel wires (Dufresne 2008, 2013, Eichhorn et al. 2010, Habibi et al. 2010, Abdul Khalil et al. 2012, Lin et al. 2012, Habibi 2014, Kim et al. 2015). Depending on the treatment fibrils with a diameter of 5 to 30 nm and length in micrometers, or rod-like crystals with a diameter of 5 to 20 nm and length of tens to hundreds of nanometers to micrometers can be obtained. The development of nanocellulose has attracted significant interest in the last few decades due to their unique characteristics, such as high crystallinity, high purity, high surface area, unique optical properties, very high tensile strength, high Young's modulus and very good reinforcing capabilities as a filler for various composite materials (Habibi et al. 2010, Eichhorn et al. 2010). Moreover, environmental awareness and the international demand for green technology, bionanocomposites have the great potential to replace present petrochemical-based materials.

In recent research, nanocellulose has been evaluated as a reinforcing phase in nanocomposites using different polymer matrices such as polyvinyl alcohol (Roohani et al. 2008), acrylic latex (Pu et al. 2007), polyethylene (Junior de Menezes et al. 2009), polyurethane (Cao et al. 2007), thermoplastic starch (Zainuddin et al. 2013) and rubber (Pasquini et al. 2010). Among them, a great number of polyesters including poly(lactic acid) (PLA) (Oksman et al. 2006), polyhydroxylalkanoate, aromatic polybutylene adipate-co-terephthalate (PBAT) (Avérous 2007), aliphatic polybutylene succinate adipate (Wollerdorfer and Herbert 1998) and unsaturated polyester (Rusmirović et al. 2017, Kargarzadeh et al. 2015c) have been investigated in association with diverse nanocellulose fillers obtained from various origins.

Most of the published studies of nanocellulose/polyester based composites are based on biodegradable polyesters (biopolyesters) (Avérous and Le Digabel 2006) matrices and non-biodegradable matrices obtained from waste polymers such as poly(ethylene terephthalate) (PET) (Rusmirović et al. 2017). From the non-biodegradable polyester matrices, unsaturated polyester resins (UPe) represent the most frequently used industrial polyester for nanocellulose-basedpolymer composites due to their high strength and modulus, high resistance to water, room temperature cure capability and transparency (Kargarzadeh et al. 2015a). Moreover, UPes can be obtained from waste PET and after mixing with the vinyl monomer in the molten state, further used for nanocellulose-based composites production. In that manner, the applying cycle of reusable and renewable materials for high-performance composites will be rounded up (Rusmirović et al. 2017). Existing literature on CNF/UPe composites primarily focuses on low content of CNFs (up to 20% CNF). Chirayil et al. (Chirayil et al. 2014, Chirayil et al. 2014) studied UPe composites with 0.5 to 5 weight percentage of CNFs and observed that the mechanical properties were favorable at 0.5 weight percentage of CNFs. The falling in mechanical properties at higher CNFs content is usually due to agglomeration of the fibrils, and the agglomerates initiate premature failure. A reduction in gel time of UPe in the presence of CNF was reported and it was suggested that CNF accelerates the curing reaction (Ansari et al. 2015).

The main problem with using nanocellulose as reinforcement in polymer composites is its polar and hydrophilic properties which prevents homogenous dispersion in hydrophobic polymer matrices, as well as moisture absorption. To overcome this problem, cellulose fibers can be modified in different ways, either by coating the cellulose nanoparticles with surfactants or by chemically modifying the cellulose surface with hydrophobic groups. The use of surfactants is the easiest method, but a very high amount of surfactants is needed to coat the surface of the fillers, which causes problems in composite applications. Chemical modifications of nanocellulose are used to improve the incorporation between nanocellulose and hydrophobic polymer matrix. Beside reducing the hydrophilic nature of nanocellulose, improving the interaction between nanocellulose filler and polymer matrix is one of the most important factors for obtaining good reinforcement in the composite (Kushwaha and Kumar 2010). Introduction of reactive cross-linkable functional group on cellulose nanoparticles surface is important to achieve direct binding with monomer chains. In recent years, many techniques for chemical surface modification of nanocellulose have been considered in order to improve dispersibility, compatibility and bonding with nonpolar polymer matrices (Cheng 2012, Taipina et al. 2013, Yu et al. 2015). One of them is the introducing of hydrophobic groups by esterification of surface hydroxyl groups of nanocellulose. Fatty acids (FA) and their derivates isolated from natural vegetable oils (VO) stand out for the "Green One-Pot" method for surface nanocellulose modification (Yoo and Youngblood 2016, Uschanov et al. 2011, Bryuzgin et al. 2014, Freire et al. 2006).

This chapter focusses on reviewing nanocellulose processing and functionalization methods and dynamic-mechanical properties of composites based on bare/functionalized cellulose nanoparticles and polyester matrices. The main objective is to evaluate/summarize the unique mechanical and dynamic-mechanical properties of functionalized cellulose-reinforced polyester-based composites, particularly the various reinforcing mechanisms of functionalized cellulose nanoparticles. Detailed overview of designing the surface and morphological properties of nanocellulose (aggregate size, crystallinity and surface energy) using appropriate drying method, and influence of applied drying method on performances of the cellulose-reinforced polymer nanocomposites is presented. Some relevant suggestions/solutions for the incorporation of reactive groups through modification procedures and making nanocellulose more suitable for free radical polymerization with polyester chain were also investigated. Additionally, the future perspectives in designing of nanocellulose/polyester composites are provided.

Cellulose

The chemical structure of cellulose was first described by the French chemist Payen in 1838 (Dufresne 2012a). Cellulose is characterized as a high molecular weight linear homopolymer formed by repeating β-D-anhydroglucopyranose units (AGU) linked together by β-1,4-glycosidic bonds in both crystalline and amorphous regions. Intra-chain hydrogen bonds between the hydroxyl groups and oxygen atoms of neighboring ring molecules stabilize the cellulose chain and they are responsible for the linear structure of the polymer chain. Every other AGU is turned 180° with respect to its neighbor and two AGU next to each other form a cellobiose unit, the smallest repeating unit in the polymer. The degree of polymerization (DP) is a measure of how many AGUs there is in the polymer and since no polymer is homogenous in length, the molecular weight distribution will have an important influence on the fibers' properties. The DP of a cellulose polymer can be as high as 10,000 in wood cellulose and even higher in, for example, native cotton plant fibers. The basic chemical structure of cellulose (Fig. 1) represents a dimer cellobiose that appears as a repeated segment. After degradation reactions and purification processes, the DP is reduced to about 300 to 1700 in wood cellulose (Börjesson and Westman 2015). The cellulose molecule contains three different kinds of AGU: a reducing end group that contains a free hemiacetal or aldehyde at the C1 position, a non-reducing end group with a free hydroxyl group at the C4 position, and internal glucose rings joined at the C1 and the C4 positions. The internal glucose units are predominant due to the long chain lengths. Each internal AGU bears three hydroxyl groups. These groups and their ability to form strong hydrogen bonds confer upon cellulose its most important properties, in particular its multi-scale microfibrillated structure, hierarchical organization (crystalline vs. amorphous regions), and highly cohesive nature (with a glass transition temperature higher than its degradation temperature). At the C6 position, there is a primary hydroxyl group, while the hydroxyl groups at the C2 and C3 positions are secondary ones. These hydroxyl groups present all possible sites for chemical modification of cellulose, where the hydroxyl group at the C6 position is the most reactive (Dufresne 2012a).

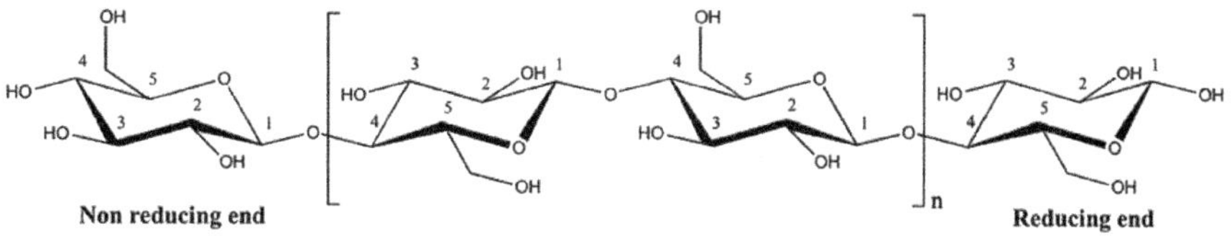

Fig. 1. Chemical structure of cellulose.

Due to the linear and regular structure of cellulose and the many hydroxyl groups in the molecule, cellulose polymers can form ordered crystalline structures held together by hydrogen bonds. These crystalline regions give important mechanical properties to the cellulose fibers. The hydroxyl groups in the cellulose polymer can form hydrogen bonds between different cellulose polymers (intermolecular hydrogen bonds) or within the polymer itself (intramolecular hydrogen bonds). The intramolecular bonds give stiffness to the polymer chain, while the intermolecular bonds allow the linear polymers to form sheet structures. The high crystallinity and the many hydrogen bonds in the cellulose fibers make cellulose insoluble in water and in most conventional organic solvents (Viet et al. 2006). The structure of a cellulose fiber can be divided into three different levels: the molecular level, the supramolecular level, and the morphological level. The molecular level was described in the beginning of this section (Fig. 1). The supramolecular level is the polymer chains ordered in crystalline and non-crystalline regions due to hydrogen bonds and the morphological level consists of the cellulose fiber and its cell walls. Cellulose can exist in six different polymorphs; I, II, III_I, III_{II}, IV_I and IV_{II}. In nature cellulose is found in the cellulose I polymorph structure, and it occurs in two allomorphs, I_α and I_β, which have different hydrogen bonding networks (Lavoine et al. 2012). The two forms coexist in cellulosic materials but the ratio varies depending on the source. Cellulose I_β is predominant in higher plants, while cellulose I_α in bacteria and tunicate. There are regions in the cellulose fibrils where the cellulose chains are arranged in a highly ordered structure (crystalline regions) and regions where the chains are disordered (amorphous regions). The ratio of crystalline-to-amorphous cellulose also varies depending on the source of cellulose.

Native cellulose can be converted irreversibly into the thermodynamically more favorable cellulose II polymorph by swelling native cellulose I (metastable form) in concentrated sodium hydroxide aqueous solutions (17 to 20% wt/vol) followed by removal of the swelling agent (named mercerization after its invention by Mercer in 1844). Mercerization is used to activate the polymer prior to the production of technical cellulose ether. The mercerization of cellulose leads only to its swelling, but not to dissolution. The insertion of chemical species induces the structural change and the transition from a structure with parallel cellulosic chains to a configuration with anti-parallel chains. Cellulose II can also be prepared by regeneration, i.e., the solubilization of cellulose I in a solvent followed by precipitation by dilution into an aqueous medium. Cellulose II is formed naturally by a mutant strain of *Gluconacetobacter xylinum* and occurs in the alga *Halicystis*. On the other hand, treatment with liquid ammonia or with certain organic amines, such as ethylene diamine (EDA), followed by washing with alcohol allows the preparation of cellulose III either from cellulose I (which leads to the cellulose allomorph III_I) or from cellulose II (which leads to the allomorph III_{II}). Cellulose III treated at high temperature in glycerol is transformed into cellulose IV, which also exists as two types, cellulose IV_I and cellulose IV_{II} obtained from cellulose III_I and III_{II}, respectively. It is generally accepted that cellulose IV_I is a disordered form of cellulose I. This could explain the reported occurrence of this form in native state in some plants (primary walls of cotton and some fungi).

Nanocellulose as novel biorenewable reinforcement in polymer composites

Nanocellulose, variously termed nanocrystals, whiskers, rods, nanofibrils, or nanofibers, refers to cellulose fibers or crystals that have at least one dimension within the nanometer size range. In 2011, TAPPI released a roadmap for the development of international standards of nanocellulose (Dufresne 2012a, Börjesson and Westman 2015), where the abbreviations for different nanocelluloses were established as: cellulose nanocrystals (CNCs), cellulose nanofibrils (CNFs), and cellulose microfibrils (CMFs). For cellulose nanocrystals, the noncrystalline regions are hydrolyzed and the remaining crystals are in nanometer size range in all dimensions. For cellulose nanofibrils some of the hydrogen bonds between fibrils will break and form fibers with micrometer size in length (noncrystalline regions still present) and nanometer size in width. Bacterial nanocellulose (BNCs), on the other hand, is synthesized by special bacteria and grown as microfibrils in a culture medium. BNC microfibrils can also be hydrolyzed into bacteria nanocrystals by an acid hydrolysis similar to CNC.

Cellulose nanocrystals (CNCs). They are also known as whiskers and consist of rod-like cellulose crystals with width of 5 to 70 nm. Nanocrystals can be several hundreds of nanometers in length and they are formed during acid hydrolysis of cellulose fibers, often followed by ultrasonic treatment, where a selective degradation of the more accessible, disordered parts takes place. Since the noncrystalline regions act as structural defects in the microfibril, it is responsible for the transverse cleavage of the microfibrils into short monocrystals under acid hydrolysis (Håkansson and Ahlgren 2005). In the early stage of the hydrolysis, the acid diffuses into the non-crystalline parts of the cellulose fiber and hydrolyzes the glycosidic bonds. Afterwards, more easily accessible glycosidic bonds in the polymer are hydrolyzed and finally hydrolysis occurs at the reducing end group and at the surface of the nanocrystals. The harder it is for the acid to access the glycosidic bonds, the slower is the reaction (Dong et al. 1998, Revol et al. 1992). The hydrolysis of the reducing end groups and the surface of the nanocrystals will make the nanocrystals charged depending on what acid is used. By using a 64 weight percentage of sulphuric acid solution, 0.5 to 2 percentage of sulfate groups will be attached to the surface of the nanocrystal (Dagang Liu et al. 2014). Due to the charged sulfate groups, CNC will form stable colloidal dispersion when diluted in water to specific concentrations. Cellulose nanocrystals have been hydrolyzed from many different cellulose sources such as hardwood pulp (Revol et al. 1992, Håkansson and Ahlgren 2005), softwood pulp (Popescu et al. 2011, Hamad and Hu 2010, Håkansson and Ahlgren 2005), microcrystalline cellulose (MCC) (Bondeson et al. 2006, Alves et al. 2014), sisal (Joseph 1999, Builes et al. 2013, Garcia de Rodriguez et al. 2006), cotton (Morais et al. 2013, Rusmirović et al. 2017), rice straw (Lu and Hsieh 2012), bacterial cellulose (Hu et al. 2009, Gatenholm and Klemm 2010), banana fibers (Abraham et al. 2011), and tunicin (Yuan et al. 2006, Berlioz et al. 2009). The different types of cellulose sources give some different structures of the nanocrystals and the aspect ratio will differ for the different sources.

Nanofibrillated cellulose (CNFs). It can be prepared from wood-based pulp material by mechanical or simultaneous enzymatic and mechanical treatment (Dufresne 2012a). CNFs material has extraordinary properties due to their nanoscale dimensions. The long and entangled fibrils have high aspect ratio and large surface area exhibiting vast amounts of reactive hydroxyl groups on their surface. They form an extensively percolating network with low weight percentages. In water CNF forms a gel, which strength is dependent on the fibril concentration. Upon drying, CNF forms hierarchical nanoporous aerogels or strong films depending on the drying process. The mechanical properties of CNFs paper films have been reported to be better than ordinary paper and the reinforcing ability in composite materials has also been shown to be superior compared to pulp fibers. This is caused by the larger surface area of fibrils exhibiting extensive hydrogen bonding compared to macroscopic fibers. Moreover these films have good gas barrier properties which can be of benefit in many coating and barrier technologies. The aerogels have been shown to have very different properties compared to traditional aerogels due to the long and entangled fibrils forming a hierarchical nanoporous structure. They open up new application areas for the use of aerogels in general. Due to their biocompatibility, CNF is also a promising material for biotechnological and pharmaceutical applications.

Bacterial nanocellulose (BNCs). It is also called bacterial cellulose, microbial cellulose, or biocellulose and formed by aerobic bacteria, such as acetic acid bacteria of the genus *Gluconacetobacter*. Opposite to CNFs and CNCs, BNC is formed as a polymer and nanomaterial by biotechnological assembly processes from low-molecular-weight carbon sources, such as D-glucose. The resulting BNC hydrogel is composed of a nanofiber network where fiber diameter is 20–100 nm enclosing up to 99 percentage of water.

Nanocellulose, including cellulose nanocrystals, cellulose nanofibril and bacterial cellulose as well, becomes an important building element in creation of novel polymer composites (Gatenholm and Klemm 2010, Habibi et al. 2010, Siró and Plackett 2010, Eichhorn 2011). Reinforcing effect of nanocellulose depends on their mechanical properties. Additionally, several factors may influence the nanocellulose mechanical properties such as: crystal structure (Iα, Iβ, II), percentage of crystallinity, anisotropy, defects, and the properties measurement methods and techniques (Moon et al. 2011). The mechanical properties of different types of cellulose particles are summarized in Table 1.

High mechanical performance composite materials can be obtained by dispersing CNC in polymer matrices even at a small mass percentage (Favier et al. 1995). Excellent mechanical properties of CNC/polymer composites result from high mechanical strength and rigidity of the CNC crystal structure that represents reinforcement in high plant skeletons (Dufresne 2013). Moreover, the optimal length/width ratio of CNC crystals and their reactivity express the positive impact on CNC mechanical properties (Dufresne 2013). Under certain conditions, the mechanically strong CNC network can be formed inside the polymer matrix, and thus improve the mechanical characteristics of polymer. Comparison of CNC mechanical properties with those of other commercial reinforcing agents, indicates that CNC has a great reinforcing potential (Durán et al. 2011).

CNCs have numerous advantages compared to other nanostructures, such as ease of formation, low cost of raw materials, diverse characteristics depending on the source of the natural substrate, and more beneficial mechanical properties (Li et al. 2016, Liu et al. 2017, Bruna et al. 2017). Cellulose nanocrystals possess higher elastic modulus in axial direction and similar mechanical properties

Table 1. Properties of cellulose-based materials (Moon et al. 2011).

Material	E_A (GPa)	E_T (GPa)	σ_f (GPa)	ε_f (%)	Technique
WF*	14–27	-	0.3–1.4	4–23	Tensile
PF*	5–45	-	0.3–0.8	1.3–8	Tensile, Raman
MCC*	25 ± 4	-	-	-	Raman
MFC & FC	N/A	-	-	-	
CNC	57, 105	-	-	-	Raman
plant wood	-	18–50	-	-	AFM indentation
t-CNC	143	-	-	-	Raman
Acid[a]	151 ± 29	-	-	-	AFM-3pt bend
TEMPO[a]	145 ± 31	-	-	-	AFM-3pt bend
BNC	78 ± 17	-	-	-	AFM indentation
	114	-	-	-	AFM-3pt bend Raman
Cellulose Iβ					
Experimental	120–138		-	-	XRD
Modeling	220 ± 50	15 ± 1	-	-	IXS
	110–173	-	-	-	
	137–168	10–50	7.5–7.7	-	
Cellulose Iα Modeling	128–155	5–8	-	-	
Cellulose II					
Experimental	9–90	-	0.2–1.0	-	Raman
Modeling	98–109	17–31	4.9–5.4	-	

E_A, E_T = elastic modulus in axial direction and transverse direction. σ_f = tensile strength, ε_f = strain to failure. [a]Treated t-CNCs.
*WF – wood fiber, PF – plant fiber, MCC – microcrystalline cellulose, Copyright © 2011, Royal Society of Chemistry.

(strength) compared to the glass and synthetic fibers. The comparative review of CNC mechanical characteristic and other commercial reinforcing agents is shown in Table 2.

Table 2. Strength and stiffness of reinforcement materials.

Materials	Tensile strength, GPa	Modulus, GPa	Reference
Cellulose nanocrystals	7.5	150	(Durán et al. 2011)
E-glass fiber*	3.4	-	(Bingham and Wallenberger 2009)
S-2 glass fiber*	4.9	-	(Bingham and Wallenberger 2009)
Steel wire	4.1	207	(Durán et al. 2011)
Carbon nanotubes	11–63	270–970	(Yu 2000)

* E-glass fiber-Electrical Resistance; alumino-lime silicate with less than 1% w/w alkali oxides, with high acid resistance; S-2 glass fiber-alumino silicate glass without CaO but with high MgO content with high tensile strength.

Cellulose Nanocrystals Preparation

The isolation of cellulose nanoparticles from cellulose source materials occurs in two stages. The first stage is a purification and homogenization pretreatment of the source material so that it reacts more consistently in subsequent treatments. The particular pretreatment is dependent on the cellulose source material and, to a lesser degree, on the desired morphology of the starting cellulose particle for the second stage treatments. Recently, numerous methods have been developed combining chemical, mechanical and enzymatic treatments to disintegrate wood fibers to produce nanocellulose (CNC) (Dufresne 2008, 2013, Eichhorn et al. 2010, Habibi et al. 2010, Lin et al. 2012, Habibi 2014, Kim et al. 2015). Depending on the treatment, fibrils with a diameter of 5–30 nm and length in micrometers, or rod-like crystals with a diameter of 5–20 nm and length of tens to hundreds of nanometers to micrometers can be obtained. Preparation of CNCs involves a chemical hydrolysis process intended to dissolve amorphous chains from the cellulose fibers and to release crystal domains (Dufresne 2012a). Apart from pure cellulosic sources such as cotton, bleached wood pulp, bacterial cellulose, and MCC, the biomass is generally first submitted to different pretreatments. A purification treatment consisting of an alkali treatment with sodium hydroxide (NaOH) or potassium hydroxide (KOH) followed by a bleaching step using acetate buffer (solution of NaOH and glacial acetic acid) and sodium chlorite ($NaClO_2$). This preliminary step to obtain pure cellulose fibers is crucial and must be done carefully. The alkali extraction is performed to solubilize most of the pectins and hemicelluloses. Although several bleaching methods exist in the wood industry, bleaching of natural fibers is typically performed with the $NaClO_2$ process over a period depending on the source of cellulose. The bleaching treatment is performed to break down phenolic compounds or molecules having chromophoric groups present in lignin and to remove the byproducts of such breakdown, to whiten the material. After removal of the non-cellulosic constituents, such as lignin, pectins and hemicelluloses, the bleached material is disintegrated in water, and the resulting suspension is submitted to the hydrolysis treatment with acid.

The cellulose amorphous regions are randomly oriented in a spaghetti-like arrangement leading to a lower density compared to nanocrystalline regions. The amorphous regions act as structural defects which are susceptible to acid attack and, under controlled conditions, they may be removed leaving crystalline regions intact (Ly et al. 2008). This transformation consists of the disruption of amorphous regions surrounding it and embedded within the cellulose microfibrils. During the acid hydrolysis process, the hydronium ions can penetrate into the cellulose chains in the amorphous domains promoting the hydrolytic cleavage of the glycosidic bonds and releasing individual crystallites. It is ascribed to the faster hydrolysis kinetics of amorphous domains compared to crystalline ones. From this process, cellulose monocrystals may be released and extracted from the cellulose substrate. It has been reported that these crystallites can grow in size because of the large freedom of motion after hydrolytic cleavage. For this reason, the crystallites can be larger in dimension than the original microfibrils (Dufresne 2012a). The preparation of cellulose nanocrystals by sulfuric acid, hydrolysis of dried and never-dried chemical pulps was reported (Kontturi et al. 2013). The average length of the nanoparticles was found to be fairly similar, but a higher number of longer crystals and a lower number of shorter crystals were found when using never-dried pulps. This was hypothetically ascribed to tensions building in individual microfibrils upon drying, resulting in irreversible supramolecular changes in the amorphous regions. The tense amorphous regions occurred to be more susceptible to acid hydrolysis.

Acid hydrolysis

Cellulose pulp treated with acid starts to degrade at the most accessible parts of the fiber, followed by the reducing end groups and the crystal surfaces. Acid concentration, reaction time, and reaction temperature are some of the most important parameters for controlling acid hydrolysis of wood pulp (Ioelovich 2012, Lu et al. 2013, Bondeson et al. 2006). A reaction time that is too long will hydrolyzes the cellulose crystals completely, and a reaction time that is too short will give a high degree of polymerization (DP) due to large undispersed fibers (Beck-Candanedo et al. 2005). The reaction temperature and time correlate to each other and a higher reaction temperature shortens the reaction time. Not only time and temperature affect the properties of nanocellulose, but also the acid concentration and acid-to-pulp ratio (Dong et al. 1998). After hydrolyzation and purification through dialysis, small crystalline rod-like particles are yielded in an aqueous suspension. The nanocrystalline cellulose that is formed through the acidic treatment is of colloidal dimensions and forms an aqueous suspension when stabilized. The critical concentration of the colloidal suspension, which is the lowest concentration where the whiskers self-organize, depends on particle size, acidic treatment, preparation conditions, aspect ratio, and ionic strength (Dong et al. 1998, Beck-Candanedo 2005).

The final properties of cellulose nanocrystals depend on the origin of the cellulosic fibers. Nanocrystals extracted from tunicates and algae sources are several micrometers in length since the cellulose microfibrils in tunicate and algae are highly crystalline. Nanocrystals from bacterial cellulose also have dimensions similar to those obtained from tunicate and algae, while nanocrystals of smaller dimensions are obtained from cotton and wood cellulose for instance. Therefore, specific hydrolysis and extraction procedures have been developed depending on the source of cellulose (Dufresne 2012a).

Sulfuric acid (H_2SO_4) is the most common acid for nanocellulose preparation through chemical hydrolysis, but it is possible to use other acids as well (Dufresne 2012a). Hydrochloric acid (HCl), hydrobromic acid (HBr), and phosphoric acid (H_3PO_4) have been used for CNC preparation, but compared to sulfuric acid, the application of hydrochloric or hydrobromic does not induce any surface charges and therefore, it is harder to form a stable colloidal dispersion when these acids are used. Use of phosphoric acid will give charged phosphate groups on the nanocrystal surface. Regarding the industrial application, sulfuric acid is a more suitable choice of acid than hydrochloric acid.

Sulfuric acid hydrolysis of cellulose pulp is a heterogeneous process where the acid diffuses into the pulp fiber and cleaves the glycosidic bonds in the cellulose polymer. Depending on reaction times, the hydrolysis could also occur in the crystalline regions and some of the hydroxyl groups on the crystalline surface will convert into sulfate groups (e.g., conversion of cellulose$-$OH to cellulose$-OSO_3^--H^+$). Other side reactions are also possible during acid hydrolysis such as dehydration and oxidation (Hamad and Hu 2010, Revol et al. 1992). In the cellulose pulp sample, hemicelluloses or pectin might be present and these polysaccharides will also undergo hydrolysis but at faster rates due to their higher reactivity. As the acid hydrolysis proceeds, the DPs are expected to decrease since the non-crystalline regions in the microfibril will be removed. Due to the loss of non-crystalline regions the crystallinity will increase and also the insolubility in water because the crystalline parts are less accessible (Hamad and Hu 2010). The sulfuric acid hydrolysis reaction has been optimized by several researchers and one general way to produce CNC from sulfuric acid is by using a 64 weight percentage of sulfuric acid solution at 45°C for 45–60 min with constant stirring, followed by quenching the suspension with 10-fold deionized water. The CNC is then concentrated through centrifugation and dialysis against deionized water until constant neutral pH is reached. To achieve the separation of crystals, the suspension has to be ultrasonicated repeatedly (Dong et al. 1998).

The sulfuric acid hydrolysis reaction conditions on softwood pulp has been evaluated (Hamad and Hu 2010, Revol et al. 1992). In the case of using 64 weight percentage of sulfuric acid solution, an acid-to-pulp ratio of 8.75 ml/g and temperature of 45°C at two different reaction times, 25 and 45 min, the longer reaction time showed a less polydisperse length distribution and a higher sulfur content than the shorter reaction time conditions. When the acid-to-pulp ratio is increased to 17.5 ml/g and the pulp is treated for 45 min at 45°C, a smaller length and polydispersity was observed but the effect was not that pronounced compared to the much higher acid-to-pulp ratio. Also, different sulfuric acid concentrations on softwood Kraft pulp and their effect on DP, crystallinity, crystal size, and yield were observed (Hamad and Hu 2010). Three different sulfuric acid concentrations (16, 40, and 64 weight percentage) with the acid-to-pulp ratio 8.75 ml/g and three different reaction temperatures (45, 65, and 85°C) were evaluated at a reaction time of 25 min. Both higher acid concentration and higher reaction temperature resulted in a lower DP, individually. The yield should be lower for CNC due to the loss of the non-crystalline regions and for the pulp samples hydrolyzed with a 64 weight percentage of sulfuric acid, the yield decreased at all three temperatures that were tested. The highest crystallinity was obtained for the highest concentration of sulfuric acid, which also gives the smallest crystals and lowest DP.

Dong et al. studied the preparation of CNC from cotton fibers where the reaction time and temperature were in focus (Dong et al. 1998). A 64-weight percentage of sulfuric acid was used and the acid-to-pulp ratio was 8.75 ml/g, the temperatures tested were 26, 45, and 65°C, and the reaction times were 15 min up to 18 hr. For low temperatures (26°C), the reaction time needed to be very long (18 hr) to produce CNC that could form an ordered suspension. At 65°C, the reaction was hard to control and already after 15 min a color change was noticed, indicating side reactions such as dehydration, while after 1 hr at 65°C, the whole suspension had turned from white to black.

Chemical Modification of Nanocellulose

The presence of hydroxyl groups within the cellulose structure makes it a unique platform for surface modification to graft a myriad of functional (macro) molecules using various chemistries, thus extending its use to a wide range of highly sophisticated applications. Naturally, most of the chemical modifications of nanocellulose represent a logical extension of reactions applied to cellulose fibers. The designation nanocellulose commonly refers to all types of nanosize cellulosic substrates including CNCs, NFCs and BNC. All these chemical transformations are applicable to every type of nanocellulose, i.e., CNCs, NFCs or BNC. They are carried out in order to improve the processability and performances of nanocellulose-based materials and to obtain nanocellulose derivatives that disperse well in a polar organic media (solvents and polymermatrices) (Abdul Khalil et al. 2014, Habibi 2014, Kontturi et al. 2013). However, preserving the integrity of the original morphology of nanocelluloses, avoiding any polymorphic conversion and maintaining their native crystalline structure is even more challenging than in the case of cellulose fibers. A wide variety of chemical modification techniques, including coupling hydrophobic small molecules, grafting polymers and oligomers, and adsorbing hydrophobic compounds to surface hydroxyl groups of cellulosic nanoparticles, can be applied. Besides the problem of dispersibility in polymeric matrices, improved nanofiller-matrix interaction is expected to enhance the stress transfer from the matrix to the dispersed phase and then improve the load bearing capability of the material (Dufresne 2012b). Moreover, the surface of the nanoparticles or nanofibers can be tailored to impart specific functionality to nanocellulose. The degree of modification affects the materials' properties. In the case of CNCs, the amount of negative groups on the surface will determine the quality of dispersion and concentrations at which they will behave as gels or liquid crystals, while chemical modification of CNFs, can reduce the amount of hemicellulose and affect the colloidal stability (epoxy modified CNFs), or it can alter the pH sensitivity of the material. The modification conditions need to be kept mild enough in order not to swell and subsequently dissolve the cellulose. It is also important to avoid flocculation in order to obtain good dispersion of fibrils and crystals during modification.

From this point, the focus will be especially on chemical modification processes of cellulose nanocrystals (CNCs), in order to be applicable as reinforcement filler in polymer nanocomposites. Because of their nanoscale dimensions, cellulose nanoparticles display a high surface area of the order of a few 100 m^2/g and their surface hydroxyl groups (2–3 mmol/g) enable targeted surface modification to introduce desired surface functionality (Eyley and Thielemans 2011). However, the surface chemistry of cellulose nanoparticles is primarily governed by the extraction procedure used to prepare these nanoparticles from the native cellulosic substrate (Moon et al. 2011). Figure 2 shows the surface functionalization provided by the most common extraction methods.

Chemical functionality of CNCs surfaces dictates CNCs suspension properties, the composite fabrication processes and the resulting composite properties. Different chemical modification of nanocellulose can be divided roughly into three major groups; preparing negatively charged, positively charged, and hydrophobic nanocelluloses (Moon et al. 2011, Dufresne 2012a, Habibi 2014). Common nanocellulose modifications include sulfuric acid treatment that provides sulfate esters, acyl-halides and acid anhydrides provide ester linkages, epoxides give ether linkages, isocyanates give urethane linkages, TEMPO-mediated hypochlorite

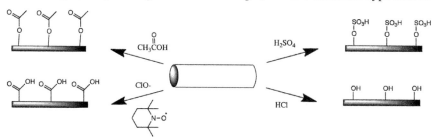

Fig. 2. Common syntheses of nanocellulose provide for distinctive surface chemistries: sulfuric acid treatment provides sulfate esters, hydrochloric acid treatment provides hydroxyl, acetic acid provides acetyl, TEMPO mediated hypochlorite treatment provides carboxylic (uronic) acid surface (Moon et al. 2011).

oxidation provides carboxylic acids, halogenated acetic acids form carboxymethyl surfaces, and chlorosilanes create an oligomeric silylated layer (Moon et al. 2011, Habibi 2014).

Esterification of nanocellulose

The hydrofobization of cellulose surface has often been achieved through the cellulose ability to undergo esterification reactions (Dufresne 2012a). The cellulose esterification process usually takes place in the presence of acid anhydrides or acyl chlorides as acetylation agents (Almasi et al. 2015, Lee et al. 2011, Kalia et al. 2014, Kim et al. 2002, Hu et al. 2011, Habibi 2014, Uschanov et al. 2011) introducing ester functional group on the nanocellulose surface. This method was also found to prevent the hornification of nanocellulose network upon drying, allowing dried nanocellulose to be re-dispersed in different solvents (Nechita and Panaitescu 2013). The modification by esterification reaction has been conducted on different nanocellulose substrates and some of these studies are summarized in Table 3.

The first procedure used to introduce acetyl groups onto the surface of nanocellulose is the esterification reaction with an excess of acetic anhydride usually in inert atmosphere since the acetylation reaction is equilibrium limited (Nechita and Panaitescu 2013, Hu et al. 2011). Before the reaction, solvent exchange step is performed, whereby the nanocellulose from water suspension was transferred to the reaction medium, usually dimethyl formamide (DMF) or toluene, through multiple solvents exchanged. The water removal by (vacuum) drying of nanocellulose results in the hornification of fibrous network of nanocellulose. As a result, the redispersion of (vacuum) dried nanocellulose in the subsequent reaction medium is no longer possible. Although freeze-dried nanocellulose could be easily redispersed in the subsequent reaction medium, a recent study (Lee and Bismarck 2012) on the susceptibility of freeze-dried and never-dried BNC toward esterification showed evidence that freeze-dried BNC would undergo severe bulk modification while solvent-exchanged BNC would undergo surface-only modification. Such results were also confirmed more recently by Žepič et al. (2015, 2014). The first stage of acetylation occurs primarily on the most accessible hydroxyl groups located on the surface of nanocellulose or in the disordered/amorphous regions of nanocellulose (Tingaut et al. 2009, Fahma et al. 2014). At this stage, the acetylation reaction is kinetically limited. The acetylation reaction then proceeds onto the less accessible hydroxyl groups, such as those located inside the cellulose crystals. These hydroxyl groups are hindered by the cellulose structure itself or the newly grafted acetyl groups. At this stage, the acetylation reaction becomes diffusion limited. Tang et al. (2013) showed that acetylated CNCs could be produced directly from wood pulp when the acetylation reaction was coupled with ultrasonication. The ultrasonication of wood pulp of up to 6 h at 40 kHz and 68–75°C was sufficient to break the hydrogen bonds between the cellulose microfibrils. Ultrasonication also influences the removal of macroscopic flaws on the cellulose fibrils, exposing more hydroxyl groups and enhancing the diffusion of acetylating reagents into

Table 3. Esterification of cellulosic nanoparticles.

Source of nanocellulose	Nanoparticle	Reagent	Objective of modification	Reference
Bacterial cellulose	BNC	Acetic anhydride	Enhancement of properties of acrylic resin	(Nogi et al. 2006, Ifuku et al. 2007)
		Palmitoyl acid (gas phase)	Hydrophobization	(Berlioz et al. 2009)
		Acetic, hexanoic and dodecanoic acid	Hydrophobization	(Lee et al. 2011)
		Acetic anhydride	Hydrophobization	(Hu et al. 2011)
		Acetic acid	Hydrophobization	(Kim et al. 2002)
		Maleic anhydride	Polyurethane solar cell encapsulation	(Yuwawech et al. 2017)
Cotton linters	CNC	Acetic and butyric acid	Dispersion in ethyl acetate and toluene	(Braun et al. 2008, Sobkowicz et al. 2009)
		Acetic anhydride	Dispersion in chloroform	(Lin et al. 2011)
		Succinic anhydride	Metal ions adsorption	(Yu et al. 2013)
	CNC	Maleic anhydride	Compatibility with unsaturated polyester resins	(Rusmirović et al. 2017)
Kenaf Bast fibers	MFC	Acetic anhydride	Dispersion in acetone and ethanol	(Jonoobi et al. 2012)
MCC	CNC	Acetic anhydride	Redispersibility in water	(Wang 2007)
Kraft pulp	MFC	Acetic anhydride	Hydrophobization	(Rodionova et al. 2011)
	CNC	Canola oil fatty acid methyl ester	Hydrophobization	(Wei et al. 2017)
	MFC	Succinic anhydride	Metal ions adsorption	(Hokkanen et al. 2013)
		Pyromellitic dianhydride (PMDA), benzophenone-3,30,4,40-tetracarboxylic dianhydride (BPDA), 1,4-phenylene diisocyanate (PPDI)	Hydrophobization	(Ly et al. 2008)

Table 3 contd. ...

...Table 3 contd.

Source of nanocellulose	Nanoparticle	Reagent	Objective of modification	Reference
Ramie	CNC	Hexanoyl, lauroyl and stearoyl chloride	Extrusion with LDPE	(Junior de Menezes et al. 2009)
Tunicin	CNC	Iso-octadecenyl succinic and n-tetradecenyl succinic anhydride	Dispersion in low polarity solvents	(Yuan et al. 2006)
		Palmitoyl acid	Hydrophobization	(Berlioz et al. 2009)
MCC	MCC	Pyromellitic dianhydride (PMDA), benzophenone-3,30,4,40-tetracarboxylic dianhydride (BPDA), 1,4-phenylene diisocyanate (PPDI)	Hydrophobization	(Ly et al. 2008)
NFC	NFC	Maleic anhydride	Polyuretane solar cell encapsulation	(Yuwawech et al. 2017)
Filter paper Whatman no. 1	CNC	2-propynoic acid, 4-pentenoic acid, 2-bromopropionic acid, or 3-mercaptopropionic acid	compatibilizing the surface of the fibers with the matrix material	(Boujemaoui et al. 2015)

the cellulosic fibers. This method allows for the production of acetylated CNCs in a single step without the need for an acid hydrolysis step prior to acetylation. The acetylation of nanocellulose with acetic anhydride could also be accelerated with the aid of lipase in supercritical carbon dioxide (scCO$_2$) (Božič et al. 2015). The acetylation reaction in scCO$_2$ was found to be three times faster than the conventional acetylation reaction that is also accompanied by an increase in the degree of acetylation of nanocellulose from 0.07 without scCO$_2$ to 0.18 with the aid of scCO$_2$. This was due to the formation of an acyl-enzyme, which acts as a catalyst greatly reducing the activation energy of the acetylation reaction.

In addition to acetic anhydrides, other hydrophobic moieties such as long-chain hydrocarbons could also be attached onto the surface of nanocellulose to render the hydrophilic surface of nanocellulose hydrophobic *via* esterification reactions with carboxylic acids derivatives. In such a reaction, *p*-toluenesulfonyl chloride is added into the reaction medium (such as pyridine) containing nanocellulose to generate anhydrides during the reaction (Lee and Bismarck 2012, Lee et al. 2011). Water from nanocellulose dispersion should be exchanged using a second solvent prior to redispersing it in the reaction medium to avoid bulk modification of nanocellulose. Using various carboxylic acids, namely, acetic, hexanoic, and dodecanoic acids showed that BNC with various degrees of surface substitution (DS) with different organic acids could be obtained (Lee and Bismarck 2012). The advancing (θa) and receding (θr) water-in-air contact angles measurements performed on cellulose nanopapers made from organic acid-esterified BNC showed that θa and θr increased with increasing chain length of the organic acids used. Furthermore, direct-wetting measurements based on generalized drop length–height method also showed that the contact angles of polylactide on a single BNC nanofiber decreased with increasing DS of (modified) BNC (Lee et al. 2009), indicating that the individual BNC nanofiber became more hydrophobic with increasing chain length of the organic acids used.

TEM observations and diffraction contrast images of the nanoparticles at various stages of the reaction showed that the acetylation proceeded by a reduction of the diameters of the nanocrystals, while their lengths were reduced to a lower extent (Dufresne 2012a) and it was supposed that the nanocrystals break down laterally but not longitudinally upon acetylation (peeling effect). A model of acetylation based on a non-swelling reaction mechanism affects only the cellulosic chains located at the surface of the nanoparticle, while for homogeneous acetylation, the partially acetylated molecules were found to be dissolved into the acetylating medium when they are sufficiently soluble. In heterogeneous conditions, the cellulose acetate remained insoluble and surrounded the crystalline core of unreacted cellulose.

Braun et al. (2008) reported the simultaneous hydrolysis and acetylation of cellulose nanocrystals in a single step in order to avoid complex surface functionalization routes using a mixture of hydrochloric acid (HCl) and organic acid (acetic and butyric). In that case, amorphous cellulose chains were hydrolyzed and cellulose nanocrystals were functionalized by the esterification process of hydroxyl groups in a one-pot reaction. It was shown that about half of the surface hydroxyl groups were esterified using this procedure. This approach is also quite versatile as the chain length of the covalently grafted surface group is easily controlled through the choice of organic acid employed in the reaction. The presence of acetate and butyrate groups affected the hydrophilicity of cellulose nanocrystals making their aqueous suspensions unstable, but they possessed better dispersibility in ethyl acetate and toluene.

Esterification of ramie cellulose nanocrystals by reacting organic fatty acid chlorides with different lengths of the aliphatic chain (C6 to C18) under reflux was also reported (Junior de Menezes et al. 2009). The crystalline core of the nanoparticles was found to be unaffected while the grafting density was high enough to allow crystallization of grafted chains when using backbones of 18 carbon atoms. Crystallization of the grafted chains was evidenced from X-ray diffraction and differential scanning calorimetry (DSC) experiments. The degree of substitution was found to decrease with increasing carbon chain length of the organic acid used. Similar results were reported for bacterial cellulose (BNC) modified using an esterification reaction with acetic acid, hexanoic acid or dodecanoic acid (Lee et al. 2011).

In order to prevent self-association of CNC particles and improve dispersibility in non-polar media, Rusmirović et al. (2017) studied two modification methods performed with fatty acids (FA) and methyl ester of fatty acids (MEFA) isolated from linseed, sunflower and soya bean oils, attached directly or *via* cross linker, ethylene diamine-maleic anhydride, bonded to nanocellulose surface (Fig. 3). Applied vacuum or scCO$_2$ drying post-treatment (SCD) on modified CNC indicates transition from crystalline to amorphous form, and this method contributes to better dimensional and thermal stability, dispersion stability in solution and polymer

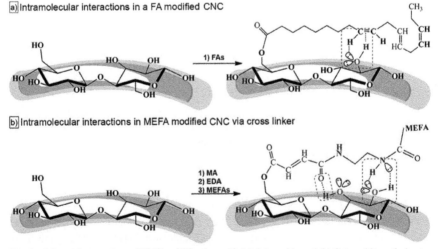

Fig. 3. Schematic overview of CNC modification with (a) fatty acids and (b) fatty acids methyl esters introduced *via* cross-linker.

matrix. SCD method also preserved nanosized characteristics and homogenous particle size distribution, while the vacuum drying method resulted in the CNC particles greater than 8 μm.

Gas phase esterification. Since the previously described esterification reactions involve numerous solvent exchange steps that are laborious and require a potentially harmful organic solvent, including DMF, toluene, and pyridine, a solvent-free gas-phase surface esterification of nanocellulose that does not require tedious solvent exchange steps has been developed. Berlioz et al. (2009) and Fumagalli et al. (2015) have also extended this esterification method to bulk esterification of nanocellulose (Fumagalli et al. 2013) and the esterification of nanocellulose crystals (Fumagalli et al. 2013). The reaction is based on the reduction of boiling temperatures of reactants following the reduction of the surrounding pressure. To esterify the nanocellulose, freeze-dried nanocellulose was first placed on a grid to avoid direct contact with the reactant, palmitoyl chloride. The pressure of the reaction vessel was then reduced to 100 mbar and temperature raised to evaporate palmitoyl chloride. The reactant diffuses onto the surface of the nanocellulose, reacting with the accessible hydroxyl groups. By varying the reaction parameters, such as time and temperature, hydrophobized nanocellulose with degree of substitutions (DS) varying from 0.15 to 2.7 can be obtained. It should be mentioned that the DS obtained using this method is comparable to the DS obtained via conventional wet-state esterification reactions.

The experimental conditions, nature and conditioning of cellulose were found to be important factors controlling the extent of esterification and morphology of the grafted nanoparticles. It was observed that the esterification proceeded from the surface of the cellulosic substrate to the crystalline core. For moderate degree of substitution, the surface was fully grafted whereas the cellulose core remained unmodified, while under certain conditions, an almost total esterification has been achieved, leading to highly substituted cellulose esters.

Reactive cross-linkable cellulose nanoparticles

Direct binding of a specific functional group of monomer with nanocellulose surface is not possible when desired monomer chain does not sufficiently react with nanocellulose hydroxyl groups. Thus, different reactive groups that show higher reactivity toward the target functional group of monomer have been attached to nanocellulose surface (Gericke et al. 2013). The introducing of vinyl functionalized molecules onto cellulose surface provides polymerizable centers suitable for copolymerization with unsaturated bonds in monomer chain. This is especially important in nanocellulose/unsaturated polyester based composites since curing of the UPe resin is a radical polymerization. In that manner, the introduction of double bonds (vinyl reactive sites) onto the CNF should make it possible to achieve covalent bonding between CNF and UPe matrix (Zadorecki and Flodin 1985). Zadorecki and Flodin (1985) used two coupling agents, based on trichloro-*s*-triazine molecule but with different terminal unsaturated (vinyl) groups for cellulose fiber modification in order to improve the bonding between cellulose fibers and an unsaturated polyester matrix. They found that vinyl treatment of cellulose fiber induces an increase of strength, as well as excellent adhesion without debonding and fractured fibers. Kanimozhi et al. (2014) developed hybrid nanocomposites based on vinyl silane-functionalized rice husk ash (lignocellulosic materials) and UPe resin. The results of the Fourier transform infrared spectroscopy confirmed that cross-linking was achieved *via* copolymerization reaction of the vinyl groups of rice husk ash with the unsaturated bonds in the UPe matrix to form the nanocomposites (Kanimozhi et al. 2014). Cross-linkable CNC particles become reactive in copolymerization with UPe matrix by introduction of unsaturated double bonds through the esterification with oleic acid or fatty acids isolated from linseed and sunflower oils through CNC amidation of ethylene diamine/maleic acid (EDA/MA) cross-linkable bridges (Rusmirović et al. 2017).

Designing of Nanocellulose Material Properties by Different Drying Methods

Considering that the unique nanomaterials properties depend on individual nanoparticles dimensions, a significant challenge occurs in applying of nanomaterials when they must be dried from a suspension of well dispersed individual particles (Beck et al. 2012). When preparing nanocellulose applying different isolation methods (acid, base hydrolysis), the nanocellulose particles are in water suspension. By removing some water, a critical concentration will be reached. Above the critical concentration, the cellulose nanoparticles will be in a stable colloidal dispersion. The critical concentration for nanocellulose is normally between 2 and 10 weight percentage of nanocellulose (Revol et al. 1992). In both cases, either nanocellulose isolation or chemical modification procedures, the majority of these methods are keeping nanocellulose in solvent medium (aqueous or organic solvents), and it is necessary to perform dehydration or desolvation of nanocellulose suspension to obtain dried nanostructures (Zimmermann et al. 2016). Moreover, nanocellulose hydrophilic nature and their propensity to agglomerate during drying are the reason for keeping nanocellulose particles in a solvent medium after isolation/modification (Peng et al. 2012). Due to that, the drying process while maintaining the cellulose nanoscale dimensions is a critical point during dehydration (Zimmermann et al. 2016). The most frequently used drying methods for CNC or CNF suspensions are oven drying (OD), freeze drying (FD), supercritical drying (SCD) and spray-drying (SD).

During conventional drying (OD or conventional vacuum drying) by evaporation of water under atmospheric or lower pressure, the interaction between the cellulose particles is promoted by the intermolecular hydrogen bonding of nanocellulose surface hydroxyl groups. This interaction promotes the agglomerates formation, called hornification, and the loss of the nanocellulose nanometric scale particles (Zimmermann et al. 2016). Additionally, it is a challenge to avoid irreversible agglomeration during conventional vacuum drying of nanocellulose which would cause the loss of properties or functionality (Rusmirović et al. 2017, Beck et al. 2012). On the other hand, Peng et al. (Peng et al. 2012) investigated agglomeration mechanisms occurring among the FD, SCD and SD drying methods applied on cellulose nanofibrils. They found that SCD and FD drying methods created highly networked structures of cellulose agglomerates with multi-scalar dimensions including the nanoscale, while SD method produced particles of dried cellulose which, range in size from nanosize to microsize (Peng et al. 2012). Prior to SCD, water or solvent CNC dispersion have to be subjected to a solvent exchange with absolute ethanol due to poor affinity of $scCO_2$ to water, as well as for providing dimensional stability of crystals (Liebner et al. 2010, Rusmirović et al. 2017). But removing water from nanocellulose suspension by solvent exchange method using absolute ethanol can cause significant increase in nanocellulose diameter (Peng et al. 2012). The characterizations of dried nanocrystalline and nanofibrillated cellulose particles (CNC and CNF) using OD, FD, SCD and SD methods are presented in Table 4.

Voronova et al. (Voronova et al. 2012) demonstrated that applying of FD method causes the decrease of CNC crystallinity and forming of CNC agglomerates with the size 200–400 nm per 10–20 nm. CNC particles dried by FD method possess high surface energy and a strong tendency to form aggregates and it is impossible to obtain a powder consisting of separate crystallites (Voronova et al. 2012). Rusmirović et al. (2017) investigated the influence of vacuum and SCD drying methods on the surface properties, morphology and thermal stability of CNC and confirmed that SCD drying method-induced transition from crystalline to amorphous form and contributed to better dimensional and thermal stability (Rusmirović et al. 2017). Representative SEM micrographs of the vacuum and SCD dried untreated and fatty acid-treated CNC particles, treated directly or *via* EDA/MA cross-linker, are shown in Fig. 4.

Besides influence of drying method on morphology and crystallinity of the dried cellulose particles, the obvious influence occurs on the thermal properties (Voronova et al. 2012). It was demonstrated (Rusmirović et al. 2017) that thermal degradation of vacuum and SCD dried CNC particles take place in a three-step reaction: process which takes place at decomposition temperature $\leq 200°C$, then between 200 and 300°C, and the last one at decomposition temperatures higher than 300°C. During conventional oven or vacuum drying the CNC films are obtained, and for that kind of sample processing, the lowest temperature at which thermal decomposition starts is 155°C and shows two pyrolysis processes which are clearly seen on derivative thermogravimetric (DTG) curves (Voronova et al. 2012). Obtained results indicated higher uniformity of CNC and difficult evaporation of water in SCD dried CNC compared to vacuum dried ones (Peng et al. 2013, Rusmirović et al. 2017). Moreover, vacuum and air drying cause stepwise dehydration that is accompanied with random agglomeration leading to broad distribution of the radius of CNC agglomerates and giving rise to multiple peaks in the DTG curves (Peng et al. 2013). Morphology of the CNCs examined by SEM after (a) air drying, (b) FD, and (c) SD drying is shown on Fig. 5 (Peng et al. 2013a).

Table 4. Comparison of CNC and CNF characteristics following different drying methods (Peng et al. 2012).

Drying method	Particle size in suspension		Dried particle size		Morphology	Advantages	Disadvantages
	CNC (nm)	CNF (nm)	CNC	CNF			
Oven drying	24–44	712–1484	Hundreds of microns to millimeters depends on drying process		Bulk network structure	Well established industrial processses, e.g., paper industry	Forms bulk material and loses the nano-91–295 scale dimensions of CNF
Freeze drying	91–295		Nano-scale in thickness and microns to millimeters in width or length		Ribbon-like structure	Maintain one dimension in nano-scale and well established process	Agglomeration of CNF and high cost
Supercritical drying			Nano-scale fibrous CNF		Fibrous and network	Maintaining nano-scale dimensions of the CNF	Complicated process using solvent replacement, high cost and impractical scale up
Spray-drying			$D_{[n,0.9]} = 6.76\ \mu m$	$D_{[n,0.9]} = 7.48\ \mu m$	CNF: irregular fibril structure CNC: spherical particles	Low cost and scalable continuous drying process, controllable particle size	Some agglomeration, the particle sizes range from nano to micron size

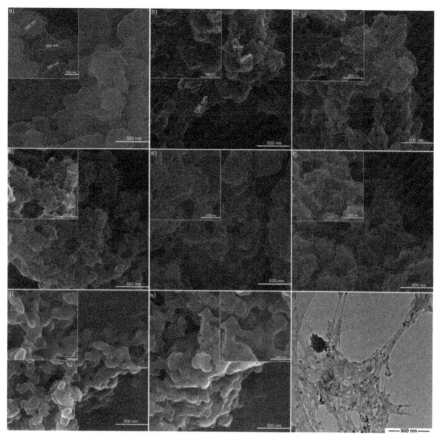

Fig. 4. Representative SEM images of the (a) vacuum dried CNC and (b) SCD dried CNC, (c) CNC-OA,* (d) CNC-FALO,* (e) CNC-FASO,* (f) CNC-MEFA/LO,* (g) CNC-MEFA/SO,* (h) CNC-MEFA/SOYA* and (i) TEM micrograph of CNC (Rusmirović et al. 2017) (*CNC-OA – Oleic acid modified CNC; CNC-FALO – Linseed oil fatty acids modified CNC, CNC-FASO – Sunflower oil fatty acids modified CNC, CNC-MEFA/LO – Linseed oil fatty acids modified CNC *via* EDA/MA cross-linker, CNC-MEFA/SO – Sunflower oil fatty acids modified CNC *via* EDA/MA cross-linker, CNC-MEFA/SOYA – Soybean oil fatty acids modified CNC *via* EDA/MA cross-linker) Copyright © 2011, Elsevier.

Moreover, mechanical and thermal properties of nanocellulose polymer composites depend on the quality of nanocellulose dispersion in the polymer matrix and the strength of adhesion at the polymer-filler interface (Peng et al. 2013a). Because of their influence on nanocellulose dispersion and the adhesion between the polymer and the cellulose nanoparticles surface, forces play an important role in the preparation and performance of polymer nanocomposites (Peng et al. 2013a). Peng et al. (Peng et al. 2013b) investigated the effect of drying method (air-drying, FD, SD and SCD) on the surface energy of nanomaterial cellulose nanofibrils that contained CNF and CNC. They found that the cellulose surface energy, determined

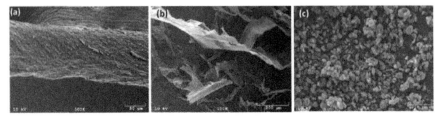

Fig. 5. SEM micrographs of dried CNCs: (a) air-dried, (b) freeze-dried, and (c) spray-dried (Peng et al. 2013a) Copyright © 2013, Tappi Press.

by inverse gas chromatography, is subjected to the influence of applied drying method and each of them produced unique cellulose nanoparticles with different surface morphologies and surface energies (Peng et al. 2013b).

According to the literature (Peng et al. 2013b), SCD method produces entangled individual CNFs with the highest dispersion component of surface energy at 30, 40, 50, 55 and 60°C, while higher temperature measurement (70, 75, and 80°C) decreases the dispersion component of surface energy (Fig. 6). For all applied drying methods, the dispersion component of surface energy decreased linearly with increasing temperatures from 30 to 60°C (Peng et al. 2013b).

By designing the surface and morphological properties of nanocellulose (aggregate size, crystallinity and surface energy) using appropriate drying method, performances of the cellulose-reinforced polymer nanocomposites can be controlled mainly by: (1) reducing agglomeration of cellulose particles, (2) improved interactions between cellulose and the polymer, and (3) improved distribution/dispersion of cellulose within the polymer matrix.

Processing and Applications of Nanocellulose in Polymer Composites

Nanocomposites are an interesting class of materials in which the reinforcing filler has at least one dimension up to 100 nm (Mondal 2017). In recent years, polymer nanocomposites (Fig. 7) have received great attention among the scientific and industrial communities because these materials can achieve significant improvements in mechanical properties, dimensional stability, and solvent/gas barrier properties. These significant improvements in material performances are achieved at very low concentration of fillers as compared to continuous phase (Mondal 2017).

Nanocomposites processing involves the incorporation of nanoparticles within the polymer matrices to produce a single composite material that has homogeneous dispersed nanoparticles (Ojijo and Sinha Ray 2013). The improvement of such properties usually arises as a result of particle organization and distribution and proper interactions between the nanoparticle and the polymer. Applied processing technique directly influences the final nanocomposite properties, and the major challenge in nanocomposite preparation is the quality of the particle dispersion

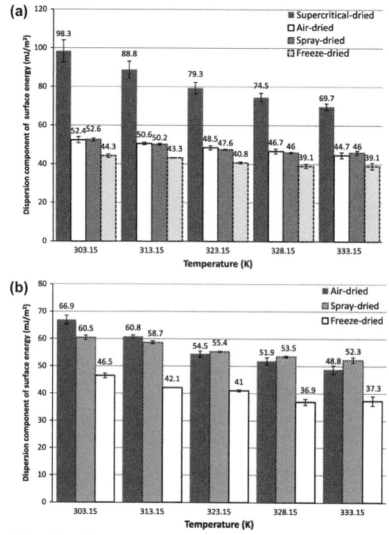

Fig. 6. Comparison of the dispersion component of surface energies of CNFs dried with different methods: (a) CNFs and (b) CNCs (Peng et al. 2013b) Copyright © 2011, Elsevier.

(Kargarzadeh et al. 2017). Overcoming the agglomeration and achievement of a completely homogeneous dispersion of individual nanoparticles have been proven to be remarkably difficult (Ober et al. 2009). The natural tendency of CNC for reorganization, and co-crystallization during drying, caused by high surface energy, leads to CNC agglomeration (Atalla et al. 2014, Kargarzadeh et al. 2017). In general, four different techniques are used for cellulose nanocomposites preparation: casting and evaporation, *in situ* polymerization, melt processing and layer-by-layer lamination among which casting and evaporation is the most widely used (Thakur

2014, Ojijo and Sinha Ray 2013, Kargarzadeh et al. 2017). Some of the most common cellulose nanocomposites processing techniques are shown in Fig. 8.

Fig. 7. TEM micrograph with schematic illustrations of possible interaction of oleic acid modified CNC and UPe based composites.

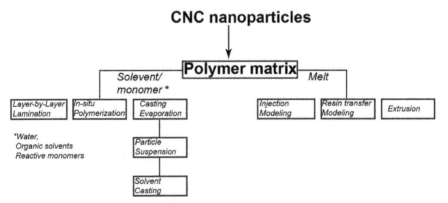

Fig. 8. Schematic illustration of cellulose nanocomposites processing techniques.

Melt processing

Melt processing method is based on the idea of using heat to soften and mold polymers and since it does not involve solvents, it represents a greener approach (Kargarzadeh et al. 2017). It includes extrusion, injection and resin transfer molding techniques (Kargarzadeh et al. 2017).

Melt extrusion method. Melt extrusion method incorporates high shear forces to promote dispersion and distribution of the nanoreinforcements in a melted thermoplastic polymer (Pöllänen et al. 2011), and it is the most prevalent one used for polymer processing in the industry (Thakur 2014). Creation of strong hydrogen bonds between the amorphous parts and forming aggregates of dried nanocellulose particles are the main difficulties observed during the use of this technique (Thakur 2014).

Injection molding. Injection molding is a suitable solid-state processing technology of materials with complex shapes that provides excellent surface smoothness. Mold and injection temperatures are important parameters for preparing high performance CNC composites because they modify the CNC capacity of crystallization and viscosity of the polymer matrix (Yousefian and Rodrigue 2015).

In situ polymerization

During *in situ* polymerization, the nanoparticles are premixed with the liquid solution of monomer followed by polymerization that is initiated by heat, radiation or suitable initiators (Ojijo and Sinha Ray 2013). In this approach, the monomer has a dual function, serving both as an effective dispersant for the nanoparticles and the matrix precursor for the *in situ* polymerization (Dufresne 2012a). Nanocomposites based on CNC and accompanying polymers poly(ε-caprolactone) (Dufresne 2012a), poly(butylene succinate) (Dufresne 2012a), UPe (Rusmirović et al. 2017) were successfully prepared by *in situ* polymerization. In some cases, the functional groups attached to nanocellulose surfaces can act as initiators and/or polymer monomers and they can be covalently bonded to polymer chains with interfacial strengths 50 times higher than those based on van der Waals interactions (Ferreira et al. 2017).

Solution casting and evaporation

Solvent solution casting method consists of mixing the nanocrystals to the previously dissolved polymer. After that, solid nanocomposite films are obtained by solvent evaporation (Ferreira et al. 2017). The polymer is dissolved in a suitable solvent, while the nanoparticles can be dispersed in the same or a different convenient solvent before mixing (Ojijo and Sinha Ray 2013). Nanobiocomposites of poly(3-hydroxybutyrate-co-3-hydroxyvalerate) (PHBV) and PLA reinforced with nanocellulose were produced by solution casting process (Wenjuan and Zhihua 2011, Benini et al. 2017). The practical limitation of the solvent solution casting method is the lack of compatibility between the hydrophilic nanocellulose and both the hydrophobic matrix and nonpolar solvents inducing inadequate dispersion combined with a weak interaction (poor wetting) (Ferreira et al. 2017).

Layer-by-layer lamination

Layer-by-layer lamination (LBL) technique provides one of the most promising tools for immobilization of the CNCs as well as for obtaining hybrids with high loading of the material in the field of preparation of thin film composites (Podsiadlo et al. 2005). It is a simple molding method intercalating layers by hand or by spray that results in superior surface and mechanical properties (Kargarzadeh et al. 2017).

Cellulose nanocomposites obtained in one of the mentioned methods, schematically illustrated in Fig. 8, have numerous applications for environmental applications (e.g., antibacterial, catalytic, pollutant sensors, and energy application), medical (Abdul Khalil et al. 2015) (cardiovascular implants, scaffolds for tissue engineering, repair of articular cartilage, vascular grafts, urethral catheters, mammary prostheses, penile prostheses, adhesion barriers, and artificial skin), or for structural reinforcement and Li-ion battery manufacturing (Wei et al. 2014). Previously published research articles primarily discussed the nanocellulose-based composites with nanocellulose added as reinforcement (Lee et al. 2014, Visakh et al. 2012, Ten et al. 2010). Addition of cellulose nanowhiskers in poly(3-hydroxybutyrate-*co*-3-hydroxyvalerate) (PHBV) polyester matrix causes the changes in crystalline structure of PHBV (Ten et al. 2012). The crystallinity changes suggest that cellulose nanowhiskers content and isothermal crystallization temperature could significantly alter crystallization kinetics of PHBV through two processes: nucleation and spherulite growth (Ten et al. 2012). The PHBV crystallization rate is increased with loading of low concentrations cellulose nanowhiskers by promoting heterogeneous nucleation. Decreasing of the crystallization rate started when the confinement effect of cellulose nanowhiskers outweighed their nucleation effect. Ten et al. (2010) investigated the influence of cellulose nanowhisker loading (1–5 weight percentage of CNW) on crystallization, thermal, dynamic mechanical and mechanical properties of bacterial PHBV polyester. The PHBV/cellulose nanowhisker composites were prepared by solution casting technique. Morphological testing results, obtained applying transmission electron microscopy (TEM) and atomic force microscopy (AFM) showed good dispersion of cellulose nanowhiskers in the PHBV polyester matrix (Fig. 9).

Moreover, the possibility of combining cellulosic nanoreinforcements with additional nanoscaled reinforcements offers the opportunity of monofunctional filler formation that can improve a single property of host polymers, broadening the field of application of nanocomposites (biological-, metallic-, ceramic- or carbon-based material) (Thakur 2014). Ternary multifunctional nanocomposites based on hybrid cellulose-based nanofillers and polymer (thermoplastic or thermoset) systems have recently attracted significant scientific and industrial interest. For this kind of systems, hybrid cellulose-based nanofillers, such as microfibrillated cellulose (MFC), cellulose nanocrystals (CNC), or bacterial cellulose (BNC), filled with

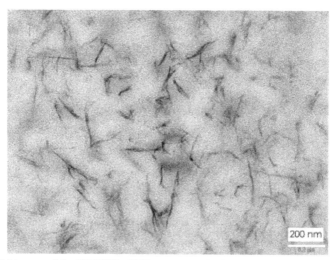

Fig. 9. TEM images of the PHBV/5% CNW nanocomposite film at 75 k magnification (Ten et al. 2010).

other types of nanoreinforcements (metallic, ceramic, carbon-based and biological) are commonly used. Fortunati et al. (2014) developed ternary nano-biocomposites based on PLA containing hybrid cellulose nanoreinforcement synthesized from CNC and silver nanoparticles (Ag). Tailor-made nano-biocomposites were prepared by the solvent casting method in order to obtain specific applications (antibacterial properties) by combination of hybrid CNC/Ag nanofillers and PLA biodegradable polymer matrix. The CNC/Ag induced PLA crystallization resulted in clear enhancement of barrier properties to oxygen and water vapor. Loading of weight percentage of Ag reduces about 40–45 percentage of barrier properties of CNC/PLA binary systems, suggesting the positive effect of silver and its combination with cellulose nanostructures (Fortunati et al. 2014). In summary, the combination of CNC particles with the antibacterial properties of Ag nanoparticles in the PLA-based composites showed high potential for development of new biodegradable materials for fresh food packaging applications.

Other ternary nanocomposites of industrial interest are based on the UPe resins. UPe resins are among the most-used thermosetting polymers with a wide range of applications due to their low cost, versatility, and ability to undergo various postpolymerization reactions (Kovačević et al. 2017, Rusmirović et al. 2016). Relatively poor mechanical and thermal properties, and low toughness limit their usage in advanced composites (Kovačević et al. 2017).

The addition of elastomeric particles such as epoxidized natural rubber (LENR) into thermoset resins (UPe) improves their low toughness, but, on the other hand, reduces the high stiffness and strength (Thomas et al. 2008, Ahmad et al. 2010, Kargarzadeh et al. 2015a). The mechanical and thermal properties of UPe may be improved by adding various types of organic/inorganic fillers such as silica (Rusmirović et al. 2017, Rusmirović et al. 2016), carbon nanotubes (Tasić et al. 2017), synthetic and natural (Lavoratti et al. 2016, Kargarzadeh et al. 2015a, Kargarzadeh et al. 2015c, Chirayil et al. 2014) polymers. The right selection of filler type is crucial for improvement of UPe characteristics. CNC particles represent a promising toughening agent for UPe nanocomposites (Kargarzadeh et al. 2015a). It was demonstrated that UPe toughness was improved by incorporating of LENR, tensile strength and modulus was significant reduced (Kargarzadeh et al. 2015b). On the other hand, forming of the ternary nanocomposite by CNC incorporation significantly improved the mechanical properties of the LENR–UPR blend (Kargarzadeh et al. 2015b).

Recent researches discussed the progress made in the development of cellulose nanofibril/inorganic nanocomposites (Wei et al. 2014). This kind of material is obtained by impregnation of a range of different nanomaterials such as metal (Au, Ag, Pd, Ni, etc.), mineral ($Ca_x(PO_4)_y$, $CaCO_3$ and montmorillonite), and carbon (carbon nanotube and graphene) nanomaterials on highly porous and mechanically strong cellulosic substrates. Resulting nanocomposites combine the advantages of both nanomaterials and perform synergistically (Wei et al. 2014).

Dynamic-Mechanical and Tensile Properties of Modified Nanocellulose Reinforced Polyester Composites

Beside of structural, surface morphology and energy characteristics of cellulose nanoparticles, mechanical properties of nanocellulose/polymer based composites are directly related to the interfacial properties and compatibility between polymer matrix and polar cellulose nanoparticles, especially if polymeric matrix is non-polar (Builes et al. 2013). The unmodified cellulose nanoparticles groups have a much smaller shear or slippage plane and there is a strong hydrophobic interaction between two independent nanoparticles, leading to agglomeration. The introduction of a coupling agent increases the shear plane boundary resulting in more stable suspensions. Moreover, bonding strength between natural fillers and polymer matrix is one of the most important factors for obtaining good reinforcement effect in the composite (Kushwaha and Kumar 2010). Good interfacial adhesion and compatibility between non-polar polyester matrix and cellulose filler, leading to outstanding improvement in mechanical properties of composites, can be reached using dispersing/coupling agents such as silane (Yu et al. 2015, Cheng 2012, Taipina et al. 2013), acid anhydrides (Cheng et al. 2012, Rančić et al. 2015), trichloro-*s*-triazine derivatives (Zadorecki and Flodin 1985), fatty acids (FAs) (Almasi et al. 2015, Kulomaa et al. 2015, Freire et al. 2006, Rusmirović et al. 2017, Almasi et al. 2015) or methyl esters of fatty acids (biodiesel) (Rusmirović et al. 2017, Yoo and Youngblood 2016). Woven bamboo fibers chemically surface modified with maleic anhydride, permanganate, benzoyl chloride, benzyl chloride, and pre-impregnation were used as reinforcements in epoxy and polyester matrices (Kushwaha and Kumar 2010). All modification types improved interfacial adhesion (interfacial bonding) between the fiber and the epoxy/polyester matrices more or less significantly. Builes et al. (2013) used PEO-*block*-PPO-*block*-PEO block copolymer (EPE20) as both nanostructuring agent for UPe resin and dispersing/coupling agent for sisal microfibrillated cellulose (MFC). It was demonstrated that the use of EPE20 helps avoiding the establishment of strong hydrogen bonding between hydroxyl groups of MFC fibers by replacing them with intermolecular interactions between fibrils and EPE20 PEO-blocks. Consequently, steric effects made more space between fibrils that can be generated allowing locations of UPe oligomers into twists/turns of MFC. Schematic illustration of this effect is shown in Fig. 10.

Interactions between polymer chains and filler nanoparticles have influence on material crystallinity, phase formation, and dynamic mechanical properties of composite materials (Rusmirović et al. 2016). More information about composite material and filler-matrix interaction can be obtained applying dynamic mechanical tests that are especially sensitive to all kinds of transitions and relaxation processes of matrix over a wide range of temperature and frequency (Saha et al. 1999). Dynamic-mechanical analysis (DMA) is a sensitive technique, which measures the modulus (stiffness) and damping properties (energy dissipation) of materials as the materials are deformed under periodic stress (Saha et al. 1999). During a DMA testing, three parameters can be obtained: (1) storage modulus (G'), which is a measure of the maximum energy, stored in a material during one cycle of oscillation; (2) loss modulus (G''), which is proportional to the amount of energy that has been dissipated as heat by the sample; and (3) mechanical damping term tanδ, which is the ratio of the G'' to the G' and is related to the degree of molecular mobility in the polymer material (Mandal and Alam 2012).

The tensile properties are the basic composite characteristics indicating material behavior under axial straining, and tensile tests are used to determine tensile strength, modulus, material deformation and Poisson coefficient. The fracture surfaces, resultant of the tests, can be used to support the failure mode analysis (Paiva 2006). Recently, cellulose nanosized fillers have received significant attention in materials science and engineering as reinforcement because of their

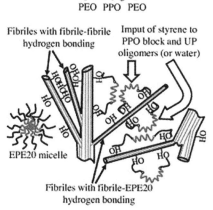

Fig. 10. Schematic description of the dispersion/adsorption phenomena of MFC/UPe resin (or water) after mixed with EPE20 block copolymer (Builes et al. 2013). Copyright © 2011, Elsevier.

high mechanical strength, ease of chemical modification and high surface to volume ratio (Thakur 2014). Moreover, the high surface area and functionality of cellulose nanofillers can also be used as template for polymerization, which then could create the ability of making highly reactive surface (Thakur 2014). The nanocellulose effect on improving tensile and dynamic-mechanical properties of composites is well documented (Zhang and Zhang 2016, Kargarzadeh et al. 2015a, Lavoratti et al. 2016, Kargarzadeh et al. 2017, Kargarzadeh et al. 2015c, Kargarzadeh et al. 2015b). The efficiency of nanocellulose as reinforcement in various thermoset polymer matrices, such as polyester, epoxies, formaldehyde resins and polyethylene terephthalate (PET) was reported in the literature (Kargarzadeh et al. 2017).

Dynamic-mechanical and tensile properties of modified CNC reinforced polyester composite

Loading of the CNCs into the maleic anhydride grafted PBAT matrix (MA-g-PBAT) matrix causes shifting of the glass transition temperature (T_g) of the matrix to higher values indicating restriction effects of CNCs on PBAT chain motion (Kashani Rahimi et al. 2017). Acetic anhydride (AA) modified CNC with lower polarity was used for preparation of composites based on poly PBAT (Zhang et al. 2016). It was reported that incorporation of the AA-modified CNC particles increases storage modulus of neat PBAT. This is apparently because the acetylation improves the dispersion of modified CNC and the compatibility between CNC and PBAT matrix (Zhang et al. 2016). The increase of moduli of AA modified CNC/PBAT composites indicates an enhanced interfacial adhesion between CNC and PBAT matrix, which should contribute to the mechanical properties of PBAT/ACNC composites.

As previously emphasized, another dynamic-mechanical parameter, tan δ is a meaningful parameter obtained from dynamic-mechanical tests to evaluate the interfacial bonding in composites since tan δ value is dependent on the filler-matrix adhesion level in composites systems (Ng et al. 2017). It was demonstrated that the incorporation of the CNCs to the MA-g-PBAT matrix causes the viscosity increase of the composites throughout the studied frequency range that is especially noticeable at low frequencies as the CNC concentration increased (Kashani et al. 2017). There is a report that the addition of CNC particles slightly reduced the tan δ value of PLA matrix due to the increase in the elastic response of the composite (Ng et al. 2017). The variations of storage modulus (G') and tan δ vs. temperature are shown in Fig. 11.

Bastiurea et al. (2015) compared the dynamic-mechanical properties of polyester/microcellulose and polyester/nanocellulose composites. They found that strongest chemical bonds between nanocellulose and polyester matrix contribute to better properties of polyester/nanocellulose composites than polyester/microcellulose (Bastiurea et al. 2015).

The incorporation of bare CNC crystals derived from cotton by acid hydrolysis method using 64 weight percentage of H_2SO_4 in UPe matrix, synthesized from the products obtained by poly(ethylene terephthalate) (PET) depolymerization with

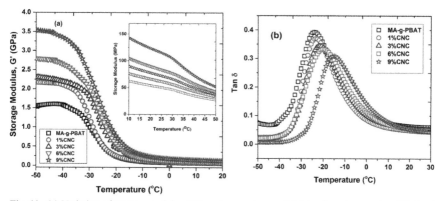

Fig. 11. (a) Variation of storage modulus (*G'*) vs. temperature and (b) tan δ vs. temperature (Kashani et al. 2017). Copyright © 2017, American Chemical Society.

propylene glycol (PG) and malic anhydride, contributed to achievement of better mechanical properties of composites (Rusmirović et al. 2017). An increase of stress at break (σ_b) from 43 MPa to 59 MPa, Young modulus (*E*) from 1.8 GPa to 1.9 GPa and elongation at break (ε_b) from 2.8 to 3.2 percentage of UPe/NC nanocomposites was obtained with 1 weight percentage of CNCs loading (Rusmirović et al. 2017). Chirayil et al. (2014) confirmed the hydrogen bonding interactions between the UPe and cellulosic filler (nanofibrils extracted from *Helicteres isora* plant) from the results of Fourier transform infrared spectroscopy of the composites. Due to network formation between the nanofibrils in UPe matrix and strong UPe chain confinement associated with and hydrogen bonding interaction between nanofibrils and UPe chains the superior mechanical and barrier properties of the nanocomposites are obtained (Chirayil et al. 2014). The water absorption of the nanocomposites is significantly lowered, although cellulose contains a number of hydrophilic OH groups, due to the hydrogen bonding between nanofibril and polyester (Chirayil et al. 2014).

Improvement of the toughness of UPe by incorporation of liquid epoxidized natural rubber (LENR) causes, on the other hand, significant loss in stiffness, yield strength, and thermal resistance (Kargarzadeh et al. 2015b). Therefore, with the aim to provide a nanocomposite based on UPe modified with LENR with better balance of material properties (i.e., a higher modulus, T_g, and impact energy), a rigid material such as cellulose CNC can be added into the LENR–UPe blend (Kargarzadeh et al. 2015b). The CNC incorporation significantly improved the mechanical properties (tensile modulus) of the LENR–UPR blend. Even the T_g of UPe was shifted to a lower temperature with addition of LENR; however, it increased with the loading of the CNCs.

Kargarzadeh et al. (2015a) investigated the effects of CNC silane surface treatment on the properties of CNC reinforced UPe have been studied. UPe-based nanocomposites reinforced with *N*-(*β*-aminoethyl)-*γ*-aminopropyltrimethoxysilane (APS) modified CNC were obtained by *in situ* polymerization method. Transmission electron microscopy results showed that more individual CNC nanoparticles were homogenously dispersed in UPe in the case of APS modified CNCs because of the reduction in the hydrophilicity of CNCs after the silane treatment due to fewer intramolecular interactions between the hydroxyl groups. The tensile strength and stiffness of the APS modified CNC/UPe composites were improved compared to the bare UPe matrix, whereas no significant changes were observed in the impact energy after composites treatment.

As one of the commercially available semicrystalline biodegradable polyesters, poly(butylene adipate-*co*-terephthalate) (PBAT) is used in a number of industrial and packaging applications, biomedical and tissue engineering (Kijchavengkul et al. 2010, Avérous and Digabel 2006). PBAT is an aromatic-aliphatic copolyester with 35–55 molar percentage of aromatic units, which possesses poor mechanical properties and low thermal stability (Shi et al. 2005, Kashani Rahimi et al. 2017). Kashani et al. (2017) demonstrated that CNC loading (1,3,6 and 9 weight percentage concentration) in maleic anhydride grafted PBAT matrix (MA-g-PBAT) improved the thermal stability and dynamic-mechanical and rheological properties of the composites, which increased with the increase of the CNC content. An improvement of 114 percentages in the elastic modulus and of 25 percentages in the ultimate tensile strength of the PBAT was obtained.

Mechanical testing results, examined by both tensile and bulge tests, confirmed that CNC can be used as an effective reinforcing agent for PHBV (Ten et al. 2010). Moreover, a 77 percentages (by tensile test) and 91 percentages (by bulge test) improvement in Young's modulus, and 35.5 percentages increase in tensile strength at 5 percentage loading of CNC have been determined (Ten et al. 2010). Aitomäki et al. (2014) confirmed CNF reinforcing efficiency by calculation of elastic modulus and strength from various nanocellulose composites using established micromechanical models: Halpin–Tsai and Rule of Mixture (ROM) model, as well. They found that the forming of CNF network is advantageous for soft matrices, while less beneficial to matrices of high modulus. Besides, they confirmed that very low fraction of CNF can have a remarkable effect on the modulus (Aitomäki and Oksman 2014).

Dynamic-mechanical and tensile properties of modified CNF reinforced polyester composite

It is well known that the incorporation of a small amount of nanoscopic fillers in thermoset matrices, such as polyester, increases thermomechanical properties (Wang et al. 2005, Chirayil et al. 2014). Chirayil et al. (2014) reported that isora nanofibrils stiffening effect causes increases of the storage modulus of CNF filled composites. Large surface area of isora nanofibrils promotes the interfacial interactions between the CNF and polyester matrix. Uniform CNF dispersion at 0.5 weight percentage concentration in polyester contributes achieving higher values of modulus; while formed CNF agglomerates at 3 weight percentage of CNF reinforced composites cause lower values than neat polyester (Chirayil et al.

2014). Such variable trends of modulus can be explained by the fact that a uniform dispersion allows transferring of stress from polymer matrix to nanoparticles during the dynamic-mechanical test (Ketabchi et al. 2016). This transfer of stress can reduce the mobility and deformation of the matrix. Increasing of temperature causes storage modulus decreasing because of the transition from glassy to rubbery state (Chirayil et al. 2014).

Lavoratti et al. (2016) evaluated the dynamic-mechanical and thermomechanical properties of the composites based on UPe resin and cellulose nanofibers obtained from dry cellulose waste of soft- and hardwood (*Pinus* sp. and *Eucalyptus* sp., respectively). They demonstrated that bleached *Eucalyptus* sp. pulp displayed higher crystallinity and cellulose content, which causes the increase of storage modulus compared to the neat UPe, while T_g of composites was not significantly altered. Increased T_g may indicate constrained molecular mobility in the UPe network due to the proximity of high specific surface area CNF fibrils. Carbonyl groups from UPe may interact with the surface hydroxyls of CNF *via* hydrogen bonding and other secondary interactions (Ansari et al. 2015). Also, the better dispersion and higher interaction in the fiber/matrix interface were also evidenced by the higher activation energy and dimensional stability of the composites based on bleached *Eucalyptus* sp. pulp. Besides, the use of unbleached *Pinus* sp. pulp waste without any purification yielded agglomerated CNFs in UPe matrix (Lavoratti et al. 2016).

Previously published study presented results related to the effect of addition of CNF modified with 3-methacryloxypropyltrimethoxysilane (MEMO) on the mechanical properties of PLA matrix (Qu et al. 2012). The final yield concentrations of MEMO silane relative to the CNF particles, during modification procedure, were 0.5, 1.0, 1.5, and 2.0 v/v percentages. The tensile results indicated that the addition of MEMO modified CNFs can promote the reinforcement of CNF/PLA composites (Qu et al. 2012). The enhanced mechanical properties of the composites were determined in comparison with the neat PLA (39 MPa). The highest tensile strength and elongation, increased by 42.3 percentages and 28.2 percentages compared to pure PLA, were determined by incorporating 1.0 percentage of MEMO modified CNFs, while lower content than 0.5 percentage did not contribute to formation of a network in the PLA matrix. However, the higher content than 1.0 percentage of MEMO modified CNF in the PLA matrix caused aggregates formation which weaken the interfacial adhesion between filler and PLA matrix, due to zones with accentuated fragility which lead to a brittleness in this material (Qu et al. 2012).

In order to maximize CNF structural and functional application potentials in poly(L-lactide) based nanocomposites, the CNF surface functionalization by dendritic polyamidoamine (PAMAM) *via* 3-aminopropyltrimethoxysilane (APTMS) represents the preferred approach to engineer the interfaces (Lu et al. 2015). The obtained results demonstrate that the amine-modified CNFs have both positive and negative effects on the reinforcement of PLLA (Lu et al. 2015). The interfacial adhesion between APTMS modified CNF fibers within the aggregates was much weaker due to an insufficient PLLA binding effect. These aggregates are thus defects with weak adhesion within the PLLA matrix. On the other hand, the larger amount of amines on the PAMAM modified CNFs was expected to decrease the molecular weight of the PLLA matrix more significantly, via the chain scission induced by aminolysis. Aminolysis resulted in the PLLA grafting and crosslinking with the matrix, which is accompanied by the degradation of the PLLA matrix. Excessive amines could compromise the mechanical properties of the nanocomposite.

Dynamic-mechanical and tensile properties of modified BNC reinforced polyester composite

Earlier published literature reviews on the production and application of bacterial nanocellulose as reinforcement in polyesters composites have pointed to disruption and/or functionalization of 3D network of BNC by mechanical and/or chemical treatments to provide fine functionalized BNC, bacterial CNF or bacterial cellulose nanowhiskers (BNCW) (Panaitescu et al. 2016). The PLA nanocomposite with 5 wt% BCNW showed a significant improvement of the tensile and compressive strength and, also, of the tensile modulus, compared to pure PLA. But for better reinforcing efficiency, BNCW were chemically modified by *in situ* polymerization of lactic acid oligomers on the BCNW (Ambrosio-Martín et al. 2015). The PLA-based nanocomposites containing lactic acid modified BCNW showed improved mechanical properties (the elastic modulus and tensile strength increased by 52 and 31 percentage, respectively) that can be explained by the better dispersion of lactic acid modified BCNWs due to the *in situ* polymerization technique which facilitated matrix–filler interactions.

PLLA-based composites filled with 2 mass percentages of BNC esterified with hexanoic acid exhibited the improvement of tensile strength in comparison to neat PLLA, 141 MPa versus 127 MPa, respectively. The improvements observed in the tensile strengths of the nanocomposites reinforced with esterified BNC could be ascribed to the good dispersibility of hexanoic acid functionalized BNC in the PLLA matrix and the intimate contact, i.e., rather good interfacial adhesion between functionalized BNC and the PLLA matrix. Moreover, the reduced draw ratio for hexanoic modified BNC was indicative of improved compatibility with PLLA and mechanical properties (Ambrosio-Martín et al. 2015).

Gao et al. (2011) proved positive effect of vinyl-triethoxy silane modified BNC on mechanical properties of UPe-based composites. They found that tensile properties (strength and modulus) and shear and flexural strengths increase with BNC volume fraction up to Vf of 20 percentage, but further increasing of filler percentage does not show improvement and tensile strength begins to decline. XPS analysis results confirmed that chemical interaction between treated BNC fiber and UPe matrix was promoted by using vinyl-triethoxy silane coupling agent (Hu et al. 2009). It was shown that both, contents of elements and chemical bonds were changed after vinyl silane treatment compared to original BNC fibers and that hydrophilic BNC fibers and hydrophobic resin were connected by silanol groups (Hu et al. 2009). Tensile strength, flexural strength, shear strength and Young's modulus of the vinyl treated BNC and UPe based composites were increased by 117.7, 38.4, 38.7 and 27.6 percentage respectively that attributed to the chemical bonding on the BNC fiber–UPe matrix interface under the function of silane coupling agent (Hu et al. 2009).

Future Perspective in Designing of Nanocellulose/Polyester Composites Properties

The development of completely renewable nanocomposites that are safer to handle and cheaper to produce, is still in its infancy, but already showing great potential to deliver sustainable materials for a variety of applications. Nanocellulose and composites based on nanocellulose and bio-based polymers are considered as a valuable replacement to non-biodegradable materials and they will be one of the alternatives to fulfill the needs of better materials for the coming future. Beside wood and non-wooden plants as the main sources for cellulose extraction, the industry bioresidues, agricultural waste, cellulose-rich municipal solid waste, corrugated cardboard and waste papers have great potential for producing nanocellulose materials. Appropriate nanocellulose extraction and modification technologies will declare nanocellulose as bio-based products for use in applications ranging from environmental, medical, rheology modifier, structural reinforcement, 3D printing to flexible electronics.

Although nanocellulose is inherently strong and has many potential applications, the industrially applicable technology is yet to develop. The problems associated with nanocellulose composites are related to agglomeration, dispersion techniques and compatibility of nanocellulose with polymer matrices, and high cost involved with surface modification. The future perspective in engineering of high performance nanocellulose based composites will be related to development of highly-efficient, low cost modification methods for introduction of reactive cross-linkable functional group on cellulose nanoparticles surface that improve direct binding with monomer chains, as well as their dispersibility and compatibility with polymer matrix. In such manner, very recently, a new kind of cellulose nanoparticle consisting of a nanocrystalline body and two-end attached amorphous chains, named hairy nanocrystalline cellulose (HNCC), has been developed (van de Ven and Sheikhi 2016). The highly reactive aldehyde groups on the amorphous regions and the unique colloidal structure endow HNCC with some superior features as compared to conventional CNC, such as high functionality and reactivity, tunable charge, and colloidal stability. Due to many aldehyde groups, which can form hemiacetal linkages with hydroxyl groups, it is expected to produce composites with strong tensile strength, and the protruding dialdehyde cellulose chains on HNCC will contribute to the flexibility of the composites. HNCC can also be used to replace some common but toxic crosslinkers, in the application as a green crosslinker for preparation of composite films, hydrogels or aerogels with PVA (polyvinyl alcohol).

Moreover, various surface and/or bulk modifications of the nanocellulose (CNC, CNF and CNB) bring up new opportunities for developing unique functional materials. The modification leads to some modifications of physico-chemical properties of materials, especially at the nanoscale. The focus of future research and development in nanocellulose/polymer composites will be on designing different types of new bio-materials based on polyester derived from cellulosic and hemicellulosic biomass, or bio-refinery residues and reactive HNCC particles or CNC/CNF/BNC nanoparticles modified with natural eco-friendly coupling agents/compatibilizers (fatty acid isolated from natural oils) with a wide range of applications as high performance bio-nanocomposites. Also, one more focus of nanocellulose/polyester research will be development of processing techniques that can be applied on industrial scale.

Conclusion

The present chapter reported advances in the application of modified cellulose nanoparticles (CNC, CNF and BNC) as reinforcement in polyester based composites. It reviewed uptodate studies in order to promote future research in processing and characterization of modified nanocellulose/polyester composites. In addition to its numerous advantages such as biorenewability, biodegradability, availability, low cost, high mechanical performances, nanocellulose shows better properties as reinforcement of polymer matrices than micro- or macrocellulose. Since nanocellulose isolation and modification techniques have a significant influence on the processing and dynamic-mechanical properties of cellulose/polyester composites, this chapter also overviewed previous research related to them. Due to numerous surface hydroxyl groups that induce high nanocellulose hydrophilicity, its use as reinforcement in polar polyester matrices is limited. However, the hydroxyl groups on the nanocellulose surface represent suitable centers for various types of modification thus introducing non-reactive or reactive cross-linkable hydrophobic groups with an emphasis on the esterification methods. Modified nanocellulose can be applied in a wide range of polymer matrices as reinforcement, including rubbers, thermoplastics, thermosets, and prevalent biodegradable polyesters obtained from biorenewable sources and agricultural wastes.

Different modification patterns provide improvement of dispersibility and better interactions/bonding of nanocellulose in the polyester matrices. Moreover, the appropriate post-treatment (drying method) has significant impact on nano-scale dimensions and aggregation of nanocellulose thus influencing the composite processing and its mechanical properties. Various preparation methods create the final products with improved inherent mechanical and dynamic-mechanical properties that promise a wide possibility of applications (medical, rheology modifier, structural reinforcement, 3D printing and flexible electronics). The future trends in nanocellulose/polyester composite production will be focused on development of new bio-materials based on polyester derived from cellulosic and hemicellulosic polymers, or bio-refinery residues and reactive cross-linkable cellulose nanoparticles. The development of new type of highly reactive nanocelluloses and new CNC, CNF and BNC surface modification techniques will bring new opportunities for developing unique functional materials and modification of physico-chemical properties of materials will be important topics for future research.

References

Abdul Khalil, H.P.S., A.H. Bhat and A.F. Ireana Yusra. 2012. Green composites from sustainable cellulose nanofibrils: a review. Carbohydrate Polymers 87(2): 963–979. http://linkinghub.elsevier. com/retrieve/pii/S0144861711007624.

Abdul Khalil, H.P.S., Y. Davoudpour, M.N. Islam, A. Mustapha, K. Sudesh, R. Dungani and M. Jawaid. 2014. Production and modification of nanofibrillated cellulose using various mechanical processes: a review. Carbohydrate Polymers 99: 649–665. http://dx.doi.org/10.1016/j.carbpol.2013.08.069.

Abdul Khalil, H.P.S., A.H. Bhat, A. Abu Bakar, P.M. Tahir, I.S.M. Zaidul and M. Jawaid. 2015. Cellulosic nanocomposites from natural fibers for medical applications: a review. pp. 475–511. *In*: J. Pandey, H. Takagi, A.N. Nakagaito and H.-J. Kim (eds.). Handbook of Polymer Nanocomposites. Processing, Performance and Application. Berlin, Heidelberg: Springer Berlin Heidelberg. http:// link.springer.com/10.1007/978-3-642-45232-1_72.

Abraham, E., B. Deepa, L.A. Pothan, M. Jacob, S. Thomas, U. Cvelbar and R. Anandjiwala. 2011. Extraction of nanocellulose fibrils from lignocellulosic fibres: a novel approach. Carbohydrate Polymers 86(4): 1468–1475.

Ahmad, I. and Farihan Mohamed Hassan. 2010. Preparation of unsaturated polyester liquid natural rubber reinforced by montmorillonite. Journal of Reinforced Plastics and Composites 29(18): 2834–2841. http://journals.sagepub.com/doi/10.1177/0731684410363167.

Aitomäki, Y. and K. Oksman. 2014. Reinforcing efficiency of nanocellulose in polymers. Reactive and Functional Polymers 85: 151–156. http://linkinghub.elsevier.com/retrieve/pii/ S1381514814001783.

Almasi, H., B. Ghanbarzadeh, J. Dehghannia, S. Pirsa and M. Zandi. 2015. Heterogeneous modification of softwoods cellulose nanofibers with oleic acid: Effect of reaction time and oleic acid concentration. Fibers and Polymers 16(8): 1715–1722. http://link.springer.com/10.1007/s12221-015-4294-1.

Almasi, H., B. Ghanbarzadeh, J. Dehghannya, A.A. Entezami and A. Khosrowshahi Asl. 2015. Novel nanocomposites based on fatty acid modified cellulose nanofibers/poly(lactic acid): Morphological and physical properties. Food Packaging and Shelf Life 5: 21–31. http://dx.doi.org/10.1016/j. fpsl.2015.04.003.

Alves, L., B. Medronho, F.E. Antunes, M.P. Fernández-García, J. Ventura, J.P. Araújo, A. Romano and B. Lindman. 2014. Unusual extraction and characterization of nanocrystalline cellulose from cellulose derivatives. Journal of Molecular Liquids 210: 106–112.

Ambrosio-Martín, J., M.J. Fabra, A. Lopez-Rubio and J.M. Lagaron. 2015. Melt polycondensation to improve the dispersion of bacterial cellulose into polylactide via melt compounding: Enhanced Barrier and Mechanical Properties. Cellulose 22(2): 1201–1226. http://link.springer.com/10.1007/ s10570-014-0523-9.

Ansari, F., M. Skrifvars and L. Berglund. 2015. Nanostructured biocomposites based on unsaturated polyester resin and a cellulose nanofiber network. Composites Science and Technology 117: 298–306.

Atalla, R.S., M.F. Crowley, M.E. Himmel and R.H. Atalla. 2014. Irreversible transformations of native celluloses, upon exposure to elevated temperatures. Carbohydrate Polymers 100: 2–8. http:// linkinghub.elsevier.com/retrieve/pii/S0144861713006024.

Avérous, L. and F. Le Digabel. 2006. Properties of biocomposites based on lignocellulosic fillers. Carbohydrate Polymers 66(4): 480–493. http://linkinghub.elsevier.com/retrieve/pii/ S014486170600186X.

Avérous, L. 2007. Cellulose-based biocomposites: Comparison of different multiphasic systems. Composite Interfaces 14(7-9): 787–805. https://www.tandfonline.com/doi/full/10.1163/15685 5407782106410.

Bastiurea, M.S., M. Bastiurea, G. Andrei and A. Materials. 2015. Dynamic mechanical properties for polyester/microcellulose and polyester/nanocellulose. In 2015 IEEE 15th International Conference on Nanotechnology (IEEE-NANO): 0–2.

Beck-Candanedo, S., M. Roman and D. Gray. 2005. Effect of conditions on the properties behavior of wood cellulose nanocrystals suspensions. Biomacromolecules 6: 1048–1054.

Beck, S., J. Bouchard and R. Berry. 2012. Dispersibility in water of dried nanocrystalline cellulose. Biomacromolecules 13(5): 1486–1494.

Benini, K.C.C. de C., M.O.H. Cioffi and H.J.C. Voorwald. 2017. PHBV/cellulose nanofibrils composites obtained by solution casting and electrospinning process. Matéria (Rio de Janeiro) 22(2) (July 20). http://www.scielo.br/scielo.php?script=sci_arttext&pid=S1517-70762017000200504&lng =en&tlng=en.

Berlioz, S., S. Molina-Boisseau, Y. Nishiyama and L. Heux. 2009. Gas-phase surface esterification of cellulose microfibrils and whiskers. Biomacromolecules 10(8): 2144–51.

Bingham, F. and P. Wallenberger. 2009. Fiberglass and Glass Technology: Energy-Friendly Compositions and Applications. Springer Berlin Heidelberg.

Bondeson, D., A. Mathew and K. Oksman. 2006. Optimization of the isolation of nanocrystals from microcrystalline cellulose by acid hydrolysis. Cellulose 13(2): 171–180.

Börjesson, M. and G. Westman. 2015. Crystalline Nanocellulose—Preparation, modification and properties. pp. 159–191. *In*: Poletto, M. and H.L. Ornaghi (eds.). Cellulose—Fundamental Aspects and Current Trends, InTech.

Boujemaoui, A., S. Mongkhontreerat, E. Malmström and A. Carlmark. 2015. Preparation and characterization of functionalized cellulose nanocrystals. Carbohydrate Polymers 115: 457–464.

Božič, M., V. Vivod, S. Kavčič, M. Leitgeb and V. Kokol. 2015. New findings about the lipase acetylation of nanofibrillated cellulose using acetic anhydride as acyl donor. Carbohydrate Polymers 125: 340–351.

Braun, B., J.R. Dorgan and J.P. Chandler. 2008. Cellulosic nanowhiskers. Theory and application of light scattering from polydisperse spheroids in the rayleigh-gans-debye regime. Biomacromolecules 9(4): 1255–1263.

Bruna, A.S.M., H. de O.R. João, S.C. Lindaiá, L.L. Ingrid, D.V.B. Josiane, B.A. Joyce and I.D. Janice. 2017. Characterization of cassava starch films plasticized with glycerol and strengthened with nanocellulose from green coconut fibers. African Journal of Biotechnology 16(28): 1567–1578. http://academicjournals.org/journal/AJB/article-abstract/7A68A4765094.

Bryuzgin, E.V., V.V. Klimov, O.V. Dvoretskaya, L.D. Man', A.V. Navrotskiy and I.A. Novakov. 2014. Hydrophobization of cellulose-containing materials with fluoroacrylic polymers and fatty carboxylic acids. Russian Journal of Applied Chemistry 87(8): 1119–1125. http://www.scopus. com/inward/record.url?eid=2-s2.0-84912006738&partnerID=tZOtx3y1.

Builes, D.H., J. Labidi, A. Eceiza, I. Mondragon and A. Tercjak. 2013. Unsaturated polyester nanocomposites modified with fibrillated cellulose and PEO-B-PPO-B-PEO block copolymer. Composites Science and Technology 89: 120–126. http://dx.doi.org/10.1016/j. compscitech.2013.09.015.

Cao, X., H. Dong and C.M. Li. 2007. New nanocomposite materials reinforced with cellulose nanocrystals in waterborne polyurethane. Biomacromolecules 8: 899–904.

Çaylı, G. and S. Küsefoğlu. 2006. Thermal and mechanical behavior of unsaturated polyesters filled with phase change material. Journal of Applied Polymer Science 100(1): 832–838. http://doi. wiley.com/10.1002/app.23181.

Cheng, Z., Q.H. Xu and Y. Gao. 2012. Research progress in nano-cellulose modification. Advanced Materials Research 627: 859–863. http://www.scientific.net/AMR.627.859.

Chirayil, C.J., J. Joy, L. Mathew, J. Koetz and S. Thomas. 2014. Nanofibril reinforced unsaturated polyester nanocomposites: Morphology, mechanical and barrier properties, viscoelastic behavior and polymer chain confinement. Industrial Crops and Products 56: 246–254. http://dx.doi. org/10.1016/j.indcrop.2014.03.005.

Chirayil, C.J., L. Mathew, P.A. Hassan, M. Mozetic and S. Thomas. 2014. Rheological behaviour of nanocellulose reinforced unsaturated polyester nanocomposites. International Journal of Biological Macromolecules 69: 274–281. http://linkinghub.elsevier.com/retrieve/pii/S0141813014003511.

Chirayil, C.J., L. Mathew and S. Thomas. 2014. Review of recent research in nano cellulose preparation from different lignocellulosic fibers. Reviews on Advanced Materials Science 37(1-2): 20–28.

Dong, X.M., J.-F. Revol and D. Gray. 1998. Effect of microcrystallite preparation conditions on the formation of colloid crystals of cellulose. Cellulose 5: 19–32.

Dufresne, A. 2008. Cellulose-based composites and nanocomposites. pp. 401–418. *In*: Monomers, Polymers and Composites from Renewable Resources. Elsevier. http://linkinghub.elsevier.com/retrieve/pii/B9780080453163000193.

Dufresne, A. 2012a. Nanocellulose: From Nature to High Performance Tailored Materials. Berlin: De Gruyter.

Dufresne, A. 2012b. Nanocellulose: Potential reinforcement in composites. *In*: Natural Polymers. Royal Society of Chemistry 2: 1–32. http://ebook.rsc.org/?DOI=10.1039/9781849735315-00001.

Dufresne, A. 2013. Nanocellulose: a new ageless bionanomaterial. Materials Today 16(6): 220–227. http://dx.doi.org/10.1016/j.mattod.2013.06.004.

Durán, N., A. Lemes, M. Durán, J. Freer and J. Baeza. 2011. A minireview of cellulose nanocrystals and its potential integration as co-products in biethanol production. Journal of the Chilean Chemical Society 56(2): 672–677. http://www.scielo.cl/scielo.php?script=sci_arttext&pid=S0717-97072011000200011&lng=en&nrm=iso&tlng=en.

Eichhorn, S.J. 2011. Cellulose nanowhiskers: Promising materials for advanced applications. Soft Matter 7(2): 303. http://xlink.rsc.org/?DOI=c0sm00142b.

Eichhorn, S.J., A. Dufresne, M. Aranguren, N.E. Marcovich, J.R. Capadona, S.J. Rowan, C. Weder, W. Thielemans, M. Roman, S. Renneckar, W. Gindl, S. Veigel, J. Keckes, H. Yano, K. Abe, M. Nogi, A.N. Nakagaito, A. Mangalam, J. Simonsen, A.S. Benight, A. Bismarck, L.A. Berglund and T. Peijs. 2010. Review: Current international research into cellulose nanofibres and nanocomposites. Journal of Materials Science 45: 1–33. https://doi.org/10.1007/s10853-009-3874-0.

Eyley, S. and W. Thielemans. 2011. Imidazolium grafted cellulose nanocrystals for ion exchange applications. Chemical Communications 47: 4177–9.

Fahma, F., A. Takemura and Y. Saito. 2014. Acetylation and stepwise solvent-exchange to modify hydrophilic cellulose whiskers to polychloroprene-compatible nanofiller. Cellulose 21(4): 2519–2527.

Favier, V., G.R. Canova, J.Y. Cavaillé, H. Chanzy, A. Dufresne and C. Gauthier. 1995. Nanocomposite materials from latex and cellulose whiskers. Polymer for Advanced Technologies 6(5): 351–355.

Ferreira, F.V., I.F. Pinheiro, R.F. Gouveia, G.P. Thim and L.M.F. Lona. 2017. Functionalized cellulose nanocrystals as reinforcement in biodegradable polymer nanocomposites. Polymer Composites 1–21. http://doi.wiley.com/10.1002/pc.24583.

Fortunati, E., S. Rinaldi, M. Peltzer, N. Bloise, L. Visai, I. Armentano, A. Jiménez, L. Latterini and J.M. Kenny. 2014. Nano-biocomposite films with modified cellulose nanocrystals and synthesized silver nanoparticles. Carbohydrate Polymers 101: 1122–1133. http://linkinghub.elsevier.com/retrieve/pii/S014486171301076X.

Freire, C.S.R., A.J.D. Silvestre, C.P. Neto, M.N. Belgacem and A. Gandini. 2006. Controlled heterogeneous modification of cellulose fibers with fatty acids: Effect of reaction conditions on the extent of esterification and fiber properties. Journal of Applied Polymer Science 100(2): 1093–1102.

Fumagalli, M., D. Ouhab, S.M. Boisseau and L. Heux. 2013. Versatile gas-phase reactions for surface to bulk esterification of cellulose micro Fibrils aerogels. Biomacromolecules 14(9): 3246–3255.

Fumagalli, M., F. Sanchez, S.M. Boisseau and L. Heux. 2013. Gas-phase esterification of cellulose nanocrystal aerogels for colloidal dispersion in apolar solvents. Soft Matter 9(47): 11309.

Fumagalli, M., F. Sanchez, S. Molina-Boisseau and L. Heux. 2015. Surface-restricted modification of nanocellulose aerogels in gas-phase esterification by di-functional fatty acid reagents. Cellulose 22(3): 1451–1457. http://link.springer.com/10.1007/s10570-015-0585-3.

Gao, C., Y. Wan, F. He, H. Liang, H. Luo and J. Han. 2011. Mechanical, moisture absorption and photodegradation behaviors of bacterial cellulose nanofiber-reinforced unsaturated polyester composites. Advances in Polymer Technology 30(4): 249–256.

Garcia de Rodriguez, N.L., W. Thielemans and A. Dufresne. 2006. Sisal cellulose whiskers reinforced polyvinyl acetate nanocomposites. Cellulose 13(3): 261–270.

Gatenholm, P. and D. Klemm. 2010. Bacterial nanocellulose as a renewable material for biomedical applications. MRS Bulletin 35(3): 208–213. http://www.journals.cambridge.org/abstract_ S0883769400001536.

Gericke, M., J. Trygg and P. Fardim. 2013. Functional cellulose beads: Preparation, characterization, and applications. Chemical Reviews 113(7): 4812–4836.

Habibi, Y., L.A. Lucia and O.J. Rojas. 2010. Cellulose nanocrystals: Chemistry, self-assembly, and applications. Chemical Reviews 110(6): 3479–3500. http://pubs.acs.org/doi/abs/10.1021/ cr900339w.

Habibi, Y. 2014. Key advances in the chemical modification of nanocelluloses. Chemical Society Reviews 43(5): 1519–42.

Håkansson, H. and P. Ahlgren. 2005. Acid hydrolysis of some industrial pulps: Effect of hydrolysis conditions and raw material. Cellulose 12(2): 177–183.

Hamad, W.Y. and T.Q. Hu. 2010. Structure-process-yield interrelations in nanocrystalline cellulose extraction. Canadian Journal of Chemical Engineering 88(3): 392–402.

Hokkanen, S., E. Repo and M. Sillanpää. 2013. Removal of heavy metals from aqueous solutions by succinic anhydride modified mercerized nanocellulose. Chemical Engineering Journal 223: 40–47.

Hu, L., Y. Wan, F. He, H.L. Luo, H. Liang, X. Li and J. Wang. 2009. Effect of coupling treatment on mechanical properties of bacterial cellulose nanofibre-reinforced UPR ecocomposites. Materials Letters 63(22): 1952–1954. http://linkinghub.elsevier.com/retrieve/pii/S0167577X09004479.

Hu, W., S. Chen, Q. Xu and H. Wang. 2011. Solvent-free acetylation of bacterial cellulose under moderate conditions. Carbohydrate Polymers 83(4): 1575–1581. http://dx.doi.org/10.1016/j. carbpol.2010.10.016.

Ifuku, S., M. Nogi, K. Abe, K. Handa, F. Nakatsubo and H. Yano. 2007. Surface modification of bacterial cellulose nanofibers for property enhancement of optically transparent composites: Dependence on acetyl-group DS. Biomacromolecules 8(6): 1973–1978. http://pubs.acs.org/doi/ abs/10.1021/bm070113b.

Ioelovich, M. 2012. Optimal conditions for isolation of nanocrystalline cellulose particles. Nanoscience and Nanotechnology 2(2): 9–13. http://article.sapub.org/10.5923.j.nn.20120202.03.html.

Jonoobi, M., A.P. Mathew, M.M. Abdi, M.D. Makinejad and K. Oksman. 2012. A comparison of modified and unmodified cellulose nanofiber reinforced polylactic acid (PLA) prepared by twin screw extrusion. Journal of Polymers and the Environment 20(4): 991–997.

Joseph, P. 1999. Effect of processing variables on the mechanical properties of sisal-fiber-reinforced polypropylene composites. Composites Science and Technology 59(11): 1625–1640. http:// linkinghub.elsevier.com/retrieve/pii/S0266353899000024X.

Junior de Menezes, A., G. Siqueira, A.A.S. Curvelo and A. Dufresne. 2009. Extrusion and characterization of functionalized cellulose whiskers reinforced polyethylene nanocomposites. Polymer 50(19): 4552–4563. http://dx.doi.org/10.1016/j.polymer.2009.07.038.

Kalia, S., S. Boufi, A. Celli and S. Kango. 2014. Nanofibrillated cellulose: Surface modification and potential applications. Colloid and Polymer Science 292(1): 5–31.

Kanimozhi, K., P. Prabunathan, V. Selvaraj and M. Alagar. 2014. Vinyl silane-functionalized rice husk ash-reinforced unsaturated polyester nanocomposites. RSC Advances 4(35): 18157. http://xlink. rsc.org/?DOI=c4ra01125b.

Kargarzadeh, H., R. Sheltami, I. Ahmad, I. Abdullah and A. Dufresne. 2015a. Cellulose nanocrystal: a promising toughening agent for unsaturated polyester nanocomposite. Polymer 56: 346–357. http://dx.doi.org/10.1016/j.polymer.2014.11.054.

Kargarzadeh, H., R. Sheltami, I. Ahmad, I. Abdullah and A. Dufresne. 2015b. Toughened polyester cellulose nanocomposites: Effects of cellulose nanocrystals and liquid epoxidized natural rubber on morphology and mechanical properties. Industrial Crops and Products 72: 125–132. http:// dx.doi.org/10.1016/j.indcrop.2014.12.052.

Kargarzadeh, H., R.M. Sheltami, I. Ahmad, I. Abdullah and A. Dufresne. 2015c. Cellulose nanocrystal reinforced liquid natural rubber toughened unsaturated polyester: Effects of filler content and surface treatment on its morphological, thermal, mechanical, and viscoelastic properties. Polymer 71: 51–59. http://dx.doi.org/10.1016/j.polymer.2015.06.045.

Kargarzadeh, H., M. Mariano, J. Huang, N. Lin, I. Ahmad, A. Dufresne and S. Thomas. 2017. Recent developments on nanocellulose reinforced polymer nanocomposites: a review. Polymer. http://linkinghub.elsevier.com/retrieve/pii/S0032386117309163.

Kashani R., S., R. Aeinehvand, K. Kim and J.U. Otaigbe. 2017. Structure and biocompatibility of bioabsorbable nanocomposites of aliphatic-aromatic copolyester and cellulose nanocrystals. Biomacromolecules 18(7): 2179–2194.

Ketabchi, M.R., M. Khalid, C.T. Ratnam and R. Walvekar. 2016. Mechanical and thermal properties of polylactic acid composites reinforced with cellulose nanoparticles extracted from kenaf fibre. Materials Research Express 3(12): 125301. http://stacks.iop.org/2053-1591/3/i=12/a=125301?key=crossref.a756daca6be7a0bb0e092703f849c0a3.

Kijchavengkul, T., R. Auras, M. Rubino, E. Alvarado, J.R. Camacho Montero and J.M. Rosales. 2010. Atmospheric and soil degradation of aliphatic–aromatic polyester films. Polymer Degradation and Stability 95(2): 99–107. http://linkinghub.elsevier.com/retrieve/pii/S014139100900411X.

Kim, D., Y. Nishiyama and S. Kuga. 2002. Surface acetylation of bacterial cellulose N. Cellulose 9: 361–367.

Kim, J.H., B.S. Shim, H.S. Kim, Y.J. Lee, S.K. Min, D. Jang, Z. Abas and J. Kim. 2015. Review of nanocellulose for sustainable future materials. International Journal of Precision Engineering and Manufacturing - Green Technology 2(2): 197–213.

Klemm, D., F. Kramer, S. Moritz, T. Lindström, M. Ankerfors, D. Gray and A. Dorris. 2011. Nanocelluloses: a new family of nature-based materials. Angewandte Chemie International Edition 50(24): 5438–5466. http://doi.wiley.com/10.1002/anie.201001273.

Kontturi, K.S., E. Kontturi and J. Laine. 2013. Specific water uptake of thin films from nanofibrillar cellulose. Journal of Materials Chemistry A 1(43): 13655. http://xlink.rsc.org/?DOI=c3ta12998c.

Kovačević, T., J. Rusmirović, N. Tomić, M. Marinović-Cincović, Ž. Kamberović, M. Tomić and A. Marinković. 2017. New composites based on waste PET and non-metallic fraction from waste printed circuit boards: Mechanical and thermal properties. Composites Part B: Engineering 127(15): 1–14. https://doi.org/10.1016/j.compositesb.2017.06.020.

Kulomaa, T., J. Matikainen, P. Karhunen, M. Heikkilä, J. Fiskari and I. Kilpeläinen. 2015. Cellulose fatty acid esters as sustainable film materials—Effect of side chain structure on barrier and mechanical properties. RSC Adv. 5(98): 80702–80708. http://xlink.rsc.org/?DOI=C5RA12671A.

Kushwaha, P. and R. Kumar. 2010. Influence of chemical treatments on the mechanical and water absorption properties of bamboo fiber composites. Journal of Reinforced Plastics and Composites 30(1): 73–85. http://jrp.sagepub.com/cgi/doi/10.1177/0731684410383064.

Lavoine, N., I. Desloges, A. Dufresne and J. Bras. 2012. Microfibrillated cellulose—its barrier properties and applications in cellulosic materials: a review. Carbohydrate Polymers 90(2): 735–764. http://linkinghub.elsevier.com/retrieve/pii/S014486171200447X.

Lavoratti, A., L.C. Scienza and A.J. Zattera. 2016. Dynamic-mechanical and thermomechanical properties of cellulose nanofiber/polyester resin composites. Carbohydrate Polymers 136: 955–963. http://dx.doi.org/10.1016/j.carbpol.2015.10.008.

Lee, K.-Y., J.J. Blaker and A. Bismarck. 2009. Surface functionalisation of bacterial cellulose as the route to produce green polylactide nanocomposites with improved properties. Composites Science and Technology 69(15-16): 2724–2733.

Lee, K.-Y., F. Quero, J. Blaker, C. Hill, S. Eichhorn and A. Bismarck. 2011. Surface only modification of bacterial cellulose nanofibres with organic acids. Cellulose 18: 595–605.

Lee, K.-Y. and A. Bismarck. 2012. Susceptibility of never-dried and freeze-dried bacterial cellulose towards esterification with organic acid. Cellulose 19(3): 891–900.

Lee, K.-Y., Y. Aitomäki, L.A. Berglund, K. Oksman and A. Bismarck. 2014. On the use of nanocellulose as reinforcement in polymer matrix composites. Composites Science and Technology 105: 15–27. http://linkinghub.elsevier.com/retrieve/pii/S0266353814003236.

Li, C., H. Liu, B. Luo, W. Wen, L. He, M. Liu and C. Zhou. 2016. Nanocomposites of poly(l-lactide) and surface-modified chitin whiskers with improved mechanical properties and cytocompatibility. European Polymer Journal 81: 266–283. http://dx.doi.org/10.1016/j.eurpolymj.2016.06.015.

Liebner, F., E. Haimer, M. Wendland, M.A. Neouze, K. Schlufter, P. Miethe, T. Heinze, A. Potthast and T. Rosenau. 2010. Aerogels from unaltered bacterial cellulose: Application of scCO$_2$ drying for the preparation of shaped, ultra-lightweight cellulosic aerogels. Macromolecular Bioscience 10(4): 349–352.

Lin, N., J. Huang, P.R. Chang, J. Feng and J. Yu. 2011. Surface acetylation of cellulose nanocrystal and its reinforcing function in poly(lactic acid). Carbohydrate Polymers 83(4): 1834–1842. http://dx.doi.org/10.1016/j.carbpol.2010.10.047.

Lin, N., J. Huang and A. Dufresne. 2012. Preparation, properties and applications of polysaccharide nanocrystals in advanced functional nanomaterials: a review. Nanoscale 4(11): 3274–94.

Liu, D., S. Wang, Z. Ma, D. Tian, M. Gu and F. Lin. 2014. Structure–color mechanism of iridescent cellulose nanocrystal films. RSC Adv. 4(74): 39322–39331.

Liu, D., Y. Dong, D. Bhattacharyya and G. Sui. 2017. Novel sandwiched structures in starch/cellulose nanowhiskers (CNWs) composite films. Composites Communications 4: 5–9. http://linkinghub.elsevier.com/retrieve/pii/S2452213916300651.

Lu, P. and Y. Lo Hsieh. 2012. Preparation and characterization of cellulose nanocrystals from rice straw. Carbohydrate Polymers 87(1): 564–573.

Lu, Y., M.C. Cueva, E. Lara-Curzio and S. Ozcan. 2015. Improved mechanical properties of polylactide nanocomposites-reinforced with cellulose nanofibrils through interfacial engineering via amine-functionalization. Carbohydrate Polymers 131: 208–217. http://dx.doi.org/10.1016/j.carbpol.2015.05.047.

Lu, Z., L. Fan, H. Zheng, Q. Lu, Y. Liao and B. Huang. 2013. Preparation, characterization and optimization of nanocellulose whiskers by simultaneously ultrasonic wave and microwave assisted. Bioresource Technology 146: 82–88.

Ly, B., W. Thielemans, A. Dufresne, D. Chaussy and M.N. Belgacem. 2008. Surface functionalization of cellulose fibres and their incorporation in renewable polymeric matrices. Composites Science and Technology 68(15-16): 3193–3201.

Mandal, S. and S. Alam. 2012. Dynamic mechanical analysis and morphological studies of glass/bamboo fiber reinforced unsaturated polyester resin-based hybrid composites. Journal of Applied Polymer Science 125(S1): E382–E387. http://doi.wiley.com/10.1002/app.36304.

Mondal, S. 2017. Review on nanocellulose polymer nanocomposites. Polymer-Plastics Technology and Engineering 1–15. https://www.tandfonline.com/doi/full/10.1080/03602559.2017.1381253.

Moon, R.J., A. Martini, J. Nairn, J. Simonsen and J. Youngblood. 2011. Cellulose nanomaterials review: Structure, properties and nanocomposites. Chemical Society Reviews 40(7): 3941. http://xlink.rsc.org/?DOI=c0cs00108b.

Morais, J.P.S., M.D.F. Rosa, M.D.S.M. de Souza Filho, L.D. Nascimento, D.M. do Nascimento and A.R. Cassales. 2013. Extraction and characterization of nanocellulose structures from raw cotton linter. Carbohydrate Polymers 91(1): 229–35.

Nechita, P. and D.M. Panaitescu. 2013. Improving the dispersibility of cellulose microfibrillated structures in polymer matrix by controlling drying conditions and chemical surface modifications. Cellulose Chemistry and Technology 47(910): 711–719.

Ng, H.-M., L.T. Sin, S.-T. Bee, T.-T. Tee and A.R. Rahmat. 2017. Review of nanocellulose polymer composite characteristics and challenges. Polymer-Plastics Technology and Engineering 56(7): 687–731. https://www.tandfonline.com/doi/full/10.1080/03602559.2016.1233277.

Nogi, M., K. Abe, K. Handa, F. Nakatsubo, S. Ifuku and H. Yano. 2006. Property enhancement of optically transparent bionanofiber composites by acetylation. Applied Physics Letters 89(23).

Ober, C.K., S.Z.D. Cheng, P.T. Hammond, M. Muthukumar, E. Reichmanis, K.L. Wooley and T.P. Lodge. 2009. Research in macromolecular science: Challenges and opportunities for the next decade. Macromolecules 42(2): 465–471. http://pubs.acs.org/doi/abs/10.1021/ma802463z.

Ojijo, V. and S. Sinha Ray. 2013. Processing strategies in bionanocomposites. Progress in Polymer Science 38(10-11): 1543–1589. http://linkinghub.elsevier.com/retrieve/pii/S007967001300052X.

Oksman, K., A.P. Mathew, D. Bondeson and I. Kvien. 2006. Manufacturing process of cellulose whiskers/polylactic acid nanocomposites. Composites Science and Technology 66(15): 2776–2784.

de Oliveira Taipina, M., M.M.F. Ferrarezi, I.V.P. Yoshida and M.C. Gonçalves. 2013. Surface modification of cotton nanocrystals with a silane agent. Cellulose 20(1): 217–226.

Paiva, J.M.F. de, S. Mayer and M.C. Rezende. 2006. Comparison of tensile strength of different carbon fabric reinforced epoxy composites. Materials Research 9(1): 83–90. http://www.scielo.br/scielo. php?script=sci_arttext&pid=S1516-14392006000100016&lng=en&tlng=en.

Panaitescu, D.M., A.N. Frone and I. Chiulan. 2016. Nanostructured biocomposites from aliphatic polyesters and bacterial cellulose. Industrial Crops and Products 93: 251–266.

Pasquini, D., E. de M. Teixeira, A.A. da S. Curvelo, M.N. Belgacem and A. Dufresne. 2010. Extraction of cellulose whiskers from cassava bagasse and their applications as reinforcing agent in natural rubber. Industrial Crops and Products 32(3): 486–490.

Peng, Y., D.J. Gardner and Y. Han. 2012. Drying cellulose nanofibrils: in search of a suitable method. Cellulose 19(1): 91–102. http://link.springer.com/10.1007/s10570-011-9630-z.

Peng, Y., D.J. Gardner, Y. Han, Z. Cai and M.A. Tshabalala. 2013a. Drying cellulose nanocrystal Suspensions. In: Postek, M.T., R.J. Moon, A.W. Rudie and M.A. Bilodeau (eds.). Production and Applications of Cellulose Nanomaterials, 321. TAPPI PRESS 15 Technology Parkway South Suite 115 Peachtree Corners, GA 30092 U.S.A.

Peng, Y., D.J. Gardner, Y. Han, Z. Cai and M.A. Tshabalala. 2013b. Influence of drying method on the surface energy of cellulose nanofibrils determined by inverse gas chromatography. Journal of Colloid and Interface Science 405: 85–95. http://dx.doi.org/10.1016/j.jcis.2013.05.033.

Peng, Y., D.J. Gardner, Y. Han, A. Kiziltas, Z. Cai and M.A. Tshabalala. 2013. Influence of drying method on the material properties of nanocellulose I: Thermostability and crystallinity. Cellulose 20(5): 2379–2392. http://link.springer.com/10.1007/s10570-013-0019-z.

Podsiadlo, P., S.-Y. Choi, B. Shim, J. Lee, M. Cuddihy and N.A. Kotov. 2005. Molecularly engineered nanocomposites: Layer-by-layer assembly of cellulose nanocrystals. Biomacromolecules 6(6): 2914–2918. http://pubs.acs.org/doi/abs/10.1021/bm050333u.

Pöllänen, M., S. Pirinen, M. Suvanto and T.T. Pakkanen. 2011. Influence of carbon nanotube–polymeric compatibilizer masterbatches on morphological, thermal, mechanical and tribological properties of polyethylene. Composites Science and Technology 71(10): 1353–1360. http://linkinghub.elsevier. com/retrieve/pii/S0266353811001710.

Popescu, M.C., M. Totolin, C.M. Tibirna, A. Sdrobis, T. Stevanovic and C. Vasile. 2011. Grafting of softwood kraft pulps fibers with fatty acids under cold plasma conditions. International Journal of Biological Macromolecules 48(2): 326–335.

Pu, Y., J. Zhang, T. Elder, Y. Deng, P. Gatenholm and A.J. Ragauskas. 2007. Investigation into nanocellulosics versus acacia reinforced acrylic films. Composites Part B: Engineering 38(3): 360–366.

Qu, P., Y. Zhou, X. Zhang, S. Yao and L. Zhang. 2012. Surface modification of cellulose nanofibrils for poly(lactic acid) composite application. Journal of Applied Polymer Science 125(4): 3084–3091. http://doi.wiley.com/10.1002/app.36360.

Rančić, M., J. Rusmirović, I. Popović and A. Marinković. 2015. Isolation and chemical modification of nanocellulose nanocrystals for reinforcement of nanocomposites. pp. 327–355. In: Wood Technology & Products Design, 2nd International Scientific Conference. Ohrid, Macedonia.

Revol, J.F., H. Bradford, J. Giasson, R.H. Marchessault and D.G. Gray. 1992. Helicoidal self ordering of cellulose microfibrils in aqueous suspension. Int. J. Biol. Macromol. 14: 170.

Rodionova, G., M. Lenes, Ø. Eriksen and Ø. Gregersen. 2011. Surface chemical modification of microfibrillated cellulose: Improvement of barrier properties for packaging applications. Cellulose 18(1): 127–134. http://link.springer.com/10.1007/s10570-010-9474-y.

Roohani, M., Y. Habibi, N.M. Belgacem, G. Ebrahim, A.N. Karimi and A. Dufresne. 2008. Cellulose whiskers reinforced polyvinyl alcohol copolymers nanocomposites. European Polymer Journal 44(8): 2489–2498.

Rusmirović, J., J. Ivanović, V. Pavlović, V. Rakić, M. Rančić, V. Djokić and A. Marinković. 2017. Novel modified nanocellulose applicable as reinforcement in high-performance nanocomposites. Carbohydrate Polymers 164: 64–74. http://linkinghub.elsevier.com/retrieve/ pii/S0144861717300978.

Rusmirović, J., T. Radoman, E.S. Džunuzović, J.V. Džunuzović, J. Markovski, P. Spasojević and A.D. Marinković. 2017. Effect of the modified silica nanofiller on the mechanical properties of unsaturated polyester resins based on recycled polyethylene terephthalate. Polymer Composites 38(3): 538–554. http://doi.wiley.com/10.1002/pc.23613.

Rusmirović, J., K. Trifković, B. Bugarski, V. Pavlović, J. Džunuzović, M. Tomić and A. Marinković. 2016. High performance unsaturated polyester based nanocomposites: Effect of vinyl modified nanosilica on mechanical properties. Express Polymer Letters 10(2).

Saha, A.K., S. Das, D. Bhatta and B.C. Mitra. 1999. Study of jute fiber reinforced polyester composites by dynamic mechanical analysis. Journal of Applied Polymer Science 71(9): 1505–1513. http://doi.wiley.com/10.1002/%28SICI%291097-4628%2819990228%2971%3A9%3C1505%3A%3AAID-APP15%3E3.0.CO%3B2-1.

Shi, X.Q., H. Ito and T. Kikutani. 2005. Characterization on mixed-crystal structure and properties of poly(butylene adipate-co-terephthalate) biodegradable fibers. Polymer 46(25): 11442–11450. http://linkinghub.elsevier.com/retrieve/pii/S0032386105015338.

Siró, I. and D. Plackett. 2010. Microfibrillated cellulose and new nanocomposite materials: a review. Cellulose 17(3): 459–494. http://link.springer.com/10.1007/s10570-010-9405-y.

Sobkowicz, M.J., B. Braun and J.R. Dorgan. 2009. Decorating in green: Surface esterification of carbon and cellulosic nanoparticles. Green Chemistry 11(5): 680. http://xlink.rsc.org/?DOI=b817223d.

Tang, L., B. Huang, Q. Lu, S. Wang, W. Ou, W. Lin and X. Chen. 2013. Bioresource technology ultrasonication-assisted manufacture of cellulose nanocrystals esterified with acetic acid. Bioresource Technology 127: 100–105.

Tasić, A., J. Rusmirović, J. Nikolić, A. Božić, V. Pavlović, A. Marinković and P. Uskoković. 2017. Effect of the vinyl modification of multi-walled carbon nanotubes on the performances of waste poly(ethylene terephthalate)-based nanocomposites. Journal of Composite Materials 51(4): 491–505. https://doi.org/10.1177/0021998316648757.

Ten, E., J. Turtle, D. Bahr, L. Jiang and M. Wolcott. 2010. Thermal and mechanical properties of poly(3-Hydroxybutyrate-co-3-hydroxyvalerate)/cellulose nanowhiskers composites. Polymer 51(12): 2652–2660. http://dx.doi.org/10.1016/j.polymer.2010.04.007.

Ten, E., L. Jiang and M.P. Wolcott. 2012. Crystallization kinetics of poly(3-hydroxybutyrate-co-3-hydroxyvalerate)/cellulose nanowhiskers composites. Carbohydrate Polymers 90(1): 541–550. http://dx.doi.org/10.1016/j.carbpol.2012.05.076.

Thakur, V.K. 2014. Nanocellulose Polymer Nanocomposites: Fundamentals and Applications. Vijay Kumar Thakur [eds.], John Wiley & Sons, USA.

Thomas, R., D. Yumei, H. Yuelong, Y. Le, P. Moldenaers, Y. Weimin, T. Czigany and S. Thomas. 2008. Miscibility, morphology, thermal, and mechanical properties of a DGEBA based epoxy resin toughened with a liquid rubber. Polymer 49(1): 278–294. http://linkinghub.elsevier.com/retrieve/pii/S0032386107010658.

Tingaut, P., T. Zimmermann and F. Lopez-Suevos. 2009. Synthesis and characterization of bionanocomposites with tunable properties from poly(lactic acid) and acetylated microfibrillated Cellulose 454–464.

Uschanov, P., L.S. Johansson, S.L. Maunu and J. Laine. 2011. Heterogeneous modification of various celluloses with fatty acids. Cellulose 18(2): 393–404.

van de Ven, T.G.M. and A. Sheikhi. 2016. Hairy cellulose nanocrystalloids: a novel class of nanocellulose. Nanoscale 8(33): 15101–15114. http://xlink.rsc.org/?DOI=C6NR01570K.

Viet, D., S. Beck-Candanedo and D.G. Gray. 2006. Dispersion of cellulose nanocrystals in polar organic solvents. Cellulose 14(2): 109–113.

Visakh, P.M., S. Thomas, K. Oksman and A.P. Mathew. 2012. Crosslinked natural rubber nanocomposites reinforced with cellulose whiskers isolated from bamboo waste: Processing and mechanical/thermal properties. Composites Part A: Applied Science and Manufacturing 43(4): 735–741. http://linkinghub.elsevier.com/retrieve/pii/S1359835X11004209.

Voronova, M.I., A.G. Zakharov, O.Y. Kuznetsov and O.V. Surov. 2012. The effect of drying technique of nanocellulose dispersions on properties of dried materials. Materials Letters 68: 164–167. http://linkinghub.elsevier.com/retrieve/pii/S0167577X11011980.

Wang, K., L. Chen, J. Wu, M.L. Toh, C. He and A.F. Yee. 2005. Epoxy nanocomposites with highly exfoliated clay: Mechanical properties and fracture mechanisms. Macromolecules 38(3): 788–800. http://pubs.acs.org/doi/abs/10.1021/ma048465n.

Wang, N., E. Ding and R. Cheng. 2007. Surface modification of cellulose nanocrystals. Frontiers of Chemical Engineering in China 1(3): 228–232. http://link.springer.com/10.1007/s11705-007-0041-5.

Wei, H., K. Rodriguez, S. Renneckar and P.J. Vikesland. 2014. Environmental science and engineering applications of nanocellulose-based nanocomposites. Environmental Science: Nano 1(4): 302–316. http://xlink.rsc.org/?DOI=C4EN00059E.

Wei, L., U.P. Agarwal, K.C. Hirth, L.M. Matuana, R.C. Sabo and N.M. Stark. 2017. Chemical modification of nanocellulose with canola oil fatty acid methyl ester. Carbohydrate Polymers 169: 108–116. http://linkinghub.elsevier.com/retrieve/pii/S0144861717303818.

Wenjuan, Z. and S. Zhihua. 2011. The preparation of nano cellulose whiskers/polylactic acid composites. pp. 123–125. *In*: 2011 International Conference on Future Computer Science and Education, IEEE. http://ieeexplore.ieee.org/document/6041675/.

Wollerdorfer, M. and B. Herbert. 1998. Influence of natural fibres on the mechanical properties of biodegradable polymers. Industrial Crops and Products 8: 105–12. http://linkinghub.elsevier.com/retrieve/pii/S1359835X1200098X%5Cnhttp://linkinghub.elsevier.com/retrieve/pii/S0926669012000623%5Cnhttp://jcm.sagepub.com/cgi/doi/10.1177/0021998313491515.

Yoo, Y. and J.P. Youngblood. 2016. Green one-pot synthesis of surface hydrophobized cellulose nanocrystals in aqueous medium. ASC Sustainable Chemistry 4: 3927–3938.

Yousefian, H. and D. Rodrigue. 2015. Nano-crystalline cellulose, chemical blowing agent, and mold temperature effect on morphological, physical/mechanical properties of polypropylene. Journal of Applied Polymer Science 132(47): 1–9. http://doi.wiley.com/10.1002/app.42845.

Yu, H.-Y., R. Chen, G.-Y. Chen, L. Liu, X.-G. Yang and J.-M. Yao. 2015. Silylation of cellulose nanocrystals and their reinforcement of commercial silicone rubber. Journal of Nanoparticle Research 17(9): 361. http://link.springer.com/10.1007/s11051-015-3165-4.

Yu, M. 2000. Strength and breaking mechanism of multiwalled carbon nanotubes under tensile load. Science 287(5453): 637–640. http://www.sciencemag.org/cgi/doi/10.1126/science.287.5453.637.

Yu, X., S. Tong, M. Ge, L. Wu, J. Zuo, C. Cao and W. Song. 2013. Adsorption of heavy metal ions from aqueous solution by carboxylated cellulose nanocrystals. Journal of Environmental Sciences 25(5): 933–943.

Yuan, H., Y. Nishiyama, M. Wada and S. Kuga. 2006. Surface acylation of cellulose whiskers by drying aqueous emulsion. Biomacromolecules 7(3): 696–700.

Yuwawech, K., J. Wootthikanokkhan, S. Wanwong and S. Tanpichai. 2017. Special issue: Green and polyurethane/esterified cellulose nanocrystal composites as a transparent moisture barrier coating for encapsulation of dye sensitized solar cells. Journal of Applied Polymer Science. 1–12.

Zadorecki, P. and P. Flodin. 1985. Surface modification of cellulose fibers I: Spectroscopic characterization of modified-cellulose fibers and their copolymerization with stirene. Journal of Applied Polymer Science 30: 2419–2429.

Zainuddin, S.Y.Z., I. Ahmad, H. Kargarzadeh, I. Abdullah and A. Dufresne. 2013. Potential of using multiscale kenaf fibers as reinforcing filler in cassava starch-kenaf biocomposites. Carbohydrate Polymers 92(2): 2299–2305.

Žepič, V. 2015. Effect of drying pretreatment on the acetylation of nanofibrillated cellulose. Bioresources 10(4): 8148–8167.

Žepič, V., E.Š. Fabjan, M. Kasunič, R.C. Korošec, A. Hančič and P. Oven. 2014. Morphological, Thermal and structural aspects of dried and redispersed nanofibrillated cellulose (NFC) Holzforschung 68(6): 657–667.

Zhang, X., P. Ma and Y. Zhang. 2016. Structure and properties of surface-acetylated cellulose nanocrystal/poly(butylene adipate-co-terephthalate) composites. Polymer Bulletin 73(7): 2073–2085. http://link.springer.com/10.1007/s00289-015-1594-y.

Zhang, X. and Y. Zhang. 2016. Reinforcement effect of poly(butylene succinate) (PBS)-grafted cellulose nanocrystal on toughened PBS/polylactic acid blends. Carbohydrate Polymers 140: 374–382. http://dx.doi.org/10.1016/j.carbpol.2015.12.073.

Zimmermann, M.V., C. Borsoi, A. Lavoratti, M. Zanini, A.J. Zattera and R.M. Santana. 2016. Drying techniques applied to cellulose nanofibers. Journal of Reinforced Plastics and Composites 35(8): 628–643. http://journals.sagepub.com/doi/10.1177/0731684415626286.

Organosolv Processes: New Opportunities for Development of High Value Products from Lignins

Georges Koumba Yoya and *Tatjana Stevanovic**

Introduction

Due to its rich carbon content, biomass constitutes an attractive alternative to petrol, both for fuel and chemicals production. In fact, the current depletion of fossil fuels imposes the diversification of sources and the development of alternative carbon content to produce energy and chemicals. Biomass fractionation technology consists of primary separation of cellulose hemicelluloses and lignin. While the transformation technologies regarding cellulose and hemicelluloses are already well advanced, lignin transformation into value-added products remains at the research and development level. The emergence of several biomass fractionation processes in the last century, as evidenced by the extensive research results confirm the renewed interest in lignins. The organosolv processes appear as the most promising fractionation technologies since they produce high grade lignins. Organosolv lignins are produced from processes involving at least an organic solvent to solubilize the lignin, while leaving the cellulose and some of the hemicelluloses in the solid pulp. Initially, the use of organic solvents and chemicals was intended to separate

Department of Wood Sciences and Forestry, Université Laval, Québec, Canada.
* Corresponding author: tatjana.stevanovic@sbf.ulaval.ca

components of wood in order to study the lignin and monosaccharides. More recently, the search for solutions to improve the quality of the cellulosic pulp, led many industries to develop new fractionation methods using organosolv process to produce pulp with qualities and some properties close to those of kraft pulp. On the other hand, one of organosolv's advantages compared to other pulping methods is the ability to obtain relatively high quality lignin in combination with a yield 4.5% higher than those recovered by kraft pulping (Pye et al. 1991). In this chapter we shall focus on the recent advances in organosolv fractionation of biomass. Indeed, several organosolv processes have explored various organic solvents in the past, as is the case with methanol in organocell process (Linder et al. 1990) or with methanol in combination with anthraquinone and alkaline sulphite in ASAM process (Kordsachia et al. 1988), ethanol in alcell (Lora et al. 2002) or lignol process, in which ethanol is applied with an acid catalyst (Pan et al. 2005), or those involving organic acids such as acetic acid on acetosolv process (Lehnen et al. 2001) or from a mixture of formic acid and hydrogen peroxide applied to transformation of wheat straw in CMIV process (Avignon et al. 2000).

Lignin Overview

Present in all vascular plants, trees and herbaceous plants, lignin biosynthesis of lignins is based on a pool of hydroxylcinnamyl alcohols which differ in degrees of methoxylation. The most abundant structural units contained in lignins derive from radical coupling of coumaryl, coniferyl and sinapyl alcohols, as presented in Fig. 1.

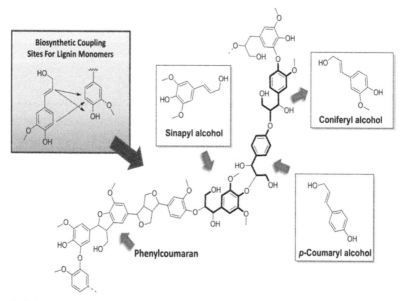

Fig. 1. Schematic representation of lignin with an example of possible biosynthetic paths (Graglia et al. 2015). © 2015, Wiley-VCH Verlag GmbH & Co. KGaA, Weinheim.

The cinnamic alcohols also named monolignols, are biosynthetized during a cascade of enzymatic reactions involving the L-phenylanine as basic phenyl-propane unit. Lignin represents between 15 and 35% of the dry masst of woods and between 5 and 15% of herbaceous plants. The amorphous nature and heterogeneous structure, the lack of well-defined primary structure, define lignins as very complex aromatic polymers, consisting of phenolic unit types, which represent about 30% of the organic carbon in the earth. Other polyphenolic polymers, such as lipid polyesters constituting suberin (mainly present in tree bark), have long been assimilated to lignin, but more recently, it has been determined that the hydroxycinnamic acides are linked to one another covalently, with a certain analogy with the monolignols of lignin (Garça et al. 2010).

From a structural point of view, lignins are composed of different structural units among which the most common types are β-O-4, α-O-4, 4-O-5, β-1, β-β, β-5 and 5-5 (Fig. 2).

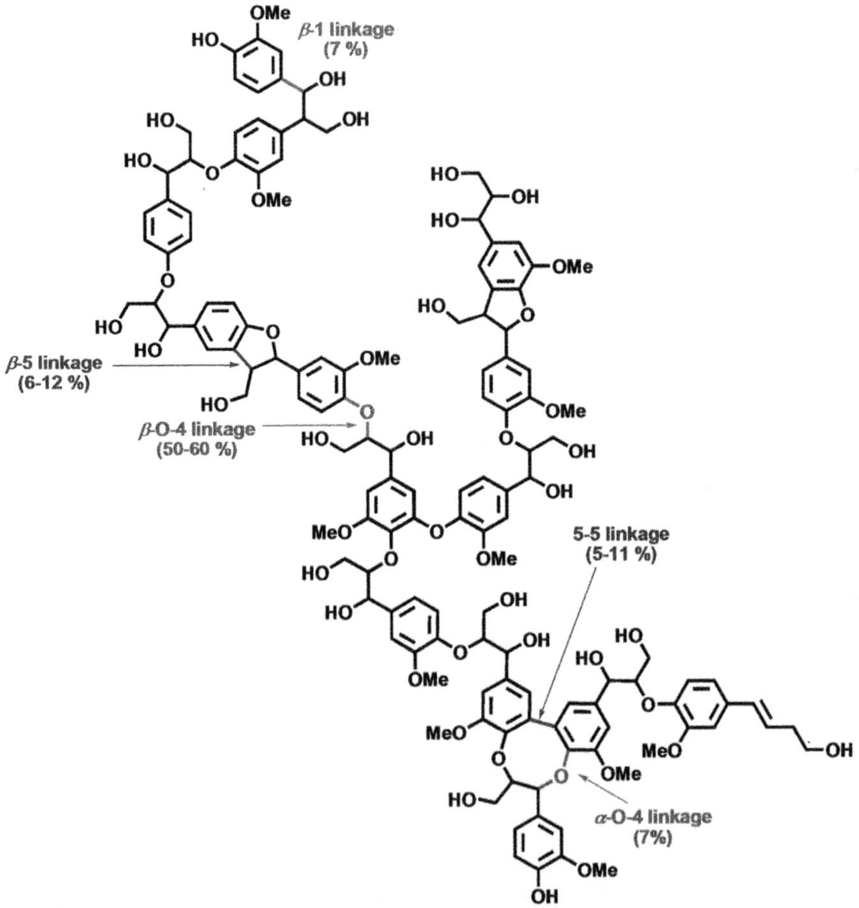

Fig. 2. Proposed lignin structure (Adapted from Crestini et al. 2011).

Thus, the proportions of the major linkages of the subunits predominantly identified in lignin are listed in Table 1 (Adler et al. 1977).

Lignins are covalently bonded to hemicelluloses, thus contributing to the formation of matrix fixing the polysaccharides in the cell walls and thus giving them strength and rigidity. Lignins are also the main constituents of middle lamellae, gluing the cells in wood and attributing the compression strength to the wood indispensable for the extraordinary growth in height of forest trees. By its polyphenolic nature lignins also play roles in protecting the plants against outside biological attacks and by their hydrophobic nature to fine regulation of water transport in plants. Lignins are attributed antimicrobial and antifungal properties and their phenolic hydroxyls are responsible for antioxidant or UV absorber when applied in polymers along with the fire retardant properties.

Table 1. Different lignin linkages and their relative abundance.

Type of linkage	Dimer	(%)	
		Softwood	Hardwood
β-O-4	Arylglycerol-β-aryl ether	50	60
α-O-4	Noncyclic benzyl aryl ether	2–8	7
β-5	Phenylcoumaran	9–12	6
5-5	Biphenyl	10–11	5
4-O-5	Diaryl ether	4	7
β-1	1,2-diaryl propane	7	7
β-β	Pinoresinol	2	3

Organosolv processes with alcohols

In the recent decades, alcohols have been extensively used in organosolv processes thus representing the most important organic solvents applied for biomass fractionation.

Methanol organosolv processes

Initially used in combination with sulfite or kraft process for improvement of cellulosic pulp quality, methanol is central to the ASAM (Alkaline Sulfite Anthraquinone Methanol) process, developed at the Federal Forest and Wood Research Center in Hamburg (Kordsachia et al. 1988). Organocell process used the same conditions and developed for implementing in an urban site. Indeed, because of its low boiling point, the use of methanol would facilitate the recovery of the solvent from the black liquor by distillation. Thus, by adding methanol and anthraquinone in alkaline sulfite (ASAM process), cellulose pulp was obtained with the advantage to become more easily bleached. Choi et al. (2001) studied the main properties of these lignins by Py-GC/MS (Choi et al. 2001). From this study, it follows that lignin extracted according to ASAM process, contains between 0.8 and 12% of polysaccharide in addition to a significant protein content that contributes to increase artificially the Klason lignin due to condensation on the phenol during acid hydrolysis. The organocell process involves some conditions of cooking soda where sodium hydroxide was add in methanol for delignification process. In the process of ASAM, wood chips were firstly impregnated between 110 and 140°C with a mixture of alcohol and water. According to Lindner and coworkers, the main idea of such a process, would be correlated to the fact that methanol has been active, confined to a methylation of aliphatic hydroxyl of lignin in benzylic positions for preventing the formation of quinone methides. This methylation allows reducing the reactions of condensation and thus blocking the release of formaldehyde (Lindner et al. 1990). Also, in this context, other methods use aqueous methanol to more than 80% of alcohol in the presence of acids hydrochloric, sulphuric, phosphoric or calcium chloride for improvement of the organosolv process. For example, with sulfuric acid (Gandolfi et al. 2014), methanol organosolv process was performed with 45–65% of MeOH using hemp as biomass source at 165°C, with 1–3% H_2SO4 as catalyst for 20–90 min of reaction time. Indeed, prior to organosolv process, the authors removed extractives in hemp with CH_2Cl_2 and acetone by soxhlet extaction, as summarized in Fig. 3.

Indeed Gandolfi and co-workers have performed the chemical characterization by FTIR and 2D 13C–1H correlation HSQC NMR spectroscopy. Results of this organosolv fractionation suggested that xylopyranose is the major carbohydrate associated with hemp lignin. Hemp organosolv lignin was examined for its antioxidant capacity, expressing the results in terms of gallic acid equivalent (GAE) and trolox equivalent antioxidant capacity (TEAC). This study revealed high values of GAE and TEAC of the hemp organosolv lignin compared to Klason lignin. However, even though there is a general consensus that short-chained alcohols are

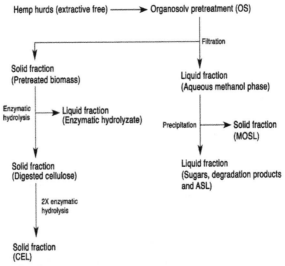

Fig. 3. Experimental methodology (Gandolfi et al. 2014). © 2014 Wiley-VCH Verlag GmbH & Co. KGaA, Weinheim.

favourable for biomass fractionation, the use of ethanol would still be preferable over methanol because of its much lower toxicity (Viell et al. 2013).

Ethanol organosolv processes

Ethanol is one of the most frequently used solvents in the organosolv processes due to its strong affinity for lignins which can be liquefied more effectively. Ethanol fractionation of lignocellulosic material can be carried out with or without the addition of catalysts. Indeed, it has been shown that the azeotropes formed between ethanol and water can promote an autohydrolysis process of the biomass without catalyst, as is the case with the Alcell process (Pye et al. 1987). Thus, some low levels of lignin depolymerization can be achieved without massive cleavage of α- and β-ether bonds, thus releasing lignin fragments of high molecular weight. Alcell process, initially named APR (Alcohol Pulping and Recovery) process, uses the ethanol-water system as solvent and exploits the acid generation from wood during organosolv pulping, acting as catalyst which promots delignification in an autohydrolysis process. In the Alcell process, several steps involving temperatures ranging between 190 and 210°C are necessary to delignify biomass for during pulping ranging between 5 and 60 minutes, depending on the raw material (Pye et al. 1987). The method proposed by Kleinert et al. (1971) involved a phase of pre-treatment with ethanol at temperatures between 100 and 110°C followed by a delignification at 180°C for 30 minutes using poplar as biomass source. An example of a more recent application of a version of Alcell process is the study of Garcia and co-workers using a mixture of ethanol-water as solvent (60:40 w/w) in a pressurized reactor at 160°C (Fig. 4) (Garcia et al. 2011).

In this process, the ethanol organosolv fractionation was carried out by application of ultra-filtration for lignin recovery. According to these authors, in this biorefinery approach when ethanol pulping is compared with soda process, the more efficient fractionation of biomass is achieved with organosolv process which allowed for the complete utilization of the raw material for high value products, while allowing for the recovery of the solvents, however, with a higher energy consumption (Garcia et al. 2011). Other organosolv processes with ethanol involve acids as catalysts, as is the case with Lignol process with sulfuric acid development and described by Pan and his co-workers (Pan et al. 2006) claiming an increase of delignification extent (Fig. 5).

Briefly, in Lignol process, hybrid poplar chips were cooked in aqueous ethanol, with sulfuric acid as the catalyst. After cooking, the pulp and liquor were separated and the spent liquor (aqueous ethanol extract) was sampled immediately for determination of furfural and HMF. The pulp was washed with aqueous ethanol and the spent liquor and ethanol washes were combined and mixed with three volumes of water to precipitate the dissolved lignin. The filtrate and water washes were combined to give a water-soluble fraction containing monomeric and oligomeric hemicelluloses-derived sugars, depolymerized lignin, and other unidentified

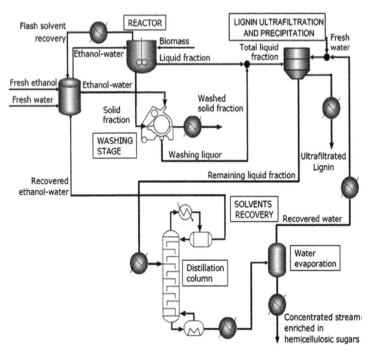

Fig. 4. Diagram of the organosolv-ethanol biorefinery process. Reprinted with permission from A. Garcia et al. (2011).

components. A variant of this process, using hydrochloric acid as catalyst during ethanol organosolv process, has been described (Akihiro et al. 2013). Typically, 0.4% of HCl was used as catalyst in the ethanol organosolv treatment. The reaction mixture solvents consisted of ethanol (50%) and water (50%), applied at 170°C for 45 minutes. In this process, the authors claimed a high glucose yield (60%) obtained by the enzymatic hydrolysis of the pulp obtained by this organosolv treatment. On the other hand, the processes performed in alkaline media usually involve use of ethanol in the presence of anthraquinone, an equivalent with ethanol for the one described for methanol, ASAE for alkali-sulphite-anthraquinone-ethanol (Kirci et al. 1994). Indeed, Kirci and co-workers have chosen *Pinus brutia* Ten. as raw material for organosolv pulping at 175°C for 30 minutes by using NaOH/Na$_2$SO$_3$ as catalyst in combination with anthraquinone in aqueous ethanol at 50%. Before organosolv cooking, an impregnation step at 115°C for 60 minutes is required. The main advantage of this approach is claimed to be the easier bleaching of ASAE pulps, requiring less bleaching agents. The ASAE pulps were demonstrated to be suitable for all common applications of sulphite pulps. More recently, a comparative study was carried out to evaluate the performance of the basic catalysts (NaOH) and acids (H$_2$SO$_4$) in aqueous ethanol (Mesa et al. 2010).

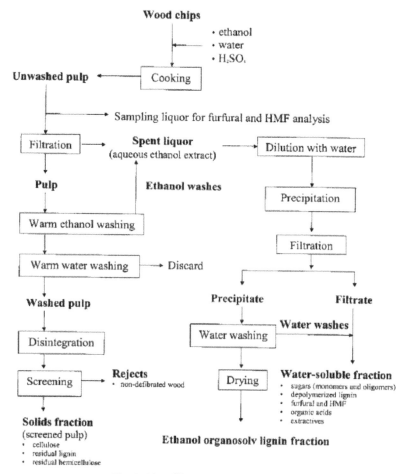

Fig. 5. Lignol™ process (Pan et al. 2006).

New Biorefinery Strategy for High Purity Lignin Production

We have recently patented the results of our studies on catalytic organosolv pulping of trembling aspen wood with ethanol water, with a Lewis acid catalyst (Stevanovic and Koumba 2016). The fractionation of the trembling aspen by that process along with the characterization of organosolv lignin and carbohydrates issued from the organosolv process (Koumba-Yoya et al. 2016).

Extractives removal

In organosolv process prior to the pulping step, the aqueous ethanol is applied in a pretreatment aimed to extractives removal. The analysis of the extractives obtained from this pre-treatment revealed a high yield of phenolic compounds, along with their glycosides and simple sugars. These extractives constituents could be valorized as antioxidants and other reactive compounds in biological system. In fact, the mains goal of this pre-extraction step is to minimize the condensation reactions between lignin and phenolic extractives during pulping. Several polyphenols, along with their glycosides and simple sugars were identified by GC-MS as the extractives constituents (Koumba-Yoya et al. 2016). After extractives removal, the ethanol can be recovered and recycled back in the organosolv process. Thus, this new organosolv process can be summarized as presented in Fig. 6.

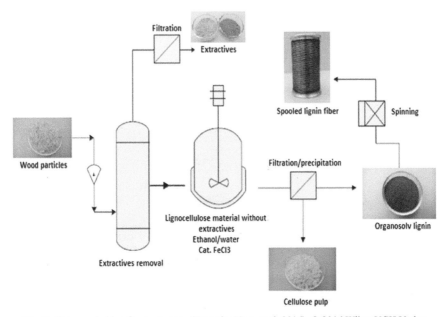

Fig. 6. Organosolv biorefinery strategy (Koumba-Yoya et al. 2016). © 2016 Wiley-VCH Verlag GmbH & Co. KGaA, Weinheim.

Lignin characterization

Lignin obtained by precipitation from spent liquor from organosolv pulping performed at 180°C without catalyst (Lignin-W), with ferric chloride (lignin-Fe) and sulfuric acid (Lignin-SA) as catalysts, were studied by physico-chemical properties for comparison (Table 2).

In order to investigate the catalyst effect on the extent of lignin degradation during the delignification process, molecular weight distributions of the organosolv lignin samples recovered were analyzed by GPC, and the results are presented in Table 3.

As shown in Table 3, the studied lignin samples are determined to have similar molecular weight distribution patterns (Mw between 1200 g/mol and 1900 g/mol). The highest polydispersity index was determined for the lignin-SA which was

Table 2. Physico-chemical indices of lignin samples.

Entry[a]	Lignin yield[b] (%)	Lignin recovery (%)	KL	ASL	Glucan	Xylan	Ash
1	15.0	69.1	89.8 ± 1.1	3.1 ± 0.1	0.45 ± 0.6	0.51 ± 0.1	0.19 ± 0.06
2	16.8	77.4	96.4 ± 0.1	2.7 ± 0.1	N.D.	N.D.	0.14 ± 0.02
3	17.1	78.8	94.4 ± 0.9	3.8 ± 0.4	0.6 ± 0.80	N.D.	0.24 ± 0.02
4	17.9	82.5	94.8 ± 0.4	3.1 ± 0.2	1.84 ± 0.12	N.D.	0.24 ± 0.02
5	15.2	70.0	93.5 ± 2.1	2.8 ± 0.1	N.D.	N.D.	0.62 ± 0.04

[a]Conditions: 1: wood delignification without catalyst; 2: Extractives-free Wood delignification with $FeCl_3$ as catalyst (3 mmol)); 3: with $FeCl_3$ as catalyst (6 mmol); 4: with $FeCl_3$ as catalyst (9 mmol); 5: with H_2SO_4 as catalyst (3 mmol).
[b]Based on oven dry mass.
KL: Klason lignin; ASL: Acid soluble lignin
N.D.: Not Detected

Table 3. Gel Permeation analysis results for lignin samples obtained after pulping at 180°C without or with different catalysts.

Entry	GPC	Lignin-W	Lignin-SA	Lignin-Fe
1	Mn (g/mol)	679	895	956
2	Mw (g/mol)	1296	1825	1642
3	PD	1.90	2.04	1.71

obtained from pulping conditions designed to simulate the Lignol process (2.04). The lowest polydispersity was determined for lignin-Fe (1.71), the process that we have designed in our laboratory, which could be yet another indication of its higher purity (Norberg et al. 2013). This result could therefore be another indication of the suitability of this organosolv lignin for electrospinning applications, as it has been shown previously that more homogeneous lignin samples gave better results in spinning tests (Brodin et al. 2009).

Lignin application for carbon fiber

Carbon fiber is regarded as a superman among materials. High initial content of carbon in lignins, along with the presence of the majority of carbons in lignin structure in sp2 hybridized form has inspired the efforts of lignin transformation into carbon fibers. We have explored the potential of lignin spinning both by electrospinning and by spinning in molten state.

Electrospinning

The studies performed in the past on the organosolv lignin spinning have demonstrated that it was impossible to transform any of the studied lignins as such, without the addition of reinforcing agents, into fiber by electrospinning. Different lignin activation procedures, such as acetylation (Zhang et al. 2014), mixing with binder agents such as polyethylene oxide (PEO) (Wang et al. 2013), polypropylene (Kadla et al. 2004), polyvinyl alcohol (PVA) (Ago et al. 2012), polylactic acid (Thunga et al. 2014) and polyethylene terephthalate (Kubo et al. 2005), or after further purification of lignin by sequential solvent extraction (Dallmeyer et al. 2014), have been tried for carbon fiber production. Also, other reinforcing agents, such as glycerin (Lallave et al. 2007) or platinum acetyl acetonate (Ruiz-Rosas et al. 2010) have been used to combine with organosolv lignin in electrospinning experiments. Figure 9 shows the nanofibers obtained by electrospinning of Lignin-Fe produced in our study (pulping at 180°C, with ferric chloride as catalyst), without any reinforcing agent addition or any other additional treatment performed on lignin sample.

The electrospinning experiments were performed using lignin solutions in DMF (50% wt in DMF), the viscosity of which ranged between 300 and 400 cPs. As shown in SEM images, only Lignin-Fe has been electrospun under the cited experimental conditions. The polymer properties of lignin-Fe seem to be more appropriate for electrospinning application than those of lignin-W or lignin-SA produced under the comparable pulping conditions. The diameters of fibers produced from lignin-Fe are determined to range between 100 and 355 nm. The diameters of lignin fibers determined in this research are also more uniform than those reported for other lignins studied in similar experiments, while also always require additives for electrospinning (Kubo et al. 2005).

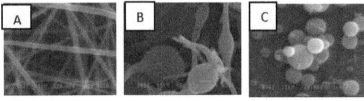

Fig. 7. SEM of electrospun lignin as collected. Scale bar is 1 µm. (A) Lignin-Fe; (B) lignin-W; (C) lignin-SA (Koumba-Yoya et al. 2016). © 2016 Wiley-VCH Verlag GmbH & Co. KGaA, Weinheim.

Carbon fiber by extrusion in melt state

Generally, carbon fibers are defined as composite materials consisting of at least 92% in weight of carbon and prepared from precursor polymers. However, the carbon fibers with almost 100% of carbon can be generated from carbon allotropes such as graphene and carbon nanotubes (Franck et al. 2014). Generally, carbon fibers are very long, with very remarkable mechanical properties and are mainly characterized by high tensile strength and high modulus of elasticity. Carbon fibers are obtained from carbonization processes mainly involving polymers such as polyacrylonitrile (PAN) and pitch from petroleum feedstock and, at a lesser extent, using the regenerated cellulose fiber (rayon). Unlike the PAN or pitch, lignin presents an advantage of being a renewable polymer, with carbon content estimated at more than 60% approximately. Lignin polymers are available in large quantities on the earth. Lignin is the second most abundant polymer in nature after cellulose. Generally, the lignin extraction from lignocellulosic materials is carried out by its gradual, but extensive fragmentation, leading to a release of fragments of lower average molecular weight, resulting in several changes in the physico-chemical properties of lignin compared to those of lignin in its native state. It is also difficult to isolate lignin by any process in a pure form, with low contents of residual polysaccharides and ashes. Since poor results on mechanical properties have been obtained with kraft lignins as carbon fiber precursors, the chemical modifications or combinations with other polymers have been required. Thus, Zhang and co-workers (Zhang et al. 2014) have conducted the esterification by acetylation of softwood kraft lignin. This acetylated lignin was tried successfully in applications related to carbon fibers, with tensile strength values higher than most of those reported previously in the literature. In 2014, Thunga and co-workers (Thunga et al. 2014) have studied the esterification of commercial karft lignin Indulin AT by butyration. This esterified lignin was then mixed with polylactic acid before spinning. These authors observed better performance of kraft lignin spinning when combined with polylactic acid at 25%. Previously, Kadla and co-workers (Kadla et al. 2002) have reported the use of a kraft lignin in combination with the PEO or PET. Indeed, taking into account the low mechanical properties with kraft lignin used alone, the authors claimed their improvement by adding the PEO to kraft lignin before spinning. With 5% of PEO or PET, the spinning of the kraft lignin had conducted to the best carbon fiber. Contrary to kraft lignins, organosolv lignins are exempt of sulfur and are the most frequently studied lignins as carbon fiber precursors. Thus in 1995, Uraki and co-workers (Uraki et al. 1995) have shown that the carbon fibers could be produced by melt spinning of Alcell lignin. Indeed, the authors observed that lignin could be used without any modification apart from those occurring during pulping (acetylated hydroxyl groups). It was found that the melt spinning had increased according to the degree of acetylation. Others lignin without sulfur, such as those from steam explosion processes were among the first to be studied for their potential as precursor of carbon fibers. Thus, Sudo et al. (1992) performed hydrogenation and thermal treatment of a steam explosion lignin (another type of lignin which is exempt of sulfur even though not an organosolv lignin) in order to obtain a material with a structure similar to pitch, thus convenient for spinning. In our case, with the high level of Tg value (143°C) for Lignin-Fe, we decided that this property could be appropriate for thermal oxidation without degradation during melt spinning. For melt spinning, different procedures using lignin in combination with others binder polymers such as polyethylene (Kubo et al. 2005), or polyacrylonitrile (Liu et al. 2015) have been explored for lignin transformation into carbon fiber. We have performed melt spinning experiments with pure organosolv lignin without adding any other polymers or additives. The results of melt spinning experiments are presented in Fig. 8.

The diameters of fibers produced by melt spinning are determined to range between 20 and 35 µm, much larger than those reported for polyacrylonitrile spun

Fig. 8. (A) Lignin fibers after melt spinning. (B) MEB image of lignin fibers. Scale bar: 10 μm.

fibers in similar experiments, the diameters of which were determined to range between 5 and 10 μm (Lin et al. 2014). In order to evaluate the best parameters for stabilization of lignin fibers, different conditions were studied by thermal oxidation under air followed by carbonization under nitrogen. In the first step, lignin samples were exposed to oxidative thermal treatment under air at different flow rates, increasing temperature from 25 to 250°C. The results of thermal stabilization are presented in Fig. 9.

As can be seen from Fig. 9, the temperature increase rate is an important factor of thermal oxidation of lignin and carbon fibers properties. Indeed, Baker et al. (2012) have studied thermal stabilization and carbonization of lignin fibers and showed that fibers could be oxidatively stabilized only at very slow heating rates (< 0.05°C/min). Stabilized lignin fibers allow the access to carbon fibers which are analyzed by XPS for carbon content evaluation. The result of XPS analysis is presented in Fig. 10 and Table 4.

The chemical analysis by XPS revealed more carbon (96.20%) in carbon fibers with C-C as major bond contrary to lignin fiber itself (79.96% of carbon) (Fig. 10). The carbon fiber is much richer in carbon than lignin fiber which indicates to the good efficiency of the stabilization process. Indeed, carbon fibers are defined as composite materials consisting of at least 92% in weight of carbon.

As for the C4 corresponding to the ester link, it is interesting to note that functional groups increase after thermal stabilization, reflecting possible oxidation of C2 bonds.

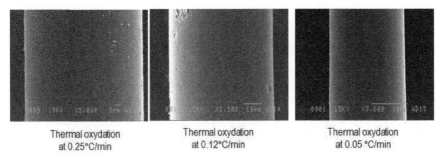

| Thermal oxydation at 0.25°C/min | Thermal oxydation at 0.12°C/min | Thermal oxydation at 0.05 °C/min |

Fig. 9. Carbon fibers from lignin under different oxidative conditions.

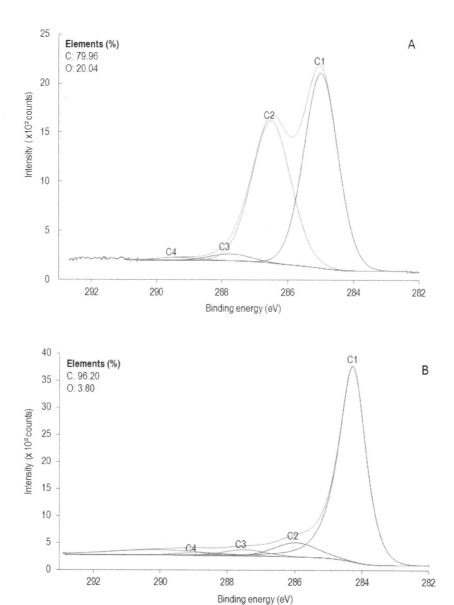

Fig. 10. Comparative XPS analysis of lignin and carbon fibers. (A) original lignin; (B) stabilized carbon fibers produced from organosolv lignin-Fe by oxidative and carbonization thermal treatments.

Mechanical Test

The mechanical evaluation of properties for carbon fiber was carried out in heterogeneous materials with different diameters which have consequently

Table 4. Comparative XPS results between original lignin and carbon fiber after stabilization (oxidation and carbonization).

Entry	Name	Bond type	Position	Atomic %	
				Original lignin	Carbon fiber
1	C_1	C–C, C–H	285	53.23	85.42
2	C_2	C–O, C–O–C	287	44.36	8.93
3	C_3	O–C–O, C = O	288	2.22	3.97
4	C_4	O–C = O	289	0.18	1.68

yielded different strengths. The results of strain at failure and elastic modulus were calculated from the strength measurements for each fiber. At this stage, these preliminary results need to be regarded simply as an indication. With strength failure between 0.5 and 1.15 GPa, the obtained elastic modulus was evaluated to be between 70 and 114 GPa. These values are much higher than those reported in literature for carbon fibers produced from other lignins. These values might be improved by further refining the temperature stabilization process. Indeed, the stabilization by carbonization of commercial carbon fiber is usually carried out at temperatures between 1000 and 2500°C under nitrogen or argon flow, obviously in that case one is speaking of industrial and not of laboratory conditions.

Conclusions

We have designed, developed and applied a new organosolv process, based on ethanol-water system and use of ferric chloride catalyst, which is allowing the access to highly pure lignin suitable for high value applications such as thermal conversion. The high purity profile of this lignin-Fe is determined according to different criteria, such as high level of Klason in addition to low acid soluble, lignin, low sugar and ash contents. These properties indicate that lignin-Fe could be appropriate for production of composites with good thermal properties. Indeed, in comparison to other organosolv lignins examined in the past in electrospinning experiments, Lignin-Fe was electrospun without addition of any binder polymer or reinforcing agent. In addition, the same lignin has been successfully spun in molten state and thermally stabilized to produce carbon fiber. The use of carbon fibers obtained by melt spinning is anticipated in applications, such as composites materials with good thermal properties along with application in construction materials after further optimization of the thermal stabilization process. The improvement of carbon fibers by increasing the carbonization temperature would result in higher strength and stiffness of fibers. In summary, we have developed a new catalytic bio-refinery concept yielding high purity lignin in combination with coproduction of extractives from pre-extraction and solid cellulosic pulp in delignification step, with carbohydrates and products of their acid transformation available from residual liquor for various applications.

References

Adler, E. 1977. Lignin chemistry-past, present and future. Wood Sci. Technol. 11: 169–218.

Ago, M., K. Okajima, J.E. Jakes, S. Park and O.J. Rojas. 2012. Lignin-based electrospun nanofibers reinforced with cellulose nanocrystals. Biomacromolecules 13: 918–926.

Avignon, G. and M. Delmas. 2000. Procédé de production de pâte à papier, lignines, sucres et acide acétique par fractionnement de matière végétale lignocellulosique en milieu acide formique/acide acétique. FR patent, WO 0068494A1.

Baker, D.A., N. Gallego and F.S. Baker. 2012. On the characterization and spinning of an organic-purified lignin toward the manufacture of low-cost carbon fiber. Journal of Applied Polymer Science 124: 227–234.

Baker, D.A. and O. Hosseinaei. 2014. High glass transition lignin and lignin derivatives for manufacture of carbon and graphite fibers. Patent US 2014, 20140271443 A1.

Choi, J.-W., O. Faix and D. Meier. 2001. Characterization of residual lignins from chemical pulps of spruce (*Picea abies* L.) and beech (*Fagus sylvatica* L.) by analytical pyrolysis-gas chromatography/mass spectrometry. Holzforschung 55: 185–192.

Crestini, C., F. Melone and R. Saladino. 2011. Novel multienzyme oxidative biocatalyst for lignin bioprocessing. Bioorg. Med. Chem. 19: 5071–5078.

Dallmeyer, I., F. Ko and J.F. Kadla. 2014. Correlation of elongational fluid properties to fiber diameter in electrospinning of softwood kraft lignin solutions. Ind. Eng. Chem. Res. 53: 2697–2705.

Frank, E., L.M. Steudle, D. Ingildeev, J.M. Sporl and M.R. Buchmeiser. 2014. Carbon fibers: Precursor systems, processing, structure, and properties. Angew. Chem. Int. Ed. 53: 5262–5298.

Gandolfi, S., G. Ottolina, R. Consonni, S. Riva and I. Patel. 2014. Fractionation of hemp hurds by organosolv pretreatment and its effect on production of lignin and sugars. ChemSusChem. 7: 1991–1999.

Garça, J. 2010. Hydroxycinnamates in suberin formation. Phytochem. Rev. 9: 85–91.

Garcıa, A., M.G. Alriols, R. Llano-Ponte and J. Labidi. 2011. Energy and economic assessment of soda and organosolv biorefinery processes. Biomass and Bioenergy 35: 516–525.

Graglia, M., N. Kanna and D. Esposito. 2015. Lignin refinery: Towards the preparation of renewable aromatic building blocks. ChemBioEng. Rev. 2: 377–392.

Hideno, A., A. Kawashima, T. Endo, K. Honda and M. Morita. 2013. Ethanol-based organosolv treatment with trace hydrochloric acid improves the enzymatic digestibility of Japanese cypress (*Chamaecyparis obtusa*) by exposing nanofibers on the surface. Bioresource Technology 132: 64–70.

Kadla, J.F., S. Kubo, R.A. Venditti, R.D. Gilbert, A.L. Compere and W. Griffith. 2002. Lignin-based carbon fibers for composite fiber applications. Carbon 40: 2913–2920.

Kadla, J.F. and S. Kubo. 2004. Lignin-based polymer blends: analysis of intermolecular interactions in lignin–synthetic polymer blends. Composites, Part A 35: 395–400.

Kirci, H., S. Bostanci and M.K. Yalinkilic. 1994. A new modified pulping process alternative to sulphate method. Alkali-Sulfite-Antraquinone-Ethanol (ASAE). Wood Sci. Technol. 28: 89–99.

Kleinert, T.N. 1971. Organosolv pulping and recovery process. US patent 3585104.

Koumba-Yoya, G. and T. Stevanovic. 2016. New biorefinery strategy for high purity lignin production. Chemistry Select. 1: 6562–6570.

Kordsachia, O. and R. Patt. 1988. Full bleaching of ASAM pulps without chlorine compounds. Holzforschung 42: 203–209.

Kubo, S. and J.F. Kadla. 2005. Lignin-based carbon fibers: Effect of synthetic polymer blending on fiber properties. J. Polym. Environ. 13: 97–105.

Lallave, M., J. Bedia, R. Ruiz-Rosas, J. Rodríguez-Mirasol, T. Cordero, J.C. Otero, M. Marquez, A. Barrero and I.G. Loscertales. 2007. Filled and hollow carbon nanofibers by coaxial electrospinning of alcell lignin without binder polymers. Adv. Mater. 19: 4292–4296.

Lehnen, R., B. Saake and H.H. Nimz. 2001. Furfural and hydroxymethylfurfural as by-products of formacell pulping. Holzforschung 55: 199–204.

Lin, J., K. Koda, S. Kubo, T. Yamada, M. Enoki and Y. Uraki. 2014. Improvement of mechanical properties of softwood lignin-based carbon fibers. Journal of Wood Chemistry and Technology 34: 111–121.

Lindner, A. and G. Wegener. 1990. Characterization of lignins from organosolv pulping according to the organocell process. Part 4. Molecular weight determination and investigation of fractions isolated by GPC. J. Wood. Chem. Technol. 10: 351–363.

Liu, H.C., A.T. Chien, B.A. Newcom, Y. Liu and S. Kumar. 2015. Processing, structure, and properties of lignin- and CNT-incorporated polyacrylonitrile-based carbon fibers. ACS Sustainable Chem. Eng. 3: 1943–1954.

Lora, J.H. and W.G. Glasser. 2002. Recent industrial applications of lignin: a sustainable alternative to nonrenewable materials. J. Polym. Environ. 10: 39–48.

Mesa, L., E. González, E. Ruiz, I. Romero, C. Cara, F. Felissia and E. Castro. 2010. Preliminary evaluation of organosolv pre-treatment of sugar cane bagasse for glucose production: Application of 2^3 experimental design. Applied Energy 87: 109–114.

Pan, X., N. Gilkes, J. Kadla, K. Pye, S. Saka, D. Gregg, K. Ehara, D. Xie, D. Lam and J. Saddler. 2006. Bioconversion of hybrid poplar to ethanol and co-products using an organosolv fractionation process: Optimization of process yields. Biotechnology and Bioengineering 94: 851–861.

Pye, E.K., W.R. Klein, J.H. Lora and M. Cronlund. 1987. The Alcell Process. Solvent Pulping: Promises and Problems Conference, Appleton, WI, 55–67.

Pye, E.K. and J.H. Lora. 1991. The alcell process. A proven alternative to kraft pulping. Tappi 74: 113–118.

Ruiz-Rosas, R., J. Bedia, M. Lallave, I.G. Loscertales, A. Barrero, J. Rodríguez-Mirasol and T. Cordero. 2010. The production of submicron diameter carbon fibers by the electrospinning of lignin. Carbon 48: 696–705.

Stevanovic, T. and G. Koumba-Yoya. 2016. Patent pending application: PCT/CA2016/000169, Organosolv process for extraction of highly pure lignin and products comprising same.

Sudo, K. and K. Shimizu. 1992. A new carbon fiber from lignin. J. Appl. Polym. Sci. 44: 127–134.

Thunga, M., K. Chen, D. Grewell and M.R. Kessler. 2014. Bio-renewable precursor fibers from lignin/polylactide blends for conversion to carbon fibers. Carbon 68: 159–166.

Uraki, Y., S. Kubo, N. Nigo, Y. Sano and T. Sasaya. 1995. Preparation of carbon fibers from organosolv lignin obtained by aqueous acetic acid pulping. Holzforschung 49: 343–350.

Viell, J., A. Harwardt, J. Seiler and W. Marquardt. 2013. Is biomass fractionation by Organosolv-like processes economically viable? A conceptual design study. Bioresource Technology 150: 89–97.

Wang, S.X., L. Yang, L.P. Stubbs, X. Li and C. He. 2013. Lignin-derived fused electrospun carbon fibrous mats as high performance anode materials for lithium ion batteries. ACS Appl. Mater. Interfaces 5: 12275–12282.

Zhang, M. and A.A. Ogale. 2014. Carbon fibers from dry-spinning of acetylated softwood kraft lignin. Carbon 69: 626–629.

Conifer Resins and Essential Oils: Chemical Composition and Applications

Nellie Francezon[1,2,3,]* and *Tatjana Stevanovic*[1,2,3]

Introduction

The ancient time, resins from trees were considered precious materials. In the Bible, fragrant frankincense and myrrh have been mentioned alongside with gold as gifts carried for infant Christ at the time of his birth. Resins also appeared to be the first raw material associated with distillation and essential oils production. Famous historians such as Herodotus (484–425 B.C.), Dioscorides (author of *De Materia Medica*) and Pliny (1st century A.D.), also mentioned the existence of turpentine oil from the distillation of pine resins. Interestingly, the word "terpene", representing the most characteristic group of essential oil components, comes from the German word "*Terpen*" which means turpentine (Urdang 1952). This chapter focuses on conifer resins and essential oils from the temperate zone and aims to highlight those little known natural products putting forward their diversity, their singularity and the wide range of their past and present applications.

[1] Renewable Materials Research Centre, Department of Wood and Forest Sciences, Université Laval, Canada.
[2] Institute of Nutrition and Functional Food (INAF), Université Laval, Canada.
[3] Centre de Recherche sur les Matériaux Avancés, Department of Chemistry, Université Laval, Canada.
* Corresponding author: nellie.francezon.1@ulaval.ca

Oleoresin from Conifers of Temperate Zone

Conifer genera from northern hemisphere forests are famous for the production of resins rich in terpenoids. The resin is composed of a volatile part (mainly monoterpenes, sesquiterpenes and their oxygenated derivatives) upon evaporation of which remains a solid residue, the rosin, constituted of heavier diterpenes and their derivatives. Highly volatile, monoterpenes are the fragrant principles of oleoresin. They can be caught by steam during the hydrodistillation process to produce essential oils. Among conifer species, pines, larches and spruces from *Pinus*, *Larix* and *Picea* genera are of particular interest as they have specific anatomical features: resin canals. Pines exhibit the larger resin ducts, which are tapped in order to collect large quantities of resins for turpentine production. Firs (genus *Abies*) belonging to the same Pinaceae family as pines, larches and spruces, are also of interest as they similarly produce resin stored in specialized blister cells in their bark (Teranishi et al. 1993). Likewise, some members of the Cupressaceae family produce terpenic resins with technological and biological properties, with a long history of various applications. Therefore, we shall limit our discussion here to those two families in order to highlight the most interesting properties and uses of the common European and North American conifers.

Resins from Pinaceae and Cupressaceae families

As mentioned, resins are composed of diterpenes, which are mainly represented by resin acids, but hydrocarbon analogs along with their alcohol and aldehyde derivatives are also often identified in conifer resins. The resin acids belonging to abietadiene derivatives, are based on tricyclic system related to phenantrene skeleton, but containing just two conjugated double bonds. The examples are laevopimaric, palustric, neoabietic and abietic acids, presented in Fig. 1. The

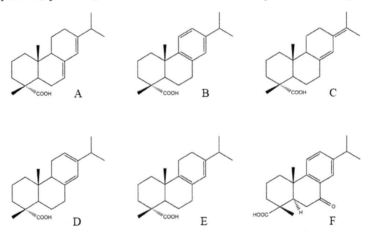

Fig. 1. Abietadiene type resins; (A) abietic acid, (B) dehydroabietic acid, (C) neoabietic acid, (D) levopimaric acid, (E) palustric acid, (F) 7-oxo-dehydroabietic acid.

dehydroabietic acid with an aromatic ring, is also present in minor quantities, while it is usually generated in higher quantities by oxidative transformation of resins (exposure to air, aging). The presence of higher quantities of dehydroabietic acid in a given resin sample may indicate its exposure to oxidative agents. Another oxidative product of abietadiene type resin acids is 7-oxo-dehydroabietic acid, which has become particularly important as a transformation product from resins originally rich in abietadiene type acids (Fig. 1).

The pimaradiene type resin acids (Fig. 2) are based on the same tricyclic skeleton, but have a vinyl and methyl group at C-7 instead of isopropyl and just one double bond in the cyclic system (Fig. 3) (which is never conjugated, by difference from the abietane type resin acids). The examples of pimaradiene skeleton-based resin acids are isopimaric, sandaracopimaric and pimaric acids (Fig. 2), which are present in variable quantities depending on the plant source.

It is interesting to discuss the chemical composition of fir resins from two firs, that from Europe, *Abies alba*, a source of resin which has traditionally been known as Strasbourg turpentine and that from balsam fir, *Abies balsamea*, a source of Canada balsam. The two resins have very similar chemical compositions, being constituted of neutral abienol in important proportions (42% *of A. alba* and 27% *of A. balsamea* resin), along with sandaracopimaric and isopimaric acids with pimaradiene skeleton and laevopimaric, abietic, dehydroabietic, abietic and neoabietic acids with abietane skeleton. While Canada balsam contains also about 6% of aldehyde analogs of the mentioned acids, those are present only in trace amounts in Strasbourg turpentine (Mills and White 1977), which could help distinguishing them.

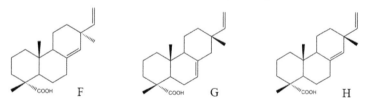

Fig. 2. Pimaradiene type resin acids; (F) pimaric acid, (G) isopimaric acid, (H) sandaracopimaric acid.

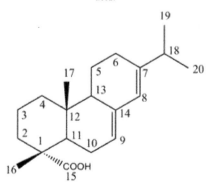

Fig. 3. Example of numbering of the tricyclic diterpene skeleton.

The resin obtained by tapping varying larch species (*Larix* spp.) yield resins rich in abietadiene and pimaradiene resin acids, but also a labdane type diterpenes among which manool appears to be present in larches of all origins. European larch *Larix decidua* has been used in the past under the name of Venice turpentine and has unique constituents, specific to this species, larixol along with its acetate (Fig. 4), the latter making 30% of Venice turpentine (Mills and White 1977). Back in the time, the term "turpentine" designated the volatile oil from resin as much as the resin itself. More recently, larixol and its derivative were identified in important quantities in a hybrid Japanese larch *Larix gmelini* var. *japonica* (Sato et al. 2009).

Yet another important type of conifer resin constituents are those based on labdane bicyclic system, such as found in communic acids, very common constituents of Cupressaceae family (Barrero et al. 2012) and notably of that from Mediterranean cypress (*Cupressus sempervirens*). Containing a conjugated system of double bonds external to bicyclic system, communic acid and similar derivatives are prone to rapid polymerization upon air exposure (Kononenko et al. 2016).

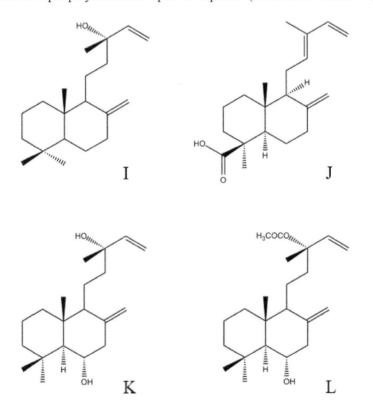

Fig. 4. Labdane type diterpenes from larches and Cupressaceae family resins; (I) 13-epi-manool, (J) communic acid, (K) larixol, (L) larixol acetate.

Speaking of Cupressaceae, it is interesting to mention that Northern white cedar *Thuja occidentalis* bark was also determined to have unusual labdane type diterpenes (Yang et al. 2014).

Essential oils from Pinaceae and Cupressaceae families

Essential oils from resin: Turpentine oils. Conifer oleoresin is composed of 10 to 20% of volatile oil which acts as a solvent of the solid part of the resin. The oil, also called gum spirit turpentine, is extracted by hydrodistillation of the oleoresin collected from tapping of trees, principally pines species. Turpentine oils yields are usually very high: between 10 and 30% of the resin, depending the pine species. Turpentine oils obtained from different raw materials are available, such as wood turpentine distilled from wood chips, sulfate and sulfite turpentines (by-products of the kraft and sulfite pulping processes) and the destructively distilled wood turpentine obtained from dry distillation of resin rich pine wood (Goldblatt 1952). Only turpentine oil from living trees tapped oleoresins will be discussed here. We shall assume that back in time the term "turpentine" was also used to name the whole resin. Turpentine oils presented here are the volatile part of resin only.

Almost exclusively composed of hydrocarbon monoterpenes such as α-pinene, β-pinene, limonene, 3-carene, camphene (Fig. 5), turpentine oils possess the particularity of containing one major compound reaching up to 90% of their whole composition. This relative purity is advantageous in the perspective of isolating one targeted compound. For example, maritime pine *Pinus pinaster* and black pine *Pinus nigra* turpentines are composed of more than 90% of α-pinene (Mirov 1948). In the same range, *Pinus strobus* and *Pinus edulis* turpentine oils can contain up to 80% of α-pinene. Longleaf pine (*Pinus palustris*) or slash pine (*Pinus caribaea*) turpentines, the most produced essential oils in the United States during the XXI century, are mixtures of α- and β-pinenes up to 90% (Goldblatt 1952). Known as

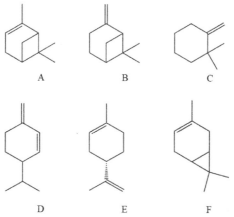

Fig. 5. Principal components of turpentine oils; (A) α-pinene, (B) β-pinene, (C) camphene, (D) β-phellandrene, (E) (–)-limonene, (F) 3-carene.

"American turpentines", they were largely used in naval stores for the maintenance of wooden naval ships. Even if pinene isomers are widely represented in conifers and are most of the time the main components of turpentine oils, some pine species turpentines, such as Jeffrey pine (*P. jeffreyi*), gray pine (*P. sabiniana*) and Torrey pine (*P. torreyana*) are devoid of pinenes (Mirov 1948). Major compounds of Jeffrey pine and gray pine turpentines are not terpenes but heptane, a very flammable paraffin hydrocarbon naturally present in gasoline, which can reach up to 90% of their turpentine composition. In these particular cases of native western North American species, short-chain alkanes make up 95 to 99% of the turpentine composition. In the late XIX century, the discovery of heptane in Jeffrey pine turpentine was quite spectacular. The first batch processed in common distillation still destroyed the plant in a dramatic explosion (Mirov 1946).

Limonene, a hydrocarbon monoterpene, is also a major compound of turpentine oils (Fig. 5). Indeed, Torrey pine and stone pine (*P. pinea*) turpentine oils are composed of 75% limonene, but the highest rate is found in pond pine (*P. serotina*) (90% of the whole composition) (Mirov 1948). However, levorotary *l*-limonene (or (−)-limonene) must be distinguished from its enantiomer, the dextrorotary *d*-limonene ((+)-limonene). The former, found in conifers, smells terpenic, piney, herbal and peppery and the latter, represented in the citrus family has a sweet citrusy scent. Another monoterpene, the very unstable β-phellandrene, is the major compound of turpentine from lodgepole pine (*P. contorta*). This turpentine has to be distilled at reduced pressure in order to avoid the polymerization of β-phellandrene into a sticky substance (Mirov 1946).

Because of its high content of a limited number of compounds, turpentine is currently regarded as a raw material for the chemical industry. It provides precursor molecules for the production of plastics, adhesives, paints, flavors and fragrances, cosmetics, and pharmaceuticals. Pinenes are by far the most important of them. α-pinene is a starting molecule in the synthesis of a number of molecules, among which borneol, camphor and terpineols. For example, α-terpineol which is lilac scented and is popular as a fragrance material, is obtained by hydration of α-pinene. The isomers α- and β-pinene can be transformed into limonene, which is widely used for perfuming household products. They are also the starting material for producing linalool, a sweet flowery scented compound. Most perfumery-grade synthetic linalool is derived from pinenes, but linalool is also an intermediate in the manufacture of artificial vitamin E.

Essential oils from needles. Most of the conifer essential oils available in the market are produced from needles. Huge amounts of foliage biomass is generated by the timber industry, representing around 13% of a tree's volume (Desrochers 2011). Far less productive than resin turpentine, needles yield only 0.4 to 1% of essential oil (Guenther 1952) and they are far more complex in composition, as they contain more compounds than turpentine oils and particularly important amount of oxygenated terpenoids. Bornyl acetate, camphor, terpinen-4-ol and α-terpineol (Fig. 6) are common oxygenated monoterpenoids found in *Pinus*, *Picea* and *Abies*

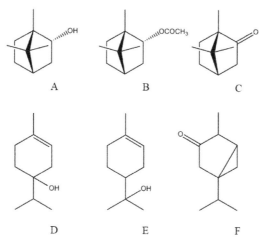

Fig. 6. Oxygenated monoterpenoids found in needle essential oils; (A) borneol, (B) bornyl acetate, (C) camphor, (D) terpinen-4-ol, (E) α-terpineol, (F) thujone.

species needle essential oils (Surburg and Panten 2006) (Fig. 6). Indeed, black spruce (*Picea mariana*) and botanically related red spruce (*Picea rubens*) foliage essential oils contain 35 to 50% of bornyl acetate (Garneau et al. 2012, Rudloff 1967, Rudloff 1975). White spruce (*Picea glauca*) and Engelmann spruce (*Picea engelmannii*) essential oils contain only 5–20% of bornyl acetate, having camphor as the major compound (Rudloff 1975). The close chemical structures of camphor and bornyl acetate and their occurrence in *Picea* spp. indicate that their biosynthesis contains a common intermediate, borneol, which by oxidation yields camphor and by esterification bornyl acetate (Fig. 6) (Sell 2010). Bornyl acetate is also a common constituent of essential oils from foliage of *Abies* species, especially of Siberian fir (*Abies sibirica*) and balsam fir (*Abies balsamea*) (Orav et al. 1995) and one pine species essential oil, that of Jack pine (*Pinus banksiana*). Otherwise, monoterpenes such as α-pinene, β-pinene, myrcene, camphene, limonene and β-phellandrene (Fig. 5) are common constituents of essential oils from Pinaceae species foliage. As for Cupressaceae family, common juniper (*Juniperus communis*) and cypress (*Cupressus sempervirens*) leaf oils are dominated by α-pinene (around 90% for juniper and 40% for cypress) (Banthorpe et al. 1973, Selim et al. 2014). But the most interesting species in this family is undoubtedly *Thuja*. Foliage essential oils from western red cedar (*Thuja plicata*) and eastern white cedar (*Thuja occidentalis*) both originating from North America, contain high levels of thujone (Fig. 6), a notorious ketone named after its discovery in thuja oil by Wallach in 1902. Western red cedar essential oil is indeed composed of around 50% of thujone isomers (α- and β-) while that of white cedar can contain up to 90% of thujone (Banthorpe et al. 1973, Kamdem et al. 1993, Rudloff 1962). α-thujone, also present in wormwood (*Artemisia absinthium*), was thought to be responsible for the toxicity of liquor absinthe which was very popular in France in the end of XIX and early XX centuries. The spirit was reported to cause fits and hallucination and sometimes led to psychoses as believed to be the case for numerous artists and writers, among whom were Vincent van Gogh and Charles Baudelaire (Höld et al. 2000). However, one should keep in mind though that alcohol content in that drink could have been a cause of many of the described troubles and not necessarily the thujone constituent itself (Pelkonen et al. 2013).

Essential oils from wood. Essential oils distilled from conifer wood possess a very distinct pattern from the ones of needles or turpentine, as they contain very few monoterpenoids but great proportion of heavier sesquiterpenoids. Their yields vary between 1 and 5%. In conifer species, the so-called "cedarwood oils" are essential oils distilled from wood, but two different cedarwood oils must be distinguished: the ones from true cedars of the *Cedrus* genus and the ones from *Juniperus* and *Thuja* genera woods, assimilated to cedarwood. Oils from true cedar such as Altas cedar (*Cedrus atlantica*), Himalayan cedar (*Cedrus deodara*) and cedar of Lebanon (*Cedrus libani*) are similar in composition as they are mainly composed of sesquiterpenoids α-, β- and γ-himachalene, and α-atlantone (Fig. 7) (Başer and Demircakmak 1995, Chalchat et al. 1994, Chaudhary et al. 2009).

Cedarwood from *Thuja* genus such as white cedar (*Thuja occidentalis*) provides, in addition to its well-known leaf essential oil, a distinct essential oil from its wood, rich in sesquiterpenes occidentalol, occidol isomers, occidenol, and α-, β-, and y-eudesmols (Andersen et al. 1995) (Fig. 7). It is interesting to indicate that occidentalol, one of the volatile constituents of *Thuja occidentalis* resin is released during the drying of its wood. Occidentalol, in the form of white crystals, envelop the kiln-drying installations, demonstrating an easy access to this pure sesquiterpene from wood drying (Roy et al. 1984).

Famous cedarwood essential oils from the wood of *Juniperus* genus are produced in the USA at a commercial scale: Virginia cedarwood (*Juniperus virginiana*) and Texas cedarwood (*Juniperus mexicana*) oils. Almost entirely composed of sesquiterpenoids, these oils contain thujopsene, cedrene isomers and

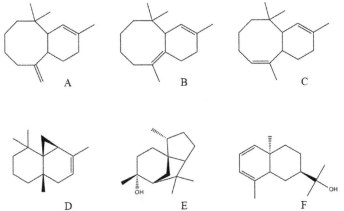

Fig. 7. Sesquiterpenes in conifer wood oils; (A) α-himachalene, (B) β-himachalene, (C) γ-himachalene, (D) (–)-thujopsene, (E) cedrol, (F) occidentalol.

its alcohol derivative, cedrol (Fig. 7). The latter, precipitates into a "cedarwax", especially when oil is distilled from fresh wood (Guenther 1952). As mentioned earlier, for occidentalol in thuja, pure cedrol crystallizes in the kilns tubing when fresh lumber from Virginia and Texas cedarwood are dried (Guenther 1952).

Essential oils from bark. Like foliage, bark is a residue of wood transformation. Huge amounts of bark produced annually by the timber industry from debarking of logs are burnt, with all their valuable molecules lost. The bark of conifers contains oleoresin, therefore an essential oil can be distilled from its bark in the same way as from wood. There is no essential oil from conifer bark currently available on the market, but researches have started paving the way for a better use of this forest biomass. Studies on pine species *Pinus brutia* and *Pinus pinaster*, reported essential oils of different composition compared to those from needle or turpentine oils (Bağci et al. 2011, Zolfaghari and Iravani 2012). Olfactory assay on needles, twigs and bark essential oils from Douglas fir (*Pseudotsuga menziesii*) revealed that the one from bark recreated the typical fir odor with a good balance of (Z)- and (E)-β-ocimene, β-pinene, sabinene and α-terpinolene. Bark from balsam fir (*Abies balsamea*) and black spruce (*Picea mariana*), the most commercially important trees in the Canadian timber industries, produced essential oils rich in α- and β-pinenes but almost devoid of bornyl acetate which is a major compound of their respective needle oils (Francezon and Stevanovic 2017a, Ross et al. 1996). As for black spruce bark oil, the absence of bornyl acetate makes its scent less pungent. However, both oils have the same fragrance pattern, as the needle oil drydown is very similar to bark oil whole odor. Thus, improvements are still necessary in order to make bark essential oil production industrially relevant (yields of distillation, reproducibility, investigations on potential applications) but it could be a good alternative for the timber industry for their post-harvesting treatment.

Hydrodistillation: Toward a better improvement of the process. Essential oils from different parts of conifer trees are extracted by steam distillation which consist of a stream of steam carrying off volatile molecules and sometimes by hydrodistillation in which raw material is immersed in boiling water. Steam with volatile compounds then condenses to yield an essential oil and a hydrosol (distillate water condensed with the essential oil floating). Hydrosol is a fragrant water, composed of low percentage of solubilized essential oil, which mainly contain oxygenated monoterpenes. A distillation water corresponding to wastewater inside the distiller tank, along with solid residues, is also produced. This distillation water can be regarded as a hot water extract of the raw material, retrieving hydrosoluble heavier molecules such as polyphenols. Known for their health benefits, polyphenols are natural powerful antioxidants exhibiting a wide range of therapeutic activities such as anti-inflammatory, anticarcinogenic or cardioprotective (Manach et al. 2005, Middleton et al. 2000). These two aqueous extracts, co-products of distillation, are most of the time discarded, considered less valuable than essential oils. However, better uses can be made of them. Hydrosols for example, can be redistilled or extracted with solvent to recover the dissolved oil, known as "secondary oil"

(Fleisher 1991, Rao et al. 2002) or even used as such in aromatherapy. Conifer hydrosols from black spruce, scots pine or balsam fir bark have demonstrated analgesic, calming, and anti-inflammatory properties among other (Price and Price 2004).

As for wastewaters drained from the distiller tank, recent studies have highlighted their potential as new sources of bioactive molecules. Rusanov et al. (2014) demonstrated that rose oil distillation wastewater was a rich source of polyphenols, among which several quercetin and kaempferol glycosides have been identified. As for conifers, wastewater from black spruce bark essential oil distillation was determined to have comparable antioxidant capacity and polyphenol content than its hot water extract (Francezon and Stevanovic 2017b). Needles, which are the main tree part used for conifer essential oil production, also contain antioxidant phenolics (Chaouche et al. 2013, Ennajar et al. 2009, Kähkönen et al. 1999, Strack et al. 1989) that can be retrieved in wastewaters after hydrodistillation. Waste treatment as source of bioactive molecules has become an important issue, especially for food and natural ingredient industries, as several studies have revealed the high concentrations of polyphenols with antioxidant activity in these residues and their potential uses as a low-cost raw material (Balasundram et al. 2006, Moure et al. 2001). Thus, this could become a new integrated approach for essential oil industries, especially those using forest biomass.

Uses of Conifer Resins and Essential Oils

Therapeutic properties of conifer essential oils

Therapeutic properties of essential oils and extracts from aromatic and medicinal plants have been known since ancient times. Ethnobotanical uses of conifers have been reported from People of First Nations. Carrier people of British Columbia, Canada, used conifer pitch and oleoresins (*Abies lasiocarpa, Picea engelmanii* x *P. glauca, Pinus contorta*) to prevent infections and treat burns, wounds, sores or dermatological disorders (Ritch-Krc et al. 1996). Eastern Canada indigenous people (Iroquoian and Algonkian) used *Abies balsamea, Larix laricina, Picea* spp., *Pinus* spp., *Tsuga canadensis*, and *Thuja* resins and gums in a variety of topical ailments to treat burns, abscesses and itching. *Picea* spp. resins were also used against tuberculosis and *Pinus* spp. and *Thuja occidentalis* as olfactory stimulants (Arnason et al. 1981). In Europe, ethnobotanical uses of *Pinus* spp. resins were referenced in Turkey for skin, respiratory affections and digestive diseases treatments and as analgesic (Kızılarslan and Sevg 2013). The first scientific experiments, conducted by Buchholtz in 1875 on antimicrobial activity of essential oils paved the way to systematic investigations of their pharmacological properties. Since 1926, conifer essential oils (from pine needle and turpentine) were referenced in the German pharmacopeia along with famous lavender, eucalyptus, thyme or rosemary oils, and then in the European pharmacopeia (Pauli and Schilcher 2010). With the emergence of alternative medicine, aromatherapy and well-being, the use of essential oils increased and consequently the number of scientific experiments on their bioactivities, even though the studies on conifer essential oils remained scarce compared to those on aromatic plants and spices.

Antimicrobial activity. Antimicrobial activity is undoubtedly the oldest empirically known essential oil bioactivity. Conifer essential oils, originating from resins, have a natural antimicrobial function to protect the tree. Indeed, resin has antibacterial and antifungal roles to fight against pathogens proliferation when the bark is wounded.

Massive screening of antimicrobial activity conducted on hundreds of essential oils and aromatic ingredients (Maruzzella and Sicurella 1960, Morris et al. 1979) revealed overall weak bioactivities of conifer essential oils against routinely tested pathogenic strains *Staphylococcus aureus*, *Escherichia coli*, *Candida albicans*, *Corynebacterium diphtheriae*, *Streptococcus fecalis*, *Salmonella typhosa*, except for pine oil and balsam fir oil. Several studies on both gram negative and gram positive bacteria and fungi have demonstrated only moderate antimicrobial activity of conifer needle oils, while no activity against *E. coli* was detected (Bağci and Diğrak 1996, Hong et al. 2004, Oh et al. 2007, Pichette et al. 2006). Nonetheless, essential oil from *Picea abies* young shoots showed good antimicrobial activity against gram positive bacteria (Kartnig et al. 1991). It was also found active against gram positive food contaminant bacteria *Listeria ivanovii*, *Listeria monocytogenes* and *S. aureus* (Canillac and Mourey 2001), while pine needle oil was found active against 18 isolated strains of *L. monocytogenes* (Lis-Balchin and Deans 1997). In addition, *Abies koreana* oil was demonstrated efficient against methicillin-resistant *S. aureus* strain, responsible for nosocomial infections (Jeong et al. 2007).

Indeed, the potential of forest essential oils might be underestimated, as demonstrated by Poaty et al. (2015) for seven essential oils from North American boreal conifers, in a study investigating their antibacterial and antifungal activities. Using an improved microdilution test with strip tubes with caps, it has been demonstrated that all studied oils were active. Moreover, jack pine and tamarack oils showed the most potent activity against bacteria *S. aureus*, *E. coli*, *S. tiphimurium* and *S. enteritidis*, which was similar to clove and lemon eucalyptus. Antifungal activity against fungi *Aspergillus niger* and *Aureobasidium pullulans* was also demonstrated, with jack pine (*Pinus banksiana*) and white cedar (*Thuja occidentalis*) being the most active.

Focusing on the active antibacterial components of essential oils, Dorman et al. (2000) tested 22 pure terpenoids on 25 bacterial strains. The phenolic constituents of oils, carvacrol, thymol and eugenol, showed the highest antimicrobial activity, being both bactericidal and bacteriostatic. None of them are, however, well represented in conifer essential oils. Despite this, other powerful antibacterial compounds, such as alcohols α-terpineol and terpinen-4-ol (Dorman and Deans 2000, Kotan et al. 2007) are widely distributed in the Pinaceae family and ketones such as thujone (Fig. 6) is a major compound in western red cedar (*Thuja plicata*) and eastern white cedar (*Thuja occidentalis*), for example. Hydrocarbon monoterpenes, for instance α- and β-pinene showed weaker activities. To illustrate this point, Monterey pine

(*Pinus radiata*) and cypress (*Cupressus sempervirens*) oils, characteristic for their monoterpene hydrocarbon pattern and representatives of resinous oils, were tested together with essential oils rich in benzyl esters (ylang-ylang, *Cananga odorata*), phenylpropanoids (curcuma, *Curcuma longa*), phenolics (thyme, *Thymus vulgaris*) and ketones (rosemary, *Rosmarius officinalis*). Even if *P. radiata* foliage oil showed a specific narrow spectrum activity against the yeast *Saccharomyces cerevisiae*, conifer oils showed an overall weaker activity against food-spoilage microorganisms (Sacchetti et al. 2005).

On the other hand, α- and β-pinene, the main constituents of turpentine oils and widely distributed in conifer oils, were reported to fight against pathogenic bacteria, especially strains responsible for jaw infections, parodontitis and periodontitis, as well as against fungi and yeasts such as *Candida albicans* and other related species (Mercier et al. 2009). Nevertheless, synergy between compounds in essential oils are more often responsible for the complete biological effect rather than a single major molecule (Burt 2004).

Testing antimicrobial activity of essential oils has become a routine in scientific papers. The huge data generated, as well as their growing use in aromatherapy and well-being therapies helped considering essential oils as potential natural antibiotics. The application of essential oils for air decontamination for buildings is also based on antimicrobial properties of conifer oils, as proposed by Hudson et al. (2011) in their study with white cedar oil (*Thuja plicata*). However, *in vivo* experiments are still required to validate their effectiveness.

Anticancer. Nature provided highly efficient drugs against cancer, such as Vinca alkaloids (vincristine, vinblastine) or diterpene-based paclitaxel. Discovered in yew bark, *Taxus brevifolia*, the latter belongs to the same terpene family as constituents of conifer resins. Being costly to extract, several semi-synthesis approaches were studied and one of them involves α-pinene as a precursor (Wender et al. 1997) which is widely available in turpentine pine oil as a major compound. Aside from being a potential precursor, α-pinene was reported to have anticancer properties itself. Significant inhibitory effect of α-pinene was demonstrated on hepatoma carcinoma cells responsible for liver cancer (Chen et al. 2015) and cytotoxicity was shown against human cancer cell lines *in vitro* for both α- and β-pinene (Setzer et al. 1999). Besides, pine needle essential oil, rich in pinenes, was used as an anticancer agent in Traditional Chinese Medicine (Chen et al. 2015). Balsam fir essential oil demonstrated cytotoxicity against tumor cancer cell lines (Legault et al. 2003). Sesquiterpene α-humulene, a minor compound was found to be the active molecule partially responsible for anticancer activity of balsam fir. Cho et al. (2014) demonstrated that Korean nut pine (*Pinus koraiensis*) needle essential oil could be a potential chemotherapeutic agent for colorectal cancer, as it reduced *in vitro* proliferation and migration of colorectal cancer cells. Moreover, *P. koraiensis* cone oil may promote anti-metastatic activity in breast cancer cells (Lee et al. 2015). Other conifer essential oils such as *Cedrus libani* (rich in α-, β-, and γ-himachalene isomers), *Pinus pinea* (composed of 65% limonene) and *Juniperus oxycedrus* (containing *cis*-thujopsene and δ-cadinene) wood oil as well as *Juniperus excelsa* fruit oil showed remarkable activity against multidrug-resistant leukemia cells, indicating that they could be used to treat drug-resistant and refractory tumors (Saab et al. 2012). *Pinus densiflora* needle essential oil was reported anti-proliferative, anti-survival and pro-apoptotic on human oral squamous cell carcinoma, an oral cavity cancer (Jo et al. 2012).

Anti-inflammatory and wound-healing. Since ancient times, cedar wood *Cedrus deodora* is known for its analgesic properties, as it was used in Ayurvedic medicine to treat inflammations and rheumatoid arthritis. Cedarwood oil was indeed proven as an effective analgesic, in writhing and hot plate tests in mice and showed anti-inflammatory activity against both exudative-proliferative and chronic phases of inflammation on the carrageenan-induced pedal edema test in rats (Shinde et al. 1999). Essential oil from another *Cedrus* genus member, *Cedrus libani* cone essential oil was found to have remarkable wound healing and anti-inflammatory activities together with *Abies cilicica* subsp. *cilicica* cone oil (Tumen et al. 2011). As reported by Gülçin et al. (2003), *Pinus nigra* turpentine had a higher analgesic effect than the reference medicine metamizol in mice writhing test.

Other Bioactivities. Several pine needle oils (*Pinus brutia, P. halepensis, P. nigra, P. pinea,* and *P. sylvestris*) tested as cholinesterase inhibitor, demonstrated moderate to high activity, making them potential therapeutic agent for Alzheimer disease (Ustun et al. 2012). Essential oil of *Pinus koraiensis* could be used against hyperlipidemia, a metabolic disorder inducing obesity, diabetes and cardiovascular diseases (Kim et al. 2012). Siberian fir needle oil was determined to affect bone metabolism *in vivo* by inhibiting bone resorption in a dose dependent manner. Therefore, it can be used to prevent bone loss in osteoporosis models (Mühlbauer et al. 2003). Cedarwood (*Cedrus libani*) oil showed antiviral activity against herpes simplex virus (Loizzo et al. 2008).

After the Second World War, studies on the use of essential oils in aromatherapy were conducted in France and Italy in order to assess their therapeutic action when inhaled. They were found to have remarkable properties in relieving symptoms of anxiety and in mood improvements (Benaim 1984, Warren and Warrenburg 1993). Inspired by the "Shinrin-yoku" (taking in the forest atmosphere or forest bathing) trend in Japan (Tsunetsugu et al. 2010), studies on therapeutic effect of essential oil by inhalation have revealed the calming and relaxing properties of conifer essential oils. α-pinene, a generally detected compound in the air of coniferous forest, was reported to induce relaxed physiological state (Tsunetsugu et al. 2010). Sesquiterpenoid cedrol inhalation was demonstrated to positively modulate cardiovascular and respiratory functions (Dayawansa et al. 2003). *Abies sachalinensis* oil showed a significant anxiolytic-like effect *in vivo* after inhalation on mouse models, providing evidences of therapeutic activity of essential oils on emotional behavior modulation (Satou et al. 2011). Siberian fir oil (*Abies sibirica*), tested by inhalation on humans, seemed to assist recovery from mental fatigue caused by visual display terminal work that could be linked to mental health disturbances and sleep-related psychiatric disorders (Matsubara et al. 2011). Forgotten since long time, healing properties of essential oils by inhalation are yet to be discovered again.

Conifer scent in the perfumer's palette

A deep invigorating breath in the forest undergrowth, the memories of luxuriant Christmas trees fragrance and the relaxing smell of traditional sauna made out of wood illustrate the wide diversity of familiar conifer's scent. While belonging to the woody olfactory family, conifer scents profile range broadly from sweet balsamic aroma of cedar woods to fresh penetrating resinous note of pines, spruces and firs. Woody scents are mostly used as base notes in perfume compositions, providing the lasting impression in the drydown period when top notes are evaporated. Associated with heart notes, base notes create the full body of a fragrance, they ground and support the composition. In addition, they can stay on the skin for hours. Very pliable and adaptable, woody scents are often associated to masculine compositions although they well match feminine fragrances as well.

In the woody note palette, alongside with the famous sandalwood, cedarwood essential oils display sweet and sensual fragrances. True cedars, Atlas cedarwood (*Cedrus atlantica*) and Himalayan cedarwood (*Cedrus deodora*) are deep, warm, woody sweet and slightly camphoraceous. True cedars must not be confused with the so-called Virginian cedarwood (*Juniperus virginiana*) which is actually a juniper. Virginian cedarwood scent is woody-dry, soft, earthy and reminiscent of wood pencils. Cedarwood is extensively used in perfumery as a base note for woody, ambery and oriental fragrances and also as a lighter alternative to precious sandalwood. Both belonging to the oldest fragrances in perfumery and among most commonly used notes, cedarwood is a classic ingredient in the perfumer's palette. Serge Lutens' line of perfume offers diverse interpretations of cedarwood scent, from cedar dedicated "Féminité du bois" and "Cèdre" to harmonies with spices in "Bois oriental" and fruity scents in "Bois et fruits" to a sensual duo in "Bois et musc" (Fragrantica 2017). An example of fragrance dominated by cedarwood would be the very famous designer's perfume "Terre d'Hermès" by Hermès.

On the other side of the woody scents, firs, spruces and pines needle essential oils embody the strong fresh resinous fragrance, characteristic of conifer boreal forest. Pine needle oils (mainly scots pine and maritime pine) are penetrating, icy-fresh and herbaceous, whereas spruces (black spruce, white spruce) are more resinous and balsamic. Fir oils evoke Christmas tree scent: balsam and Siberian firs have a woodier note whereas grand fir and silver fir possess a lemony note. Conifer needle essential oils are less frequently used than sweet cedarwood as a major note for mass perfume compositions. Niche perfumery however would fairly create more committed fragrances, using conifer scent as overreaching structure. Indeed, conifer oils can be employed as top to middle note, and spread to base note. Evocative examples from niche perfumer are "Fille en aiguilles" by Serge Lutens or "Nuit étoilée" by Annick Goutal (Fragrantica 2017), which united pine and fir oils to offer their own representations of pine forests in perfumes. Pine and fir oils are used as a main note in many fragrances from niche perfumers, "In the woods" by Esentielle, "Boisé" by Patyka but also in famous perfume such as "Eau des merveilles" by Hermès or "Kenzo for Men" by Kenzo. Unfortunately, only few compositions contain spruce scents. Natural niche perfumers however created dedicated fragrances like "Black spruce" by Teone Reinthal or "Rainforest" and "Orcas" by Ayala Moriel. Spruce note can also be found in "Tommy Boy" by Tommy Hilfiger as a base note (Fragrantica 2017).

Conifer resin in archeological and art objects

The traditional use of diterpenoid resins in the past was as components of adhesive formulations applied in fabrication of various objects, the application in objects of art being of particular importance for world cultural heritage. The resins most readily available in Europe and especially in the Mediterranean region are of the specific interest for overview here. The resins obtained from conifer species belonging to Pinaceae family are of particular interest in this context. Various pines and larches were the most important sources of resins in terms of quantities produced, mainly due to their important distribution in the region and particularly to large diameters of their resin ducts (especially so in case of pines). Even though the resins from common European conifer species are composed of similar diterpenoid resin acids, they still contain some marker constituents, some of which may influence their rheological properties and thus could have defined their use by artisans and artists from Antiquity through Renaissance and modern times.

Use of resins in adhesive formulations. Pine resins have been most commonly used in Italy and elsewhere in Europe as finishing adhesives in combination with crushed and finely dispersed colored marbles in oils, among which linseed oil was particularly often used in paintings. The study performed by use of gas chromatographic analysis of a sample taken from the altar place from Santo Stefano Church (from XVth century) in Venice provided interesting results in comparison with the results of analysis of the fresh resin from typical pine species belonging to subdivision Diploxylon. Namely, the resins from all Diploxylon pines contain considerable amounts of pimaric/sandaracopimaric acids, the only exception to that pattern being the Aleppo pine (*Pinus halepensis*) growing in the coastal region of Mediterranean. The resin from Aleppo pine is distinguishable by the absence of pimaric/sandaracopimaric acids. Thus, Mills and White (1977) described the analysis and authentication of resinous adhesive from altar of church Santo Stefano in Venice (constructed between XIIth and XVth centuries). The gas chromatogram revealed no trace of pimaric/sandaracopimaric acids which brought up the conclusion that the resin used was effectively from Aleppo pine growing in coastal region.

Yet another example of diterpene resin use as adhesive is mosaic gluing applied in East Corinth in coastal Greece at Aegean Sea. The absence of any other resin acid apart from dehydroabietic acid brought to conclusion that the original resin was used in molten form obtained by intensive heating, probably in order to improve the resin malleability. Indeed, the chromatogram obtained from Diploxylon pine type resin analyzed after heating to 300°C exhibited the same features: major peak for dehydroabietic acid, important concentrations of terpenic hydrocarbons and the absence of the other oxidation product of abietic acid, the 7-oxo-dehydroabietic acid. Its absence is ascribed to non-oxidizing conditions in which the mosaic had dwelled, since the panel of mosaic were submerged in water. They explain the dark color of resin rather by process of heating the resin, excluding the use of tar as such, available otherwise from pine wood dry-distillation (used in Antiquity for charcoal and tar production) since the latter would not have the properties required for slow gluing. The mosaics were dated by ^{14}C analysis to mid-4th century A.D.

Another remarkable example of the importance of use of conifer resins as adhesives is the analysis of a fabric wrapping the detachable battering ram at prove of the warship sunk and found underwater near Marsala in Sicily, dated from Punic wars (3rd century B.C.). It is amazing to what extent the original composition of pine resin was preserved: with abietic acid and isopimaric/palustric acid peaks larger than dehydroabietic acid peak! The anaerobic conditions under the sea, provided by decaying plants (producing methane) could have contributed to protection against oxidation. In this case, the use of conifer tar (from dry distillation of wood) is also a likely material as it was used to provide preservation of wood from which the ships were constructed in the Antiquity (naval stores).

Resin/wax mixtures used as thin protective films in renaissance paintings. The interest for conifer resins use in the past stems not only from a need to understand the ways of their application in various materials and objects of art but also from the need to design the good conservation practices to be applied to the archeological and art objects. The understanding of the chemical composition of the materials used to surface treatment of paintings for example, provides also an insight into the general practices of an artist.

Renaissance artists used very often the Venice turpentine from European larch *Larix decidua* for moldering and waxing the painting surfaces. The name "Venice turpentine" comes from the city where the resin was marketed while it was commonly tapped from larches (*Larix decidua*) growing in the Tirol region of Austria. Wax component was usually a beeswax, constituted by stearic and palmitic acids. A wax relief from Victoria and Albert Museum dating from 16th century was sampled and the methanol extract was analyzed after methylation by gas chromatography. The typical larch resin constituents, epi-manool, larixol and larixol acetate (Fig. 4) were determined, along with the oxidation products of abietane type acids, dehydroabietic and 7-oxo-dehydroabietic acids (Fig. 1). The conclusion of Mills and White was that the Venice turpentine had probably been mixed with cheaper pine resin for relief molding used by Giovanni Bologna (Mills and White 1977).

Also often used resin in Renaissance was that from cypresses, rich in communic acid derivatives. It is amazing to discover how Leonardo da Vinci cautioned about time and mode of sampling of cypress resin which "had to be collected in April-May and used quickly after collection" (Kononenko et al. 2016). Even in relation to this material uses of resin, Leonardo da Vinci's extraordinary capacity of observation is again evidenced. The commercially available sandarac resin, which is provided by tapping of Atlas thuja (*Tetraclinis articulata*) growing in Northern Africa, a member of Cupressaceae family as well, was examined to study the polymerization of its constituent communic acid.

Russo and Avino (2012) have examined the terpenoid resin used as binder in the XV century painting of Antonello da Messina from Uffizi Gallery collection. Antonello da Messina is the Renaissance master very much inspired by Flemish painting. He is credited to be the first artist who introduced oil painting in Italy. The oil painting is applying the protection layer consisting of a mixture of a terpenic resin and siccative oil, most commonly linseed oil (mixture of palmitic and stearic acid and unsaturated fatty acids). The sample from the da Messina painting "Madona con bambino e angelli" was examined by micro FT-IR and GC-MS (Russo and Avino 2012). The authors declared that the resin used was Venice turpentine from European larch even though they were unable to detect larixol or its acetate, which is curious as these constituents are usually found in Venice turpentine which was the most appreciated at the time of Antonello da Messina work. It is also interesting that the larch resin (Venice turpentine) has been otherwise proven to survive centuries. Thus, the results of gas-chromatographic analysis of methylated resin fraction of the material used by Giovanni da Bologna (known also as Gianbologna) to create the relief "Christ Rejected by the Jews" (relief dated 1579, ECCE HOMO, now exposed in Victoria and Albert Museum in London), clearly confirmed the presence of epi-manool, larixol and larixyl acetate, marker constituents of Venice turpentine. The very important concentration of dehydroabietic acid in examined sample is attributed to the probable use of less expensive pine resin (Mills and White 1977) which could have been also the case with the painting of da Messina discussed previously.

The wax reliefs of Gianbologna served for casting of bronze reliefs ordered for the Grimaldi Chapel in San Francesco di Castelletto, Genoa, are now conserved at the University of Genoa. It is curious to read that the Victoria and Albert Museum panel describes the object as "red wax on wooden ground", without a mention of the resin ingredients which are certainly of primary importance for its conservation. We cite "Wax is a delicate medium and therefore the physical survival of these pieces is remarkable, suggesting they were especially valued". They certainly have been valued which explains their conserving after casting in bronze, which would not have been a common practice for less valuable work. The study of Mills and White adds to our understanding of the survival of these objects for over five centuries and crucial to understanding of the sculpturing work of Gianbologna, for which use of conifer resin was of prime importance.

Conclusion

Conifer species belonging to Pinaceae and Cupressaceae family are widely distributed in forests of Europe and North America. They are well-known for production of terpenoid resins which are composed of volatile part containing mono- and sesquiterpenoids and solid part constituted of various diterpenoids. The extraordinary properties of these complex molecules have been exploited in the past in various fields, for design of different materials, but also in health and art products. They are continually being explored for their fragrance properties in perfumery and their biological properties are continuing to interest researchers for applications in health products and aromatherapy. These fascinating molecules have helped preserve numerous objects of art which make part of world heritage. It is therefore extremely important to share the knowledge about these precious forest substances and continue the research for finding their new and innovative applications.

References

Andersen, A., H. Gagnon, G. Collin and R.P. Adams. 1995. Essential oil of the wood of *Thuja occidentalis* L. J. Essent. Oil Res. 7: 489–495.

Arnason, T., R.J. Hebda and T. Johns. 1981. Use of plants for food and medicine by native peoples of eastern Canada. Can. J. Bot. 59: 2189–2325.

Bağci, E. and M. Diğrak. 1996. Antimicrobial activity of essential oils of some *Abies* (Fir) species from Turkey. Flavour. Frag J. 11: 251–256.

Bağci, E., S. Hayta and G. Dogan. 2011. Chemical composition of essential oils from bark and leaves of *Pinus brutia* ten. from Turkey. Asian J. Chem. 23: 2782.

Balasundram, N., K. Sundram and S. Samman. 2006. Phenolic compounds in plants and agri-industrial by-products: Antioxidant activity, occurrence, and potential uses. Food Chem. 99: 191–203.

Banthorpe, D., C. Gatford and S. Williams. 1973. Monoterpene patterns in *Juniperus* and *Thuja* species. Planta Med. 23: 64–69.

Barrero, A.F., M. Herrador, P. Arteaga, J.F. Arteaga and A.F. Arteaga. 2012. Communic acids: Occurrence, properties and use as chirons for the synthesis of bioactive compounds. Molecules 17: 1448–1467.

Başer, K. and B. Demircakmak. 1995. The essential oil of taurus cedar (*Cedrus libani* A. rich): Recent results. Chem. Nat. Compd. 31: 16–20.

Benaim, C. 1984. Fragrances & moods: New perspectives. Perfumer & Flavorist 9: 13–16.

Burt, S. 2004. Essential oils: their antibacterial properties and potential applications in foods—a review. Int. J. Food Microbiol. 94: 223–253.

Canillac, N. and A. Mourey. 2001. Antibacterial activity of the essential oil of *Picea excelsa* on *Listeria*, *Staphylococcus aureus* and coliform bacteria. Food Microbiol. 18: 261–268.

Chalchat, J.-C., R.-P. Garry, A. Michet and B. Benjilali. 1994. Essential oil components in sawdust of *Cedrus atlantica* from morocco. J. Essent. Oil Res. 6: 323–325.

Chaouche, T.M., F. Haddouchi, R. Ksouri, F. Medini and F. Atik-Bekara. 2013. *In vitro* evaluation of antioxidant activity of the hydro-methanolic extracts of *Juniperus oxycedrus* subsp. *oxycedrus*. Phytothérapie 11: 244–249.

Chaudhary, A., P. Kaur, B. Singh and V. Pathania. 2009. Chemical composition of hydrodistilled and solvent volatiles extracted from woodchips of Himalayan Cedrus: *Cedrus deodara* (Roxb.) Loud. Nat. Prod. Commun. 4: 1257–1260.

Chen, W., Y. Liu, M. Li, J. Mao, L. Zhang, R. Huang, X. Jin and L. Ye. 2015. Anti-tumor effect of alpha-pinene on human hepatoma cell lines through inducing G2/M cell cycle arrest. J. Pharmacol. Sci. 127: 332–338.

Cho, S.-M., E.-O. Lee, S.-H. Kim and H.-J. Lee. 2014. Essential oil of *Pinus koraiensis* inhibits cell proliferation and migration via inhibition of p21-activated kinase 1 pathway in HCT116 colorectal cancer cells. BMC Complement Altern. Med. 14: 275.

Dayawansa, S., K. Umeno, H. Takakura, E. Hori, E. Tabuchi, Y. Nagashima, H. Oosu, Y. Yada, T. Suzuki and T. Ono. 2003. Autonomic responses during inhalation of natural fragrance of "Cedrol" in humans. Auton. Neurosci. 108: 79–86.

Desrochers, L. 2011. Récolte et ségrégation de la biomasse ligneuse en forêt (Harvest and segregation of lineous biomass in forest), QWEB - Les extractibles forestiers (Forest extractibles), FPInnovations (https://fpinnovations.ca/).

Dorman, H.J.D. and S.G. Deans. 2000. Antimicrobial agents from plants: antibacterial activity of plant volatile oils. J. Appl. Microbiol. 88: 308–316.

Ennajar, M., J. Bouajila, A. Lebrihi, F. Mathieu, M. Abderraba, A. Raies and M. Romdhane. 2009. Chemical composition and antimicrobial and antioxidant activities of essential oils and various extracts of *Juniperus phoenicea* L. (Cupressacees). J. Food Sci. 74: M364–371.

Fleisher, A. 1991. Water-soluble fractions of the essential oils. Perfumer and Flavorist 16: 37–41.

Fragrantica. 2017. Retrieved online from https://www.fragrantica.com (perfumery database).

Francezon, N., T. Stevanovic. 2017a. Chemical composition of essential oil and hydrosol from *Picea mariana* bark residue. BioResources 12: 2635–2645.

Francezon, N. and T. Stevanovic. 2017b. Integrated process for the production of natural extracts from black spruce bark. Industrial Crops and Products 108: 348–354.

Garneau, F.-X., G. Collin, H. Gagnon and A. Pichette. 2012. Chemical composition of the hydrosol and the essential oil of three different species of the Pinaceae family: *Picea glauca* (Moench) Voss., *Picea mariana* (Mill.) B.S.P., and *Abies balsamea* (L.) Mill. J. Essent. Oil Bear. Pl 15: 227–236.

Goldblatt, L. 1952. American turpentines. pp. 253–308. *In*: Guenther, E. (ed.). The Essential Oils. D. Van Nostrand Company, Inc, New York, USA.

Gülçin, İ., M.E. Büyükokuroğlu, M. Oktay and Ö.İ. Küfrevioğlu. 2003. Antioxidant and analgesic activities of turpentine of *Pinus nigra* Arn. subsp. *pallasiana* (Lamb.) Holmboe. J. Ethnopharmacol. 86: 51–58.

Höld, K.M., N.S. Sirisoma, T. Ikeda, T. Narahashi and J.E. Casida. 2000. α-thujone (the active component of absinthe): γ-aminobutyric acid type A receptor modulation and metabolic detoxification. Proc. Natl. Acad. Sci. U.S.A. 97: 3826–3831.

Hong, E.-J., K.-J. Na, I.-G. Choi, K.-C. Choi and E.-B. Jeung. 2004. Antibacterial and antifungal effects of essential oils from coniferous trees. Biol. Pharm. Bull. 27: 863–866.

Hudson, J., M. Kuo and S. Vimalanathan. 2011. The antimicrobial properties of cedar leaf (*Thuja plicata*) oil; a safe and efficient decontamination agent for buildings. Int. J. Environ. Res. Public. Health 8: 4477–4487.

Jeong, S.I., J.P. Lim and H. Jeon. 2007. Chemical composition and antibacterial activities of the essential oil from *Abies koreana*. Phytother. Res. 21: 1246–1250.

Jo, J.R., J.S. Park, Y.K. Park, Y.Z. Chae, G.H. Lee, G.Y. Park and B.C. Jang. 2012. *Pinus densiflora* leaf essential oil induces apoptosis via ROS generation and activation of caspases in YD-8 human oral cancer cells. Int. J. Oncol. 40: 1238–1245.

Kähkönen, M.P., A.I. Hopia, H.J. Vuorela, J.-P. Rauha, K. Pihlaja, T.S. Kujala and M. Heinonen. 1999. Antioxidant activity of plant extracts containing phenolic compounds. J. Agric. Food Chem. 47: 3954–3962.

Kamdem, P.D., J.W. Hanover and D.A. Gage. 1993. Contribution to the study of the essential oil of *Thuja occidentalis* L., 1992. J. Essent. Oil Res. 5: 117–122.

Kartnig, T., F. Still and F. Reinthaler. 1991. Antimicrobial activity of the essential oil of young pine shoots (*Picea abies* L.). J. Ethnopharmacol. 35: 155–157.

Kim, J.H., H.J. Lee, S.J. Jeong, M.H. Lee and S.H. Kim. 2012. Essential oil of *Pinus koraiensis* leaves exerts antihyperlipidemic effects via up-regulation of low-density lipoprotein receptor and inhibition of acyl-coenzyme A: cholesterol acyltransferase. Phytother. Res. 26: 1314–1319.

Kızılarslan, Ç. and E. Sevg. 2013. Ethnobotanical uses of genus *Pinus* L. (Pinaceae) in Turkey. Indian J. Tradit. Know. 12(2): 209–220.

Kononenko, I., L. de Viguerie, S. Rochut and P. Walter. 2016. Qualitative and quantitative studies of chemical composition of sandarac resin by GC-MS. Environ. Sci. Pollut. Res. 24(3): 2160–2165.

Kotan, R., S. Kordali and A. Cakir. 2007. Screening of antibacterial activities of twenty-one oxygenated monoterpenes. Z Naturforsch C 62: 507–513.

Lee, J.-H., K. Lee, D.H. Lee, S.Y. Shin, Y. Yong and Y.H. Lee. 2015. Anti-invasive effect of β-myrcene, a component of the essential oil from *Pinus koraiensis* cones, in metastatic MDA-MB-231 human breast cancer cells. J. Korean Soc. Appl. Biol. Chem. 58: 563–569.

Legault, J., W. Dahl, E. Debiton, A. Pichette and J.-C. Madelmont. 2003. Antitumor activity of balsam fir oil: production of reactive oxygen species induced by α-humulene as possible mechanism of action. Planta Med. 69: 402–407.

Lis-Balchin, M. and S. Deans. 1997. Bioactivity of selected plant essential oils against *Listeria monocytogenes*. J. Appl. Microbiol. 82: 759–762.

Loizzo, M.R., A. Saab, R. Tundis, G.A. Statti, I. Lampronti, F. Menichini, R. Gambari, J. Cinatl and H.W. Doerr. 2008. Phytochemical analysis and *in vitro* evaluation of the biological activity against herpes simplex virus type 1 (HSV-1) of *Cedrus libani* A. Rich. Phytomedicine 15: 79–83.

Manach, C., A. Mazur and A. Scalbert. 2005. Polyphenols and prevention of cardiovascular diseases. Curr. Opin. Lipidol. 16: 77–84.

Maruzzella, J.C. and N.A. Sicurella. 1960. Antibacterial activity of essential oil vapors. J. Pharm. Sci. 49: 692–694.

Matsubara, E., M. Fukagawa, T. Okamoto, K. Ohnuki, K. Shimizu and R. Kondo. 2011. The essential oil of *Abies sibirica* (Pinaceae) reduces arousal levels after visual display terminal work. Flavour. Frag. J. 26: 204–210.

Mercier, B., J. Prost and M. Prost. 2009. The essential oil of turpentine and its major volatile fraction (α- and β-pinenes): a review. Int. J. Occup. Med. Environ. Health 22: 331–342.

Middleton, E., C. Kandaswami and T.C. Theoharides. 2000. The effects of plant flavonoids on mammalian cells: implications for inflammation, heart disease, and cancer. Pharmacol. Rev. 52: 673–751.

Mills, J.S. and R. White. 1977. Natural resins of art and archaeology their sources, chemistry, and identification. Stud. Conserv. 22: 12–31.

Mirov, N. 1946. *Pinus*: a contribution of turpentine chemistry to dendrology and forest genetics. J. Forest. 44: 13–16.

Mirov, N. 1948. The terpenes (in relation to the biology of genus *Pinus*). Annu. Rev. Biochem. 17: 521–540.

Morris, J., A. Khettry and E. Seitz. 1979. Antimicrobial activity of aroma chemicals and essential oils. J. Am. Oil Chem. Soc. 56: 595–603.

Moure, A., J.M. Cruz, D. Franco, J.M. Domínguez, J. Sineiro, H. Domínguez, M.a. José Núñez and J.C. Parajó. 2001. Natural antioxidants from residual sources. Food Chem. 72: 145–171.

Mühlbauer, R.C., A. Lozano, S. Palacio, A. Reinli and R. Felix. 2003. Common herbs, essential oils, and monoterpenes potently modulate bone metabolism. Bone 32: 372–380.

Oh, H.-J., H.-M. Ahn, K.-H. So, S.-S. Kim, P.-Y. Yun, G.-L. Jeon and K.-Z. Riu. 2007. Chemical and antimicrobial properties of essential oils from three coniferous trees *Abies koreana*, *Cryptomeria japonica*, and *Torreya nucifera*. J. Appl. Biol. Chem. 50: 164–169.

Orav, A., K. Kuningas and T. Kailas. 1995. Computerized capillary gas chromatographic identification and determination of Siberian fir oil constituents. J. Chromatogr A. 697: 495–499.

Pauli, A. and H. Schilcher. 2010. *In vitro* antimicrobial activities of essential oils monographed in the European Pharmacopoeia 6th edition. pp. 353–540. *In*: Başer, K.H.C. and G. Buchbauer (eds.). Handbook of Essential Oils : Science, Technology, and Applications. Taylor & Francis, Boca Raton.

Pelkonen, O., K. Abass and J. Wiesner. 2013. Thujone and thujone-containing herbal medicinal and botanical products: Toxicological assessment. Regul. Toxicol. Pharmacol. 65: 100–107.

Pichette, A., P.L. Larouche, M. Lebrun and J. Legault. 2006. Composition and antibacterial activity of *Abies balsamea* essential oil. Phytother. Res. 20: 371–373.

Poaty, B., J. Lahlah, F. Porqueres and H. Bouafif. 2015. Composition, antimicrobial and antioxidant activities of seven essential oils from the North American boreal forest. World J. Microbiol. Biotechnol. 31: 907–919.

Price, L. and S. Price. 2004. Understanding Hydrolats: The Specific Hydrosols for Aromatherapy: A Guide for Health Professionals. Churchill Livingstone, Edinburgh.

Rao, B.R., P. Kaul, K. Syamasundar and S. Ramesh. 2002. Water soluble fractions of rose-scented geranium (*Pelargonium* species) essential oil. Bioresour. Technol. 84: 243–246.

Ritch-Krc, E., N. Turner and G. Towers. 1996. Carrier herbal medicine: an evaluation of the antimicrobial and anticancer activity in some frequently used remedies. J. Ethnopharmacol. 52: 151–156.

Ross, J., H. Gagnon, D. Girard and J.-M. Hachey. 1996. Chemical composition of the bark oil of balsam fir *Abies balsamea* (L.) Mill. J. Essent. Oil Res. 8: 343–346.

Roy, D., S.K. Konar and J.R. Purdy. 1984. Sublimation of occidentalol, a sesquiterpene alcohol, from eastern white cedar (*Thuja occidentalis*) in a drying kiln. Can. J. Forest. Res. 14: 401–403.

Rudloff, E. 1962. Gas–liquid chromatography of terpenes VI. The volatile oil of *Thuja plicata* Donn. Phytochemistry 1: 195–202.

Rudloff, E. 1967. Chemosystematic studies in the genus *Picea* (Pinaceae): II. the leaf oil of *Picea glauca* and *P. mariana*. Can. J. Bot. 45: 1703–1714.

Rudloff, E. 1975. Seasonal variation in the terpenes of the foliage of black spruce. Phytochemistry 14: 1695–1699.

Rusanov, K., E. Garo, M. Rusanova, O. Fertig, M. Hamburger, I. Atanassov and V. Butterweck. 2014. Recovery of polyphenols from rose oil distillation wastewater using adsorption resins—a pilot study. Planta Med. 80: 1657–1664.

Russo, M.V. and P. Avino. 2012. Characterization and identification of natural terpenic resins employed in "Madonna con Bambino e Angeli" by Antonello da Messina using gas chromatography–mass spectrometry. Chem. Cent. J. 6: 1–10.

Saab, A.M., A. Guerrini, G. Sacchetti, S. Maietti, M. Zeino, J. Arend, R. Gambari, F. Bernardi and T. Efferth. 2012. Phytochemical analysis and cytotoxicity towards multidrug-resistant leukemia cells of essential oils derived from Lebanese medicinal plants. Planta Med. 78: 1927–1931.

Sacchetti, G., S. Maietti, M. Muzzoli, M. Scaglianti, S. Manfredini, M. Radice and R. Bruni. 2005. Comparative evaluation of 11 essential oils of different origin as functional antioxidants, antiradicals and antimicrobials in foods. Food Chem. 91: 621–632.

Sato, M., K. Seki, K. Kita, Y. Moriguchi, M. Hashimoto, K. Yunoki and M. Ohnishi. 2009. Comparative analysis of diterpene composition in the bark of the hybrid larch F 1, *Larix gmelinii* var. *japonica* × *L. kaempferi* and their parent trees. J. Wood Sci. 55: 32–40.

Satou, T., M. Matsuura, M. Takahashi, T. Umezu, S. Hayashi, K. Sadamoto and K. Koike. 2011. Anxiolytic-like effect of essential oil extracted from *Abies sachalinensis*. Flavour Frag. J. 26: 416–420.

Selim, S.A., M.E. Adam, S.M. Hassan and A.R. Albalawi. 2014. Chemical composition, antimicrobial and antibiofilm activity of the essential oil and methanol extract of the Mediterranean cypress (*Cupressus sempervirens* L.). BMC Complement Altern. Med. 14: 179.

Sell, C. 2010. Chemistry of essential oils. *In*: Baser, K.H.C. and G. Buchbauer (eds.). Handbook of Essential Oils: Science, Technology, and Applications. CRC Press, Boca Raton, USA.

Setzer, W.N., M.C. Setzer, D.M. Moriarity, R.B. Bates and W.A. Haber. 1999. Biological activity of the essential oil of *Myrcianthes* sp. nov. "black fruit" from Monteverde, Costa Rica. Planta Med. 65: 468–469.

Shinde, U., A. Phadke, A. Nair, A. Mungantiwar, V. Dikshit and M. Saraf. 1999. Studies on the anti-inflammatory and analgesic activity of *Cedrus deodara* (Roxb.) Loud. wood oil. J. Ethnopharmacol. 65: 21–27.

Strack, D., J. Heilemann, V. Wray and H. Dirks. 1989. Structures and accumulation patterns of soluble and insoluble phenolics from Norway spruce needles. Phytochemistry 28: 2071–2078.

Surburg, H. and J. Panten. 2006. Natural raw materials in the flavor and fragrance industry. pp. 177–238. *In*: Surburg, H. and J. Panten (eds.). Common Fragrance and Flavor Materials. Wiley-VCH Verlag GmbH & Co. KGaA, Weinheim.

Teranishi, R., R.G. Buttery and H. Sugisawa. 1993. Bioactive volatile compounds from plants. American Chemical Society, Washington.

Tsunetsugu, Y., B.J. Park and Y. Miyazaki. 2010. Trends in research related to "Shinrin-yoku" (taking in the forest atmosphere or forest bathing) in Japan. Environ. Health Prev. Med. 15: 27–37.

Tumen, I., E.K. Akkol, I. Suntar and H. Keles. 2011. Wound repair and anti-inflammatory potential of essential oils from cones of Pinaceae: preclinical experimental research in animal models. J. Ethnopharmacol. 137: 1215–1220.

Urdang, G. 1952. History, origin in plants, production, analysis. p. 1. *In*: Guenther, E. (ed.). The Essential Oils. D. Van Nostrand Company, Inc, New York, USA.

Ustun, O., F.S. Senol, M. Kurkcuoglu, I.E. Orhan, M. Kartal and K.H.C. Baser. 2012. Investigation on chemical composition, anticholinesterase and antioxidant activities of extracts and essential oils of Turkish *Pinus* species and pycnogenol. Ind. Crops Prod. 38: 115–123.

Warren, C. and S. Warrenburg. 1993. Mood benefits of fragrance. Perfumer & Flavorist 18: 9–16.

Wender, P.A., N.F. Badham, S.P. Conway, P.E. Floreancig, T.E. Glass, C. Granicher, J.B. Houze, J. Janichen, D. Lee and D.G. Marquess. 1997. The pinene path to taxanes. 5. Stereocontrolled synthesis of a versatile taxane precursor. J. Am. Chem. Soc. 119: 2755–2756.

Yang, J., Q. Liang, M. Wang, C. Jeffries, D. Smithson, Y. Tu, N. Boulos, M.R. Jacob, A.A. Shelat and Y. Wu. 2014. UPLC-MS-ELSD-PDA as a powerful dereplication tool to facilitate compound identification from small molecule natural product libraries. J. Nat. Prod. 77: 902.

Zolfaghari, B. and S. Iravani. 2012. Essential oil constituents of the bark of *Pinus pinaster* from Iran. J. Essent. Oil Bear Pl 15: 348–351.

Flavonoids from *Quercus* Genus: Applications in Melasma and Psoriasis

Esquivel-García Roberto,[1,2] Velázquez-Hernández
María-Elena,[2] Valentín-Escalera Josué,[1] Valencia-Avilés
Eréndira,[1] Rodríguez-Orozco Alain-Raimundo[3] and
Martha-Estrella García-Pérez[1,*]

Introduction

Quercus trees, commonly known as oaks, belong to the Fagaceae family (Kubitzki et al. 1993, Nixon 1993, Larios et al. 2013, Bellarosa et al. 2005, Luna-José et al. 2003, Sánchez-Burgos et al. 2013). The *Quercus* genus consists of around 400–500 species and is one of the most important tree genera being distributed in the Northern Hemisphere, in America, Europe, Asia and Africa (Iftikhar et al. 2009, Lesur et al. 2015, Nixon 1993).

Oaks constitute important forest resources that are harvested for a wide range of uses (biomass, fiber, wood products, and food). Besides their contributions to the economy, they are keystone species in forest ecosystems and are major drivers of

[1] Facultad de Químico Farmacobiología. Universidad Michoacana de San Nicolás de Hidalgo. Morelia, Mich., México.
[2] Centro Multidisciplinario de Estudios en Biotecnología-FMVZ, Universidad Michoacana de San Nicolás de Hidalgo. Tarímbaro, Mich., México.
[3] Facultad de Ciencias Médicas y Biológicas. Universidad Michoacana de San Nicolás de Hidalgo. Morelia, Mich., México.
* Corresponding author: margarc@live.ca

terrestrial biodiversity due to their longevity and their large geographic distribution (Brändle and Brandl 2001). In most countries, they are also considered as major patrimonial and cultural resources despite their economic, ecological, and cultural importance (Curtu et al. 2007, Kremer et al. 2012).

Oaks have been also valued for their ecological role to restrain desertification and soil erosion because of the mineral composition of their leaves that form stable aggregates of good size (SIRE 2012). Additionally, oaks have contributed to biodiversity maintenance as they harbor large communities of rodents, birds, reptiles, arachnid insects and epiphytes such as orchids, bromeliads, ferns, fungi and bryophytes (Brändle and Brandl 2001, Correia et al. 2014, Valencia-A 2004). Oaks represent a great benefit from the ongoing climate warming (Sallé et al. 2014) due to their drought resistance (*Q. petraea* (Matt.) Liebl. and *Q. pubescens* Willd.) and preference for higher temperature (Rellstab et al. 2016).

Oaks are traditionally used in cooperage, mainly the species *Q. alba* L. (American oak) and two European species: *Q. robur* L. (pedunculate oak) and *Q. petraea* (sessile oak) (Alañón et al. 2011b). The cork oak is often used for production of wine stoppers, insulation materials, tannin extraction and fungi cultivation (Şen et al. 2012, Şen et al. 2010). However in recent decades, overexploitation has led to the destruction and degradation of cork habitat (Wu et al. 2013).

The oak wood of some species has been used mainly for industrial uses, firewood and charcoal (Alañón et al. 2011c) while other species has linked its uses with indigenous groups, who have used it for various purposes such as medicinal, to treat gastrointestinal disorders (diarrhea, gastritis and ulcer) and skin. It is also used to treat throat infection, hemorrhages and dysentery (Berahou et al. 2007). Other uses involve consumption of fresh or processed food; crafts and the elaboration of different articles like rosaries or toys. Oaks are also used as fodder, mainly for swine and goat feeding; for skin tanning, as a mordant and for dyeing (Luna-José et al. 2003).

Oaks are considered a rich source of bioactive molecules mainly polyphenols. Indeed, these species constitute classic model systems in elucidating the role of polyphenols for the plant defense (Salminen et al. 2004). Within polyphenols, flavonoids are very important, since their significant effects on human health. Several studies have been performed about the flavonoid composition in oaks. Considering that Mexico is the second most important center of diversity for *Quercus* genus worldwide, this chapter presents the distribution and characteristics of these species with a special focus on the description of flavonoids present in oaks. Furthermore, we analyze the effect of these molecules for the treatment of melasma and psoriasis, two dermatological diseases of high prevalence. Finally, the challenges that need to be overcome for future research with these molecules as therapeutic candidates for both diseases are discussed.

Quercus Genus: Distribution and Characteristics

The distribution of *Quercus* genus encompasses strong ecological and climatic gradients in the Eurasiatic as in the American continents, in an almost continuous pattern. They cover very large continuous forest from the tropical to the boreal regions, and from sea level up to 4,000 m in the Himalayas, and expressed in very different life history traits (Axelrod 1983).

Traditionally in the *Quercus* genus, two subgenera have been recognized— *Cyclobalanopsis* and *Quercus*, the latter comprising three sections: *Lobatae* (red oaks), *Protobalanus* (golden cup or intermediate oaks), and *Quercus* (white oaks) (Manos et al. 1999, Nixon 1993). However, using molecular phylogenetic two major classes have been identified (Bellarosa et al. 2005, Denk and Grimm 2010, Manos et al. 1999), emphasizing a putative origin and subsequent diversification in America followed by at least one dispersal of an ancestral group of white oaks to Eurasia that gave rise to a small radiation of approximately 20 species placed in section *Quercus* spp. The larger of the two clades has been referred as the New World oaks with sections *Lobatae* (red and black oaks), *Protobalanus* (golden cup oaks), and *Quercus* spp. (white oaks). The smaller clades are strictly Eurasian and englobes two groups: the *Cerris* oaks (including species of the *Ilex* group), with approximately 50 species of temperate and semiarid regions and the tropical, evergreen *Cyclobalanopsis* or cycle cup oaks with about 90 species (Bellarosa et al. 2005, Kremer 2016, Kremer et al. 2012, Kremer et al. 2010, Manos et al. 1999).

Despite the general characteristics, the taxonomic classification of the genus *Quercus* (the oaks) is a hard task due to the unusually high frequency of interspecific hybridization and introgression as a consequence of the adaptive evolution and diversification of many plant groups (Sullivan et al. 2016, Tovar-Sánchez and Oyama 2004). Due to oaks have been taken as model species, they have been studied in terms of their genomic delineation, ecological speciation and plant adaptation (Curtu et al. 2007, Grivet et al. 2008, Kremer et al. 2012, Valencia-Cuevas et al. 2014). Hybridization in the oaks has been used to argue for alternatives to the biological species concept, as well as to suggest a mechanism by which species can adapt genotypically to a changing ecological landscape (Tovar-Sánchez et al. 2012, Valencia-Cuevas et al. 2014).

The level of hybridization and the introgression phenomena of oaks depends on the species characteristics but also could be influenced by the environmental context like their habitat, geographical localization (Tovar-Sánchez and Oyama 2004), relative abundance, rates of gene flow (Curtu et al. 2007), the survivorship of hybrid individuals, reproductive system and the density of individuals available to mating (Valencia-Cuevas et al. 2014).

Although morphological patterns of variation that support or are consistent with hybridization between many oak species pairs have been reported (González-Rodríguez et al. 2004, Sallé et al. 2014), two centers of diversity for gender are recognized. The first is in Southeast Asia with about 125 species; the second one is presented in Mexico, with about 150 species, within which 86 are endemics

(Nixon 1993). Paleobotanical evidence suggests that the cooler, drier and more variable climates developed after the Eocene-Oligocene transition in North America encouraged the evolution and migration of *Quercus* (González-Rodríguez et al. 2004, Valencia-A 2004). The overall number of species in the New World, including Latin America, the United States and Canada, is probably around 220 species. Figure 1 shows the number of endemic species in some American countries (Nixon 1993, Nixon 2006).

While in Europe, the oak tree-ring chronologies are amongst the longest in the world, the genus gathers in this continent at least 17 indigenous species plus numerous sub-species, hybrids and several introduced species (Gričar et al. 2015, Sallé et al. 2014).

All oaks share a series of common biological characteristics: woody stems, leaves with leather-like consistency (leathery or hard) and presence of a circular single fruit or acorn surrounded by the conspicuous accessory referred to as the cupule or involucres (Kremer et al. 2012). The *Quercus* seeds are good sources of oil and contain high amounts of unsaturated fatty acids (Charef et al. 2008).

Quercus growth is commonly as a tree (with a height of 3 to 40 m) and some as shrub (with heights of 10 to 60 cm). Their development is slow, because they are long-lived. In evergreen oaks, the foliage always stays green and is gradually replacing. In contrast, deciduous oaks lose all their leaves during the dry season. The leaves can have different types of apex (leaf tip), leaf base, number of ribs, margin (or leaf border), texture, sizes and colors, morphological properties that are used in the taxonomy for classification. An important characteristic is that they are hermaphrodite because it has both, male and female flowers (Arizaga et al. 2009).

Subtropical oaks, including those of North America and Europe, have relatively consistent patterns of flowering and fruit production. Most temperate species have a characteristic flowering time in the spring months (usually between February and

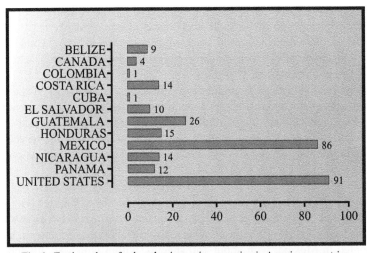

Fig. 1. Total number of oak endemic species occurring in American countries.

June, depending on the species and latitude), and a fairly fixed fall fruit production period, mostly in September-November, with the greatest production in October (Nixon 2006).

Flavonoid Composition in *Quercus* Genus

Oaks contain a variety of compounds which are bioactive molecules. Among them polyphenols, which are ubiquitous plants constituents and important for human diet. Indeed, these compounds are common components of vegetables and fruits being of considerable interest due to their bioactive properties. Polyphenols have shown protective effects against degenerative diseases such as cardiovascular diseases, cancers and infections (bacterial and viral) (Kumar and Pandey 2013). These medicinal effects have been attributed to their antioxidant properties and the capacity to interfere with pathways involved in disease pathogenesis (Ignat et al. 2011, Stevanovic et al. 2009, Sun et al. 2011).

Flavonoids constitute the largest group of plant phenolic and are very effective antioxidants (Kuppusamy et al. 2016). They possess a variable phenolic structure, being synthesized by phenyl-propanoid pathway in response to stress such as microbial infection and strong light exposure (Brossa et al. 2009). Their activities are structure-dependent.

The flavonoid classification depends on their degree of hydroxylation, presence of substitutions, conjugations and polymerization degree (Kumar and Pandey 2013). Flavonoid chemical structures are based upon a fifteen-carbon skeleton consisting of two benzene rings linked via a heterocyclic pyrane ring (C6C3C6). This structure forms an oxygenated heterocycle except in the cases of chalcones and dihydrochalcones (Kumar and Pandey 2013, Routray and Orsat 2012, Stalikas 2007). The various classes of flavonoids differ in the level of oxidation and pattern of substitution of the C ring, while individual compounds within a class differ in the pattern of substitution of the A and B rings (Kumar and Pandey 2013). Major classes of flavonoids based on abundance in food are: flavonols, flavones, isoflavones, flavanones, flavandiols, anthocyanins, proanthocyanidins, and catechins (Bravo 1998). Other classes of flavonoids include flavan-3-ols, anthocyanidins, chalcones and other biosynthetic intermediates of the flavonoid biosynthesis such as aurones, biflavonoids, and dihydrochalcones (Rice-Evans et al. 1996, Routray and Orsat 2012, Stalikas 2007). The first flavonoid isolated in 1930 from oranges was the rutin and now between 4,000 and 6,000 varieties of these compounds have been identified (Kumar and Pandey 2013, Stevanovic et al. 2009).

Many classes of flavonoids are described in *Quercus* genus which have been correlated with their antioxidant activity (Alañón et al. 2011a). Several studies have pointed out a strong variability in oak flavonoid composition depending on the species, tissues used during determinations, stave, stand, type of forest and altitude (Alañón et al. 2011c, Castro-Vázquez et al. 2013).

Oak acorns are a rich source of carbohydrates, amino acids, proteins, lipids and various sterols (Rakić et al. 2007). Several scientific studies have demonstrated a high concentration of phenolic compounds and flavonoids in various oak species (Tejerina et al. 2011). Gallic acid, digallic acid and gallotannins were identified in the ethyl-acetate fraction of *Q. acutissima* Carruth. while the methanol extracts of acorns of *Q. ilex* L., *Q. rotundifolia* Lam., and *Q. suber* L. contain hydrolysable tannins (Cantos et al. 2003, Custódio et al. 2015, Rakić et al. 2007). Caffeic acid, gallic acid, (+)-catechin, theophylline, gallotannins, ellagitannins, condensed tannins and catechol were found in acorns of *Q. suber* (Tejerina et al. 2011).

Hydrolysable tannins and gallic acid were identified in extracts of native and thermally treated kernels of *Q. robur* and *Q. cerris* L. Non-tannin phenolics, including gallic acid were present in significantly higher quantities in thermally treated samples, whilst tannin content decreased. Flavonoids were present in very small quantities in all investigated samples, with higher content in *Q. cerris* kernels (Rakić et al. 2007). In *Q. robur* bark extracts, p-coumaric, syringic, sinapic acids, (+)-catechin, (–)-epicatechin, (–)-epigallocatechin and naringenin have been identified (Bouras et al. 2015).

Rosales-Castro et al. determined that *Q. sideroxyla* Bonpl. bark is as a good source of therapeutic health products or nutraceutical ingredients, due to the presence of polyphenols like gallic acid, catechin, epicatechin, gallocatechin, catechin gallate, dimeric procyanidins, galloylated dimeric proanthocyanidins, trimeric procyanidins and tetrameric proanthocyanidins (Rosales-Castro et al. 2012).

Catechin is the major compound in *Q. durifolia* Seemen and *Q. eduardi* Trel. while epicatechin and gallocatechin gallate were detected at low concentrations. In both species was determined the existence of procyanidin monogallates, procyanidin digallate, procyanidin dimer aglycone, procyanidin trimer aglycones, procyanidin tetramers (Rosales-Castro et al. 2011). Catechin, epicatechin, procyanidin monogallate, procyanidin dimer, procyanidin dimer monogallate, procyanidin trimer, procyanidin tetramer have been described in *Q. durifolia*, while in *Q. eduardii* barks, procyanidin monogallate, procyanidin dimer, procyanidin dimer monogallate, procyanidin trimer, procyanidin tetramer are present in high concentrations (Rosales-Castro et al. 2011).

In a methanolic extract from *Q. ilex* cork (–)-epicatechin, catechol, gallic acid, myricetin and theophylline have been described, while in the *Q. suber* cork distillate (–)-epicatechin, catechol, gallic acid, (+)-catechin, (–)-catechin, ferulic acid, methyl gallate and theophylline have been identified (Custódio et al. 2013). In *Q. resinosa* leaves, phenolic acids and flavonoids were identified and quantified by Rocha-Guzmán et al. being the most prominent the gallic acid, protocatechuic acid, 4-hydroxybenzoic, catechin, vanillic acid, syringic acid, epigallocatechin gallate, epicatechin, vanillin, benzoic acid, salicylic acid, chlorogenic acid and caffeic acid (Rocha-Guzmán et al. 2012).

Brossa et al. characterized the phenolic compounds present in *Q. ilex* leaves, among which it was possible to find many flavonoids (Brossa et al. 2009). These include cyanidin-3-glucoside, galloyl-hexoside, epigallocatechin gallate, digalloyl-hexoside, (epi)gallocatechin, (epi)catechin-dihexoside, procyanidin B1, epicatechin, (epi)catechin-hexoside, trigalloyl-hexoside, miricetin-O-hexoside, quercetin-

(galloyl)-hexoside, epicatechin gallate, catechin gallate, quercetin-hexoside, quercetin-3-O-glc, quercetin glucuronide, quercetin-pentoside, rhamnetin-hexoside, kaempferol-hexoside, rhamnetin-hexoside, rhamnetin glucuronide, luteolin-malonyl-hexoside, dimeric (epi)gallocatechin.

Q. ilex collected at Mount Parnes was studied to determine the presence of phenolic compounds. The results revealed the occurrence of acylated flavonol glycosides: trans-tiliroside, cis-tiliroside, kaempferol-3-O-(2'',6''-di-*E*-p-coumaroyl)-glucoside, kaempferol-3-O-(2''-*Z*-6''-*E*-di-p-coumaroyl)-glucoside, kaempferol-3-O-(2'',6''-di-*Z*-p-coumaroyl)-glucoside, kaempferol-3-O-(3''-acetyl, 2'',6''-di-*E*-p-coumaroyl)-glucoside, kaempferol-3-O-(3''-acetyl, 2''-*E*, 6''-*Z*-di-p-coumaroyl)-glucoside, kaempferol-3-O-(3''-acetyl, 2'',6''-di-*Z*-p-coumaroyl)-glucoside, kaempferol-3-O-(3'',4''-diacetyl-2'',6''-di-*E*-p-coumaroyl)-glucoside and kaempferol-3-O-(3'',4''-diacetyl-2'',6''-di-*Z*-p-coumaroyl)-glucoside (Karioti et al. 2010). These results were confirmed by Karioti et al. who continued research about *Q. ilex* leaves extracts insolating many phenolic compounds including flavonoids, proanthocyanidins, gallocatechins, catechins and phenolic acids (Karioti et al. 2011). *Q. aucheri* Jaub. & Spach is one of the 18 endemic oak plants that grow in Turkey. Quercetin 3-O-α-L-arabinopyranoside, quercetin 3-O-β-D-galactopyranoside, tannin precursors (+)-catechin, gallocatechin and a procyanidin epicatechin-(4β→8)-catechin have been identified in the ethanolic extract from leaves of that oak (Sakar et al. 2005).

High and low molecular weight phenols have been identified in the wood of *Q. robur*, like ellagitannins (castalagin, roburins A-E, grandinin and vescalagin), ellagic acid, gallic acid, protocatechuic acid, and p-coumaric acid (Alañón et al. 2011b). *Q. rubra* L. stem presents higher concentration of flavonoids compared with *Pinus ponderosa* P. Lawson & C. Lawson and *Pseudotsuga menziesii* (Mirb.) Franco (Warren et al. 2002).

The presence of tannic acid, gallic acid, syringic acid, ellagic acid, amentoflavone hexamethyl ether, isocryptomerin, anthocyanins, methyl-betulate, methyl-oleanate, hexagalloyl-glucose and polygalloyl-glucose have been confirmed in *Q. infectoria* Olivier galls (Kaur et al. 2008). The whole plant of *Q. incana* W. Bartram was studied by Iftikhat et al.; quercusides A and B, two new flavonoid glucosides, were isolated from its chloroform soluble fractions (Iftikhar et al. 2009). Lupeol, β-sitosterol and ursolic acid have also been reported from this species.

Overall, results here described show that only few species from this important genus have been phytochemically analyzed as to their flavonoid content. Most important flavonoid aglycones from *Quercus* as per scientific literature are: catechin, epicatechin, epigallocatechin, epigallocatechin gallate, gallocatechin, kaempferol, luteolin, myricetin, naringenin and quercetin (Fig. 2). These flavonoids can be used for drug development oriented to the treatment of skin conditions of high prevalence, such as psoriasis and melasma.

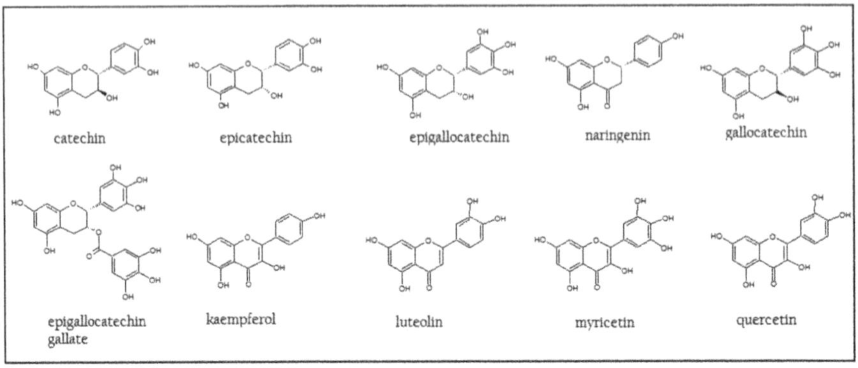

Fig. 2. Most important flavonoid aglycones from *Quercus* as per scientific literature.

Factors Influencing Oak Composition of Flavonoids

Several reports reveal the factors influencing the quantities and types of flavonoids present in the *Quercus* genus. For one instance, Rivas-Ubach et al. reported the effect of drought trees on the synthesis of flavonoids (Rivas-Ubach et al. 2014). It was concluded that plants of *Q. ilex* which underwent experimental drought had higher concentrations of flavonoids than the control group. It was also concluded that folivory activity can be a factor for the upregulation of flavonoid synthesis. This is to say, herbivores may be responsible for the increased concentrations of flavonoids found in the leaves of *Q. ilex*.

Additionally, it has been reported that patterns in the composition of flavonoids are specific to each specie (McDougal and Parks 1984). Altitude is another factor affecting composition of flavonoids is oak. Indeed, it was reported that *Q. rubra* at different altitudes yielded different type of flavonoids. A higher concentration of myricetin glycosides were found in samples of *Q. rubra* that were at an elevation between 900–1,500 m. As the altitude lowered, an increased concentration of quercetin and kaempferol was seen. Also, hydroxylation in flavonoid aglycones varied in sites below 900 m (McDougal and Parks 1984).

Other environmental factors including precipitation and variation in temperature also influence phenolic composition in oak populations (Ebrahimi et al. 2016). In another study, the variation of leaf flavonoid content of *Q. ilex* during different seasons was determined. In younger leaves, anthocyanin cyaniding-3-O-glucoside was more abundant while almost nonexistent in completely developed leaves. Meanwhile, epicathechin and epicatechin gallate showed no variation despite seasonal change, and a variation in flavonol-hexosides from winter to summer was seen. Flavonoid production is also influenced by low temperatures as the mRNA of phenylpropanoid pathway enzymes increases in a cold environment (Brossa et al. 2009). Lastly, factors such as nutrient availability and CO_2 influence plant metabolism. In a study conducted by Brossa et al., reduction in nitrogen availability resulted in higher concentrations of epicatechin gallate in leaves of *Q. ilex* (Brossa et al. 2009).

Etiology of Melasma and *Psoriasis vulgaris*

Psoriasis and melasma are complex diseases. Designing drugs to treat patients who suffer from such diseases is complicated because the cells and the solubility factors involved in their pathogenesis may vary during the development of these. Moreover, these ailments exhibit different clinical presentations and the compromised skin area also varies (Kim and Krueger 2015, Kwon et al. 2016).

Melasma, from the greek "melas" which means black, is a chronic acquired pigmentary disorder characterized by symmetrical, irregular, hyperpigmented (light to dark brown) macules on sun exposed skin areas, mainly face and neck (Ferreira Cestari et al. 2014, Kwon et al. 2016). The pathogenesis of melasma is complex, therefore has not been fully elucidated. However, chronic ultraviolet (UV) light exposure, hormone consumption, pregnancy, inflammation, use of cosmetics and photosensitizing drugs are recognized as triggering factors for its occurrence (Ortonne et al. 2009, Ferreira Cestari et al. 2014).

Melasma prevalence varies per sex, ethnic composition, skin phototype, and intensity of sun exposure. This pathology is common in women of reproductive age, particularly in their thirties and forties, though the onset of the disease is earlier in light skin types than in dark skin types. Epidemiological studies have reported higher prevalence among more pigmented phenotypes, such as East Asians, Indian, Pakistani and African, and it is also common among Americans and Brazilians who live in intertropical areas. Besides, a genetic predisposition has been observed, since almost half of the patients have relatives affected with the disease (Passeron 2013, Handel et al. 2014).

This pathology is one of the main causes of dermatological consultation. It has a significant impact in the patient quality of life, because it affects their physical appearance and consequently their emotional state. Some patients present psychological disorders such as anxiety, depression, low self-esteem and poor body image derived from melasma (Fatma et al. 2016, Sofen et al. 2016).

Histological studies have shown that melasma skin shows characteristics of solar damaged skin such as solar elastosis, that is an accumulation of abnormal elastic tissues in the dermis due to prolonged sun exposure. The increased solar elastosis in melasma suggest that accumulated sun exposure is necessary for its development (Kang et al. 2002, Kwon et al. 2016).

Face is the principal area affected. Therefore, melasma is classified in three types per their distribution on the face: (1) *centrofacial*, affecting the forehead, cheeks, upper lip, nose and chin; (2) *malar*, affecting cheeks and nose; and (3) *mandibular*, affecting the ramus of the mandible. There is another classification of melasma, according to the location of the melanin pigment: (a) *epidermal*, in which the melanin is deposited in the basal or suprabasal layer; (b) *dermal*, in which pigment laden macrophages (melanophages) are in the superficial and mid dermis; and (c) *mixed*, characterized by a combination of epidermal and dermal features (Rokhsar and Fitzpatrick 2005, Moin et al. 2006).

Melanin is responsible for cutaneous pigmentation and plays an important role in protection against sun induced skin damage (Fig. 3). Epidermal melanocytes are the melanin producing cells and skin color depends on the amount of eumelanin (the black to brown type of melanin) and pheomelanin (the yellow to reddish type of melanin) produced. Both types of melanin are synthesized from the amino acid L-tyrosine with the participation of tyrosinase, the key enzyme of melanogenesis.

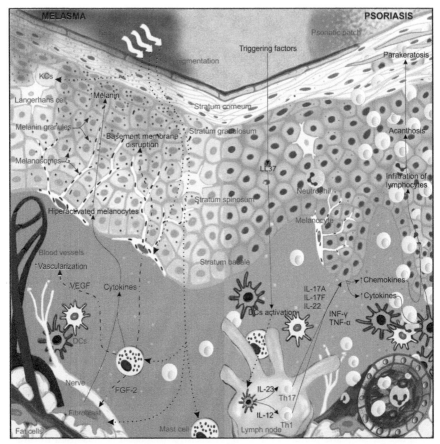

Fig. 3. Pathogenesis of melasma and psoriasis. Prolonged ultraviolet light exposure results in melanocyte hyperactivity. The increase in melanogenesis and melanosome transfer from melanocytes to epidermal keratinocytes (KCs) results in the hyperpigmentation of melasma. KCs, mast cells and fibroblasts also contribute to the development and persistence of melasma by releasing cytokines that promote melanogenesis and contribute to the histological changes. Mast cells can also induce vascular proliferation by secreting angiogenic factors (VEGF, FGF-2, etc.). In psoriasis, there is an increased production of antimicrobial peptides (LL37) in the KCs, thereby activating dendritic cells (DCs). DCs release cytokines such as IL-12 and IL-23 and stimulate the activation of Th1 and Th17 lymphocytes, activating the pathogenic pathway of the disease in which there is an increased concentration of chemokines (CXCLs) and cytokines (IL- 17A, IL-17F, IFN-γ, TNF-α, etc.), that generate an hyperproliferation of KCs and infiltration of lymphocytes in psoriatic lesions.

Melanin synthesis is restricted to vesicular organelles denominated melanosomes, which travel along melanocyte dendrites and are transferred to neighboring keratinocytes (KCs) in order to form DNA protector perinuclear melanin caps (Ando et al. 2012, Solano 2014).

Melanocyte hyperactivity caused by prolonged UV light exposure results in the increased pigmentation of melasma. It has been found that melanocytes within the affected skin are larger, with prominent dendrites and contain more melanosomes than melanocytes in unaffected skin. Also protein levels of some melanogenesis-associated factors are increased, indicating higher melanogenic activity (Kang and Ortonne 2010, Kang et al. 2011). An increase in the number of melanocytes and in melanosome transfer to epidermal KCs in lesioned skin have also been confirmed (Kang et al. 2002, Arora et al. 2014). Recently, it has been shown that the visible light is also able to induce an increase of skin pigmentation after multiple exposure (Randhawa et al. 2015).

Melanocytes are not the only cells involved in melasma pathogenesis. Melasma is caused by a network of cellular interactions among melanocytes, KCs, mast cells and fibroblasts. Dermal inflammation caused by UV radiation plays an important role in the hyperpigmentation through the production of melanogenic cytokines and growth factors. In malar melasma, lymphocytic infiltrates composed of CD4+ T cells, mast cells, and macrophages were observed in lesional skin, accompanied with elevated levels of the cytokine interleukin (IL)-17 and the proinflammatory mediator cyclooxygenase (COX-2) (Rodríguez-Arámbula et al. 2015).

UV radiation (UVA, UVB, etc.), promotes KCs to induce melanocyte proliferation and melanogenesis by secreting fibroblast growth factor (bFGF), nerve growth factor (NGF), endothelin-1 (ET-1), IL-1, inducible nitric oxide synthase (iNOS) and proopiomelanocortin (POMC)-derived peptides such as melanocyte-stimulating hormone (MSH) and adrenocorticotrophic hormone (ACTH). POMC-derived peptides and MSH receptors participate in the paracrine interaction between KCs and melanocytes. The binding of melanocortin to the MC-1 receptor (MC1R) expressed in melanocytes activates adenylate cyclase and increases cyclic adenosine monophosphate (cAMP) formation. cAMP signal transduction pathway, which results in the activation of protein kinase A (PKA) induces melanogenesis. Another signal transduction pathways involved in melanogenesis are protein kinase C (PKC) activation due to the formation of diacylglycerols, and nitric oxide (NO) production accompanied by cyclic guanylate monophosphate (cGMP) synthesis (Lee 2015, Kwon et al. 2016).

Melanogenic cytokines including stem cell factor (SCF) and hepatocyte growth factor are released from the dermal fibroblasts. A decreased expression of the Frizzled related protein WIF-1 in fibroblasts and KCs from melasma lesions increases melanogenesis and melanosome transfer through upregulation of the canonical and the noncanonical Wnt signaling pathway (Lee 2015, Kim et al. 2013). Another Wnt related gene has shown altered patterns in melasma, but its roles in pigmentation is unknown. Also, lipid metabolism associated genes are downregulated in lesional skin, nevertheless this may be related to the impaired barrier function (Kang et al. 2011).

An increase in the number of mast cells has been detected in pigmented lesional dermis of melasma patients but its role in melasma has not been completely examined. However, as in other dermatological diseases, histamine released by mast cells could rise melanin synthesis by binding to histamine receptor in melanocytes, particularly H2, through intracellular cAMP signaling pathway activation. Mast cells can also induce vascular proliferation by secreting angiogenic factors such as vascular endothelial growth factor (VEGF). There is an increase of this cytokine and in blood vessel of melasma lesion that produce altered dermal vasculature (Kim et al. 2007, Lee 2015, Kwon et al. 2016). Evidence of damage to basal membrane (BM), which could facilitate the migration of active melanocytes and melanin into the dermis has also been reported. This damage could be produced through an increase in the activity of mast cell tryptase and matrix metalloproteinases (MMP) that cause the degradation of collagen in the skin during chronic UV exposure (Torres-Álvarez et al. 2011).

There is also evidence of hormonal involvement in melasma. High levels of estrogen, progesterone and melanocortin are considered as possible triggering factors of the disease and are related to the high melasma prevalence during pregnancy. These hormones increase the level of melanogenic enzymes and augmented expression of its receptors have been found in the melasma lesional skin (Ortonne et al. 2009, Jang et al. 2010).

Psoriasis is a very common inflammatory skin disease characterized by scaly erythematous plaques predominantly in extensor body surfaces (Nestle et al. 2009). It has been estimated that psoriasis affects 2–3% of population worldwide (Goff et al. 2015), with a higher prevalence in adults than children. Indeed, prevalence in children varies from 0% (Taiwan) to 2.1% (Italy), and in adults from 0.91% (United States: African Americans) to 8.5% (Norway) (Parisi et al. 2013). Differences between regions show that the occurrence of psoriasis is higher in countries more distant from the equator, which probably reflects a genetic and environmental influence (Chandran and Raychaudhuri 2010).

Psoriasis affects the quality of life of patients, primarily because it causes not only physical but also psychological damage that makes it a pathology as disabling as cancer, diabetes or depression (Schwartz et al. 2016, Kim et al. 2015). Its etiology, although not fully elucidated, involves genetic susceptibility factors such as altered immunity-related genes (*HLA-Cw6, IL23R, STAT3*, etc.) (Harden et al. 2015) and environmental triggering factors that include physical traumas, infections, and different stressors such as smoking (Ghosh and Panda 2016, Lønnberg et al. 2016). Additionally, psoriasis has been associated with other pathologies such as metabolic syndrome and psoriatic arthritis (Ni and Chiu 2014, Michalek et al. 2016).

It was believed that psoriasis was only caused by the aberrant proliferation of KCs, however, recent research and translational medicine have demonstrated that psoriasis is a complex disease in which immunological factors play key roles and therefore is now considered as an autoimmune disease (Fig. 3) (Kim and Krueger 2015, Guttman-Yassky and Krueger 2007).

It has been described that in the initial phase of the disease, KCs trigger the activation of the innate immune system through the production of antimicrobial peptides such as cathelicidin LL-37 (Lande et al. 2014). Consequently, signaling pathways that involve transcriptional factors such as NF kappa B (NF-κB) are activated in dendritic cells (DCs) of the dermis, resulting in an increased production of type I interferons (IFN α/β) and in the release of cytokines IL-12 and IL-23 (Di Cesare et al. 2009).

IL-12 and IL-23, which share a common p40 subunit, constitute an axis of regulation and activation of lymphocytes Th1 and Th17, which are characterized by the production of pro-inflammatory cytokines such as IL-17A, IL-17F, IFN-γ and tumor necrosis factor alpha (TNF-α) (Croxford et al. 2014). These cytokines stimulate the production and release of chemokines and adhesion molecules by KCs (CCL20, CXCL1, CXCL2, CXCL8, CXCL9, CXCL10, and CXCL11), facilitating lymphocyte recruitment and infiltration into the psoriatic plaque (Kim and Krueger 2015). Psoriasis is now recognized as a disease mediated by polarized Th17 cells that produce IL-17 which originates a continuous circle responsible for the perpetuation of the disease (Fig. 3) (Harden et al. 2015).

Flavonoids for Melasma Treatment

Melasma treatment is challenging because of the multiple factors implied in its pathogenesis. To get a successful outcome it is necessary: (a) to prevent the melanocyte hyperactivity; (b) to break down deposited pigment for internal removal or external release; (c) to decrease inflammation and (d) to promote exfoliation of cells to enhance turnover. Hydroquinone (a dihydric phenol that inhibits the tyrosinase enzyme) and triple combination creams (fluocinolone acetonide, hydroquinone and tretinoin) remain the gold standard for melasma treatment. Alternative options include kojic acid, azelaic acid, arbutin, ascorbic acid, retinoids, chemical peels and lasers (Arora et al. 2014, Sofen et al. 2016).

The limited efficacy of currently available treatments has motivated the search for novel agents. Flavonoids, due to their relatively low toxicity are attractive candidates for melasma treatment, because of their anti-inflammatory, photoprotective, antioxidant and antityrosinase activities (An et al. 2008, Sarkar et al. 2012). As previously mentioned, the *Quercus* genus can be considered as a natural source of these compounds. Table 1 shows the effects of most important flavonoids found in *Quercus* genus for melasma treatment.

The antimelanogenic effect of catechins has been extensively investigated *in vitro* and *in vivo*. Treatment with 5–20 μM of catechins: (–)-epigallocatechin-3-gallate (EGCG), (–)-epigallocatechin (EGC) and (–)-catechin, during five days on B16 murine melanoma cells stimulated with α-MSH resulted in a dose-dependent inhibition of melanin formation and inhibition of tyrosinase expression, with EGCG having the higher antimelanogenic effect (Sato and Toriyama 2009). EGCG

(12.5–200 µg/ml) also reduced the melanin level, tyrosinase activity and melanosome maduration in UVA irradiated B16 cells (Liang et al. 2014) and protected human skin fibroblasts from damage induced by UVA, through relieving oxidative stress by an increase in glutathione peroxidase activity (Shin et al. 2014). In mouse melanocyte cell line Mel-Ab, EGCG reduced melanin synthesis by decreasing the protein levels of the transcription factor MITF an important transcriptional regulator of melanogenesis (Kim et al. 2004). A water extract containing EGC, epicatechin (EC), EGCG, epicatechin gallate (ECG) and theaflavins showed higher levels of antimelanogenic activity than EGCG alone. The extract reduced melanin synthesis deposition and dendrite formation in immortalized mouse melanocytes A-melan, and decreased tyrosinase protein levels without affecting mRNA expression (Kim et al. 2015). EGCG has also been shown to induce differentiation and inhibition of proliferation in KCs (Balasubramanian and Eckert 2007) with an antiinflammatory activity on immune cells (Peairs et al. 2010). A significant decrease in the melanin level was determined in 50 melasma patients who used a topical emulsion containing a plant extract enriched with catechins (Khan et al. 2013). Oral administration of low dose catechins (540 mg) for 12 weeks resulted in the incorporation of catechin metabolites into the human skin associated with reduction of proinflammatory eicosanoid 12-hydroxyeicosatetraenoic acid induced by UV irradiation, which may

Table 1. Flavonoids found in *Quercus* genus as therapeutic candidates for melasma treatment.

Flavonoid	Biological model	Pharmacological effects related to melasma treatment	References
Catechins	Healthy human subjects	Protection against UV mediated damage by reduction of proinflammatory eicosanoid 12-hydroxyeicosatetraenoic acid	(Rhodes et al. 2013)
Epigallocatechin gallate	Mel-Ab cells, B16 cells	Reduction in the melanin level, tyrosinase level and activity, and melanosome maduration	(Kim et al. 2004, Liang et al. 2014)
Kaempferol	Mice	Reduction of epidermal thickness in UV damaged skin	(Yao et al. 2014)
Luteolin	Healthy human subjects	Reduction of UVB-induced erythema	(Casetti et al. 2009)
Myricetin	Mice	Reduction of UVB-induced epidermal thickening of skin and suppression of matrix metalloproteinase-9 protein expression and enzyme activity	(Jung et al. 2010).
Naringenin	Mice	Inhibition of edema and cytokine production	(Martinez et al. 2016)
Quercetin	Mice	Reduction in photoaging by increased skin moisture content and diminution of wrinkles in number and depth	(Joshan and Singh 2013)

contribute to protection against sunburn inflammation and potentially UV mediated damage (Rhodes et al. 2013).

Luteolin is another important flavone present in *Quercus* genus with the capacity to inhibit the cellular melanogenesis as effectively as arbutin, one of the most widely used hypopigmenting agent. As opposed to catechins, luteolin increased the tyrosinase protein levels in the absence and presence of the stimulus but inhibited tyrosinase-catalysed oxidation of L-dihydroxyphenylalanine (L-DOPA) in cell-free extracts and in living cells. Therefore, the antimelanogenic effects of luteolin are attributed to their inhibitory effects on tyrosinase enzymatic activity, despite their effects on increasing tyrosinase protein levels (An et al. 2008). It has been also demonstrated that its derivate luteolin 7-sulfate attenuated cellular melanin synthesis more effectively (Kwak et al. 2016). Luteolin inhibited IL-6 secretion, matrix metalloproteinase (MMP-1) expression, and hyaluronidase activity in fibroblasts stimulated with UVA light via interference with the mitogen-activated protein kinases (MAPKs) pathway. Luteolin also inhibited the production of IL-20 in KCs irradiated with solar simulated radiation, and the treatment of KCs with luteolin before UV radiation reduced IL-6 and MMP-1 expression in fibroblasts on skin explants, suggesting that modulation of KCs by luteolin may attenuate photoaging in dermal fibroblasts (Wölfle et al. 2012). Luteolin protects the epidermis from UV induced damage by preventing the entrance of photons into the skin, reducing the formation of UV-induced cyclobutane pyrimidine dimers and preventing reactive oxygen species (ROS) generation. Topical application of a luteolin enriched extract dose-dependently reduced UVB-induced erythema on the back of healthy volunteers without causing irritation on non-irradiated skin sites. A stronger effect was observed when applied before than after irradiation, and the extract 2.5% was as effective as 1% hydrocortisone. This result support the potential use of luteolin for the prevention and treatment of inflammatory skin (Casetti et al. 2009).

Naringenin is a potent antioxidant flavanone. As luteolin, exposure of cells to naringenin resulted in the induction of melanin synthesis, tyrosinase activity, and the expression of tyrosinase and MITF in B16F10 melanocytes. Melanogenesis induced by naringenin was related to the activation of the Wnt-β-catenin-signalling pathway since it increased the intracellular accumulation of β-catenin as well as the phosphorylation of glycogen synthase kinase-3β and up-regulated the activity of phosphatidylinositol 3-kinase (Huang et al. 2011, Bouzaiene et al. 2016), however naringenin also has a tyrosinase inhibitory activity. Its derivated I3-naringenin-II8-eriodictyol inhibited the melanin content, without reducing cell viability of B16F10 cells and showed a strong antityrosinase activity exerting a non-competitive inhibition for the L-tyrosine substrate and uncompetitive inhibition for L-DOPA substrate (Yamauchi et al. 2011, Campos et al. 2015). Naringenin is a photoprotective molecule that exhibited a sun factor protection similar to that of the commercially available tinosorb (Stevanato et al. 2014), and topical formulations of naringenin-loaded elastic liposomes have demonstrated a good deposition of naringenin in different skin layers, particularly in the epidermis without causing edema or erythema in rat skin (Tsai et al. 2015). Topical application of different formulations containing naringenin in mice exhibited protection from UVB-induced skin damage by inhibiting edema and cytokine production (TNF-α, IL-1β, IL-6, and IL-10). Additionally, they inhibit superoxide anion, lipid hydroperoxides production by inducing mRNA expression of different cellular antioxidants (Martinez et al. 2016).

Flavonoids quercetin and kaempferol showed competitive inhibition towards both enzymatic activities of the tyrosinase in a cell-free system. Inhibition involved binding of their hydroxyl groups to the active site on the enzyme, resulting in steric hindrance or changed conformation (Taherkhani and Gheibi 2014). Quercetin is an aglycone with melanogenesis inhibitory activity in B16F10 cells, that reduced melanin production in a dose-dependent manner (10–50 μM). Nevertheless, low concentrations of quercetin (20 μM) induced mild stimulation of the tyrosinase activity and melanogenesis in human HMVII melanoma cells and in normal human epidermal melanocytes in the absence of UV radiation (Nagata et al. 2004). Quercetin antimelanogenic concentrations have showed little toxicity, therefore, quercetin glycosides have been isolated or synthesized for antimelanogenesis purposes with lower cytotoxicity (Choi and Shin 2016). Quercetin-3-O-β-D-glucopyranosyl-($1\rightarrow6$)-β-D-glucopyranosid reduced melanin synthesis in α-MSH-stimulated B16F10 cells by reducing expression of tyrosine and tyrosine-related proteins via extracellular signal-related protein kinase (ERK) activation, followed by down-regulation of CREB, p38, and MITF without affecting cell viability (10–100 μg/ml) (Jung et al. 2015). Topical formulations containing quercetin have showed to inhibit the UVB irradiation-induced skin damage in mice by inhibiting the myeloperoxidase activity increase, GSH depletion and proteinases secretion (Casagrande et al. 2006). Quercetin also showed to reduce photoaging. Topical application of 1% quercetin on female albino mice exposed to an UV dose over a twelve-week increased skin moisture content, reduced thiobarbituric acid reacting substances, reduced glutathione increased with a diminution of wrinkles in number and depth (Singh Joshan and Singh 2013).

Kaempferol showed inhibition of tyrosinase in B16 cells (IC$_{50}$= 171.4 μM) and inhibition of the nitric oxide production induced by lipopolysaccharide (LPS) in RAW264.7 cells by the suppression of iNOS mRNA. Its rhamnosides α-rhamnoisorobin, afzelin and kaempferitrin showed weaker activity suggesting that the 3-hydroxyl group of kaempferol is an important pharmacophore and that additional rhamnose moieties at 3 of kaempferol have a negative effect on the biological activity (Rho et al. 2011). Kaempferol is a potent inhibitor of solar UV-induced mouse skin damage. Skin from kaempferol-treated mice (1.0 mg topical kaempferol) exposed to solar UV irradiation exhibited a substantial reduction in phosphorylation of cAMP response element-binding protein (CREB), c-Fos and histone H3, and a reduction of epidermal thickness (Yao et al. 2014). In a clinical trial conducted to evaluate the efficacy and safety of whey associated with dodder (*Cuscuta campestris* Yunck.) seed extract in the treatment of moderate-to-severe atopic dermatitis in adults, a significant decrease in skin pigmentation was observed in the treated group (15 g of whey powder and four extract capsules of 500 mg) that may be attributable to the presence of the flavonoids quercetin and kaempferol in the extract (Mehrbani et al. 2015).

Myricetin is a flavonoid structurally similar to quercetin with photoprotective activity. Myricetin attenuated UVB-induced HaCaT keratinocyte death (1–30 μM), reduced the UVB-induced malondialdehyde level, the UVB-induced peroxide and the UVB-induced activation of c-jun-NH2 terminal kinase (JNK) (Huang et al. 2010). Topical application of the compound reduced epidermal thickening resulting from UVB irradiation and suppressed matrix metalloproteinase-9 (MMP-9) protein expression and activity. It was found to inhibit chronic UVB irradiation-induced wrinkles by suppressing UVB-induced Raf-kinase activity and subsequent attenuation of phosphorylation of MEK and ERK (Jung et al. 2010). As other flavonols, myricetin reduces melanin synthesis through competitive inhibition for the oxidation of L-DOPA by tyrosinase. In terms of flavonol-inhibitor strength, quercetin has the stronger activity followed by myricetin and kaempferol (Chang 2009).

Flavonoids for *Psoriasis vulgaris* Treatment

The antipsoriatic treatment is based on the severity of the disease and is mainly symptomatic, the latter discourage the adherence of the patient to the therapy (García-Pérez et al. 2012). Moreover, concerns exist about the long-term toxicities of biotech products, leading to pharmaceutical industry to search new alternatives for the treatment of this disease. In the search for new treatments, molecules recognized by its antioxidant, antiproliferative and antiinflammatory characteristics such as flavonoids, could be good candidates for treating psoriasis (Schmidt et al. 2015). Particularly, *Quercus* genus contains high amounts of flavonoids that could be used for psoriasis treatment. Table 2 shows the effects of most important flavonoids found in *Quercus* genus for psoriasis management.

Catechins, in particular EGCG, have been shown to significantly suppress proliferation and activates the differentiation pathways of normal human KCs (Balasubramanian and Eckert 2007). EGCG induce caspase 14 expression in epidermal KCs whose nuclear translocation is diminished in the psoriatic skin, and reduces psoriatic lesions in the flaky skin mouse model (Hsu et al. 2007). Psoriatic mice treated with EGCG, quercetin or curcumin showed a statistically significant differences of macrophage spreading index in the peritoneal cavity (P < 0.05) as compared to untreated psoriatic mice (Skurić et al. 2011). In a psoriasis mouse model was demonstrated that EGCG had anti-inflammatory, immune regulatory and antioxidant effects. Treatment with EGCG reduced the infiltration of T cells, the percentages of DCs in spleen and the plasma levels of IL-17A, IL-17F, IL22 e IL-23 (Zhang et al. 2016). EC suppresses the production of pro-inflammatory cytokines IL6 and IL-8 showing also an enhance of anti-inflammatory cytokine IL-10 production in whole blood culture (Al-Hanbali et al. 2009). High levels of IL-10 improves the control over Th1 cells activation. EC down-regulated the protein expression of iNOS, COX-2 and pro-inflammatory cytokines (TNF-α, IL-1β and

Table 2. Flavonoids found in *Quercus* genus as therapeutic candidates for psoriasis treatment.

Flavonoid	Biological model	Pharmacological effects related to psoriasis treatment	References
Epigallocatechin gallate	Mice	Reduction in the infiltration of T cells, the percentages of DCs in spleen and the plasma levels of cytokines	(Zhang et al. 2016)
Epicatechin	RAW264.7 cells	Inhibition of the activation of NF-κB, MAPKs, and JAK2/STAT3	(Yang et al. 2015)
Kaempferol, formononetin and luteolin	Rat	Reduction in the thickness of epidermis, retention of the stratum granulosum and absence of movement of neutrophils	(Vijayalakshmi et al. 2014)
Luteolin	HaCaT cells	Inhibition of the production of cytokines by decreasing NF-κB activation	(Weng et al. 2014)
Myricetin	HaCaT cells	Suppression of Akt, mTOR and NF-κB pathways and reduction of TNF-α production	(Lee and Lee 2016)
Naringenin	hPBMC cells	Down regulation of TNF-α	(Chlapanidas et al. 2014, Gaggeri et al. 2013)
Quercetin	Mice	Inhibition of leukocyte migration and keratinocyte proliferation	(Vijayalakshmi et al. 2012)

IL-6) in the LPS-induced macrophages through NF-κB, MAPKs and Janus kinase 2 (JAK2)/signal transducer and activator of transcription 3 (STAT3) inactivation (Yang et al. 2015). These results suggest that the use of EC and EGCG alone or in combination with current therapies could be advantageous for the treatment of psoriasis. However, more studies must be performed to confirm the effectiveness of catechins for psoriasis treatment.

Luteolin inhibits the production of IL-6, IL-8 and VEGF by decreasing NF-κB activation in HaCaT cells (Weng et al. 2014). Luteolin-7-glucoside showed reduction in proliferation, acanthosis, and inflammation by inhibition of IL-22/STAT3 pathway in KCs and mouse psoriatic model (Palombo et al. 2016). The flavonoids glycosides luteolin, formononetin and kaempferol also showed significant antiproliferative activity in HaCaT cells. Formulations of tablets containing combinations of these bioactive flavonoids were developed and their effect were evaluated in a rat model. Animals which received the tablets exhibited a remarkable inhibition of IL-17 and TNF-α. Its antipsoriatic activity was confirmed by the reduction in the thickness of epidermis, the significant retention of the stratum granulosum and the absence of movement of neutrophils (Vijayalakshmi et al. 2014).

Myricetin reduce TNF-α-stimulated inflammatory mediator production in HaCaT cells by suppressing the activation of the Akt, mTOR and NF-κB pathways (Lee and Lee 2016). Naringenin exhibit inhibitory effect on the expressions of iNOS, COX-2 and ERK, and reduce the production of IL-1β, IL-6, IL-8, TNF-α and NF-κB activation in macrophages (Hsu et al. 2013, Bodet et al. 2008). The effect of combined use of naringenin and sericin as TNF-α blockers was evaluated. The combination in microspheres was more potent than the inhibition produced alone by naringenin on cytokine blockade, suggesting this combination could be used for the development of new topical formulations suitable for the treatment of middle-stage psoriasis (Chlapanidas et al. 2014, Gaggeri et al. 2013).

The flavonoid quercetin decreased the activation of NF-κB and inhibited the mRNA of IL-1β, IL-6, IL-8 and TNF-α in primary human KCs (Vicentini et al. 2011). Additionally, a quercetin derivate suppressed the activation of the NF-κB pathway in HaCaT cells (Lee et al. 2013). The isolated quercetin from the rhizome of *Smilax china* L. was assessed for antipsoriatic activity using the mouse tail model and a significant inhibition in leukocyte migration (p < 0.001) and antiproliferative activity were observed (IC$_{50}$: 62.42 ± 10.20 μg/mL) (Vijayalakshmi et al. 2012). Pharmaceutical formulation for the topical application of quercetin was further investigated, and showed higher quercetin deposition in the upper skin layers of the epidermis, however a clinical study in humans is required to demonstrate the antipsoriatic potential of flavonoid (Hatahet et al. 2016).

Future Research

As examined in this chapter, the *Quercus* genus is an important source of flavonoids mainly of catechin, epicatechin, epigallocatechin, epigallocatechin gallate, gallocatechin, kaempferol, luteolin, myricetin, naringenin and quercetin. That is extremely important considering that these compounds are present in tissues such as barks and twigs, considered residues of the forest exploitation. The use of these wastes as a source of flavonoids, could contribute to the development of new pharmaceutical products for the treatment of melasma and psoriasis, two dermatological diseases of high prevalence.

Flavonoids can exert their positive effects for melasma management, mainly by: (1) inhibition of melanin formation; (2) inhibition of tyrosinase expression; (3) blockage of tyrosinase activity; (4) inhibition of melanosome maduration; (5) Reduction of tyrosine expression and (6) protection of epidermis against the UV-induced damage. These compounds have also been demonstrated to have antipsoriatic properties related to their capacity to (1) suppress keratinocyte proliferation; (2) activate the keratinocyte differentiation; (3) inhibit the production of proinflammatory cytokines and chemokines; (4) inactivate transcription factors such as NF-κB and STAT3, and (5) block the leukocyte migration. This polyvalence in their mechanism of action make them excellent therapeutic candidates for these multi-causal diseases.

However, some limitations related to the use of these compounds for melasma and psoriasis are appreciated. First of one, there is a lack of long-term controlled preclinical and clinical studies, which contrasts with the long treatment required for these diseases. Additionally, safety of flavonoids in preclinical studies considering the oral and topical administration should be carefully analyzed to minimize side effects of future pharmaceutical products. Moreover, their pharmacokinetic profile could offer important information about their activation/inactivation under metabolism influence. It would also be important to study their effects alone and combined with other existing drugs to know their contribution to the therapeutic armamentarium for both diseases. Overall, a greater amount of scientific evidence, standardized formulations and reliable clinical trials are necessary for the use of flavonoids in the development of new pharmaceutical products for psoriasis and melasma.

References

Alañón, M.E., L. Castro-Vázquez, M.C. Díaz-Maroto, M.H. Gordon and M.S. Pérez-Coello. 2011a. A study of the antioxidant capacity of oak wood used in wine ageing and the correlation with polyphenol composition. Food Chem. 128(4): 997–1002.

Alañón, M.E., L. Castro-Vázquez, M.C. Díaz-Maroto, I. Hermosín-Gutiérrez, M.H. Gordon and M.S. Pérez-Coello. 2011b. Antioxidant capacity and phenolic composition of different woods used in cooperage. Food Chem. 129(4): 1584–1590.

Alañón, M.E., M.S. Pérez-Coello, I.J. Díaz-Maroto, P.J. Martín-Álvarez, P. Vila-Lameiro and M.C. Díaz-Maroto. 2011c. Influence of geographical location, site and silvicultural parameters, on volatile composition of *Quercus pyrenaica* Willd. wood used in wine aging. For. Ecol. Manag. 262(2): 124–130.

Al-Hanbali, M., D. Ali, M. Bustami, S. Abdel-Malek, R. Al-Hanbali, T. Alhussainy, F. Qadan and K.Z. Matalka. 2009. Epicatechin suppresses IL-6, IL-8 and enhances IL-10 production with NF-κB nuclear translocation in whole blood stimulated system. Neuro. Endocrinol. Lett. 30(1): 131–138.

An, S.M., H.J. Kim, J.-E. Kim and Y.C. Boo. 2008. Flavonoids, taxifolin and luteolin attenuate cellular melanogenesis despite increasing tyrosinase protein levels. Phytother. Res. 22(9): 1200–1207.

Ando, H., Y. Niki, M. Ito, K. Akiyama, M.S. Matsui, D.B. Yarosh and M. Ichihashi. 2012. Melanosomes are transferred from melanocytes to keratinocytes through the processes of packaging, release, uptake, and dispersion. J. Invest. Dermatol. 132(4): 1222–1229.

Arizaga, S., J. Martínez-Cruz, M. Salcedo-Cabrales and M.Á. Bello-González. 2009. Manual de la biodiversidad de encinos michoacanos. 1st Edn. INE-SEMARNAT. D.F., Mexico. 147 pp.

Arora, P., V. Garg, S. Sonthalia, N. Gokhale and R. Sarkar. 2014. Melasma update. Indian Dermatol. Online J. 5(4): 426–435.

Axelrod, D.I. 1983. Biogeography of oaks in the arcto-tertiary province. Ann. Mo. Bot. Gard. 70: 629–657.

Balasubramanian, S. and R. Eckert. 2007. Keratinocyte proliferation, differentiation, and apoptosis-differential mechanisms of regulation by curcumin, EGCG and apigenin. Toxicol. Appl. Pharmacol. 224(3): 214–219.

Bellarosa, R., M.C. Simeone, A. Papini and B. Schirone. 2005. Utility of ITS sequence data for phylogenetic reconstruction of Italian *Quercus* spp. Mol. Phylogenet. Evol. 34(2): 355–370.

Berahou, A., A. Auhmani, N. Fdil, A. Benharref, M. Jana and C.A. Gadhi. 2007. Antibacterial activity of *Quercus ilex* bark's extracts. J. Ethnopharmacol. 112(3): 426–429.

Bodet, C., V.D. La, F. Epifano and D. Grenier. 2008. Naringenin has anti-inflammatory properties in macrophage and *ex vivo* human whole-blood models. J. Periodont. Res. 43(4): 400–407.

Bouras, M., M. Chadni, F.J. Barba, N. Grimi, O. Bals and E. Vorobiev. 2015. Optimization of microwave-assisted extraction of polyphenols from *Quercus* bark. Ind. Crops. Prod. 77: 590–601.

Brändle, M. and R. Brandl. 2001. Species richness of insects and mites on trees: expanding Southwood. J. Anim. Ecol. 70(3): 491–504.

Bravo, L. 1998. Polyphenols: chemistry, dietary sources, metabolism, and nutritional significance. Nutr. Rev. 56(11): 317–333.

Brossa, R., I. Casals, M. Pintó-Marijuan and I. Fleck. 2009. Leaf flavonoid content in *Quercus ilex* L. resprouts and its seasonal variation. Trees 23(2): 401–408.

Campos, P.M., A.S. Prudente, C.D.S. Horinouchi, V. Cechinel-Filho, G.M. Fávero, D.A. Cabrini and M.F. Otuki. 2015. Inhibitory effect of GB-2a (I3-naringenin-II8-eriodictyol) on melanogenesis. J. Ethnopharmacol. 174: 224–229.

Cantos, E., J.C. Espín, C. López-Bote, L. de la Hoz, J.A. Ordóñez and F.A. Tomás-Barberán. 2003. Phenolic compounds and fatty acids from acorns (*Quercus* spp.), the main dietary constituent of free-ranged Iberian pigs. J. Agric. Food Chem. 51(21): 6248–6255.

Casagrande, R., S.R. Georgetti, W.A. V.J., D.J. Dorta, A.C. dos Santos and M.J.V. Fonseca. 2006. Protective effect of topical formulations containing quercetin against UVB-induced oxidative stress in hairless mice. J. Photochem. Photobiol. B. 84(1): 21–27.

Casetti, F., W. Jung, U. Wölfle, J. Reuter, K. Neumann, B. Gilb, A. Wähling, S. Wagner, I. Merfort and C.M. Schempp. 2009. Topical application of solubilized *Reseda luteola* extract reduces ultraviolet B-induced inflammation *in vivo*. J. Photochem. Photobiol. B. 96(3): 260–265.

Castro-Vázquez, L., M.E. Alañón, J.M. Ricardo-da-Silva, M.S. Pérez-Coello and O. Laureano. 2013. Evaluation of Portuguese and Spanish *Quercus pyrenaica* and *Castanea sativa* species used in cooperage as natural source of phenolic compounds. Eur. Food Res. Technol. 237(3): 367–375.

Chandran, V. and S.P. Raychaudhuri. 2010. Geoepidemiology and environmental factors of psoriasis and psoriatic arthritis. J. Autoimmun. 34(3): J314–321.

Chang, T.-S. 2009. An updated review of tyrosinase inhibitors. Int. J. Mol. Sci. 10(6): 2440–2475.

Charef, M., M. Yousfi, M. Saidi and P. Stocker. 2008. Determination of the fatty acid composition of acorn (*Quercus*), *Pistacia lentiscus* seeds growing in Algeria. J. Am. Oil Chem. Soc. 85(10): 921–924.

Chlapanidas, T., S. Perteghella, F. Leoni, S. Faragò, M. Marazzi, D. Rossi, E. Martino, R. Gaggeri and S. Collina. 2014. TNF-α blocker effect of naringenin-loaded sericin microparticles that are potentially useful in the treatment of psoriasis. Int. J. Mol. Sci. 15(8): 13624–13636.

Choi, M.-H. and H.-J. Shin. 2016. Anti-melanogenesis effect of quercetin. Cosmetics 3(2): 18.

Correia, B., J.L. Rodriguez, L. Valledor, T. Almeida, C. Santos, M.J. Cañal and G. Pinto. 2014. Analysis of the expression of putative heat-stress related genes in relation to thermotolerance of cork oak. Journal of Plant Physiology 171(6): 399–406.

Croxford, A.L., P. Kulig and B. Becher. 2014. IL-12-and IL-23 in health and disease. Cytokine Growth Factor Rev. 25(4): 415–421.

Curtu, A.L, O. Gailing and R. Finkeldey. 2007. Evidence for hybridization and introgression within a species-rich oak (*Quercus* spp.) community. BMC Evol. Biol. 7: 218.

Custódio, L., J. Patarra, F. Albericio, N.R. Neng, J.M.F. Nogueira and A. Romano. 2015. Phenolic composition, antioxidant potential and *in vitro* inhibitory activity of leaves and acorns of *Quercus suber* on key enzymes relevant for hyperglycemia and Alzheimer's disease. Ind. Crops Prod. 64: 45–51.

Custódio, L., J. Patarra, F. Alberício, N.R. Neng, J.M.F. Nogueira and A. Romano. 2013. Extracts from *Quercus* sp. acorns exhibit *in vitro* neuroprotective features through inhibition of cholinesterase and protection of the human dopaminergic cell line SH-SY5Y from hydrogen peroxide-induced cytotoxicity. Ind. Crops Prod. 45: 114–120.

Denk, T. and G.W. Grimm. 2010. The oaks of western Eurasia: traditional classifications and evidence from two nuclear markers. Taxon. 59(2): 351–366.

Di Cesare, A., P. Di Meglio and F.O. Nestle. 2009. The IL-23/Th17 axis in the immunopathogenesis of psoriasis. J. Invest. Dermatol. 129(6): 1339–1350.

Ebrahimi, A., I. Mehregan, T. Nejad Sattari, M. Assadi and K Larijani. 2016. Population variability in *Quercus brantii* Lindl. based on the acorn morphometry and chomposition of phenolic compounds. Appl. Ecol. Environ. Res. 14(3): 215–231.

Fatma, F., I. Baati, M. Mseddi, R. Sallemi, H. Turki and J. Masmoudi. 2016. The psychological impact of melasma. A report of 30 Tunisian women. Eur. Psychiat. 33(Supplement): S396.

Ferreira Cestari, T., L. Pinheiro Dantas and J. Catucci Boza. 2014. Acquired hyperpigmentations. An. Bras. Dermatol. 89(1): 11–25.

Gaggeri, R., D. Rossi, M. Daglia, F. Leoni, M.A. Avanzini, M. Mantelli, M. Juza and S. Collina. 2013. An eco-friendly enantioselective access to (R)-naringenin as inhibitor of proinflammatory cytokine release. Chem. Biodivers. 10(8): 1531–1538.

García-Pérez, M.E., J. Jean and R. Pouliot. 2012. Antipsoriatic drug development: challenges and new emerging therapies. Recent Pat. Inflamm. Allergy Drug Discov. 6(1): 3–21.

Ghosh, A. and S. Panda. 2016. Recent understanding of the etiopathogenesis of psoriasis. Indian J. Paediatr. Dermatol. 18(1): 1–8.

Goff, K.L., C. Karimkhani, L.N. Boyers, M.A. Weinstock, J.P. Lott, R.J. Hay, L.E. Coffeng et al. 2015. The global burden of psoriatic skin disease. Br. J. Dermatol. 172(6): 1665–1668.

González-Rodríguez, A., D.M. Arias, S. Valencia and K. Oyama. 2004. Morphological and RAPD analysis of hybridization between *Quercus affinis* and *Q. laurina* (Fagaceae), two Mexican red oaks. Am. J. Bot. 91(3): 401–409.

Gričar, J., Š. Jagodic and P. Prislan. 2015. Structure and subsequent seasonal changes in the bark of sessile oak (*Quercus petraea*). Trees 29(3): 747–757.

Grivet, D., V.L. Sork, R.D. Westfall and F.W. Davis. 2008. Conserving the evolutionary potential of California Valley oak (*Quercus lobata* Née): a multivariate genetic approach to conservation planning. Mol. Ecol. 17(1): 139–156.

Guttman-Yassky, E. and J.G. Krueger. 2007. Psoriasis: evolution of pathogenic concepts and new therapies through phases of translational research. Br. J. Dermatol. 157(6): 1103–1115.

Handel, A.C., L.D.B. Miot and H.A. Miot. 2014. Melasma: a clinical and epidemiological review. An. Bras. Dermatol. 89(5): 771–782.

Harden, J.L., J.G. Krueger and A. Bowcock. 2015. The immunogenetics of psoriasis: a comprehensive review. J. Autoimmun. 64: 66–73.

Hatahet, T., M. Morille, A. Hommoss, J.M. Devoisselle, R.H. Müller and S. Bégu. 2016. Quercetin topical application, from conventional dosage forms to nanodosage forms. Eur. J. Pharm. Biopharm. 108: 41–53.

Hsu, C.-L., S.-C. Fang and G.-C. Yen. 2013. Anti-inflammatory effects of phenolic compounds isolated from the flowers of *Nymphaea mexicana* Zucc. Food Funct. 4(8): 1216–1222.

Hsu, S., D. Dickinson, J. Borke, D.S. Walsh, J. Wood, H. Qin, J. Winger, H. Pearl, G. Schuster and W.B. Bollag. 2007. Green tea polyphenol induces caspase 14 in epidermal keratinocytes via MAPK pathways and reduces psoriasiform lesions in the flaky skin mouse model. Exp. Dermatol. 16(8): 678–684.

Huang, J.-H., C.-C. Huang, J.-Y. Fang, C. Yang, C.-M. Chan, N.-L. Wu, S.-W. Kang and C.-F. Hung. 2010. Protective effects of myricetin against ultraviolet-B-induced damage in human keratinocytes. Toxicol. *In Vitro.* 24(1): 21–28.

Huang, Y.-C., C.-H. Yang and Y.-L. Chiou. 2011. Citrus flavanone naringenin enhances melanogenesis through the activation of Wnt/β-catenin signalling in mouse melanoma cells. Phytomedicine. 18(14): 1244–1249.

Iftikhar, B., S. Perveen, A. Malik, N. Sultana, S. Arayne and P. Muhammad. 2009. Structural determination of quercusides A and B, new flavonoid glucosides from *Quercus incana*, by 1D and 2D NMR spectroscopy. Magn. Reson. Chem. 47(7): 605–608.

Ignat, I., I. Volf and V.I. Popa. 2011. A critical review of methods for characterisation of polyphenolic compounds in fruits and vegetables. Food Chem. 126(4): 1821–1835.

Jang, Y., J. Lee, H. Kang, E.-S. Lee and Y. Kim. 2010. Oestrogen and progesterone receptor expression in melasma: an immunohistochemical analysis. J. Eur. Acad. Dermatol. Venereo. 24(11): 1312–1316.

Jung, H.G., H.H. Kim, S. Paul, J.Y. Jang, Y.H. Cho, H.J. Kim, J.M. Yu et al. 2015. Quercetin-3-O-β-D-Glucopyranosyl-(1→6)-β-D-Glucopyranoside suppresses melanin synthesis by augmenting p38 MAPK and CREB signaling pathways and subsequent cAMP down-regulation in murine melanoma cells. Saudi J. Biol. Sci. 22(6): 706–713.

Jung, S.K., K.W. Lee, H.Y. Kim, M.H. Oh, S. Byun, S.H. Lim, Y.-S. Heo et al. 2010. Myricetin suppresses UVB-induced wrinkle formation and MMP-9 expression by inhibiting Raf. Biochem. Pharmacol. 79(10): 1455–1461.

Kang, H.Y. and J.-P. Ortonne. 2010. What should be considered in treatment of melasma. Ann. Dermatol. 22(4): 373–378.

Kang, H.Y., I. Suzuki, D.J. Lee, J. Ha, P. Reiniche, J. Aubert, S. Deret, D. Zugaj, J.J. Voegel and J.-P. Ortonne. 2011. Transcriptional profiling shows altered expression of Wnt Pathway–and lipid metabolism-related genes as well as melanogenesis-related genes in melasma. J. Invest. Dermatol. 131(8): 1692–1700.

Kang, W.H., K.H. Yoon, E.-S. Lee, J. Kim, K.B. Lee, H.Yim, S. Sohn and S. Im. 2002. Melasma: histopathological characteristics in 56 Korean patients. Br. J. Dermatol. 146(2): 228–237.

Karioti, A., A.R. Bilia and H. Skaltsa. 2010. *Quercus ilex* L.: a rich source of polyacylated flavonoid glucosides. Food Chem. 123(1): 131–142.

Karioti, A., M. Sokovic, A. Ciric, C. Koukoulitsa, A.R. Bilia and H. Skaltsa. 2011. Antimicrobial properties of *Quercus ilex* L. proanthocyanidin dimers and simple phenolics: evaluation of their synergistic activity with conventional antimicrobials and prediction of their pharmacokinetic profile. J. Agric. Food Chem. 59(12): 6412–6422.

Kaur, G., M. Athar and M.S. Alam. 2008. *Quercus infectoria* galls possess antioxidant activity and abrogates oxidative stress-induced functional alterations in murine macrophages. Chem. Biol. Interact. 171(3): 272–282.

Khan, B.A., N. Akhtar, I. Hussain, K.A. Abbas and A. Rasul. 2013. Whitening efficacy of plant extracts including *Hippophae rhamnoides* and *Cassia fistula* extracts on the skin of Asian patients with melasma. Adv. Dermatol. Allergol. 4: 226–232.

Kim, D.-S., S.-H. Park, S.-B. Kwon, K. Li, S.-W. Youn and K.-C. Park. 2004. (–)-Epigallocatechin-3-gallate and hinokitiol reduce melanin synthesis via decreased MITF production. Arch. Pharm. Res. 27(3): 334–339.

Kim, E.H., Y.C. Kim, E.-S. Lee and H.Y. Kang. 2007. The vascular characteristics of melasma. J. Dermatol. Sci. 46(2): 111–116.

Kim, G.E., E. Seidler and A.B. Kimball. 2015. Effect of age at diagnosis on chronic quality of life and long-term outcomes of individuals with psoriasis. Pediatr. Dermatol. 32(5): 656–662.

Kim, J. and J.G. Krueger. 2015. The immunopathogenesis of psoriasis. Dermatol. Clin. 33(1): 13–23.

Kim, J.-Y., T.-R. Lee and A.-Y. Lee. 2013. Reduced WIF-1 expression stimulates skin hyperpigmentation in patients with melasma. J. Invest. Dermatol. 133(1): 191–200.

Kim, Y.C., S.Y. Choi and E.Y. Park. 2015. Anti-melanogenic effects of black, green, and white tea extracts on immortalized melanocytes. J. Vet. Sci. 16(2): 135.

Kremer, A., R. Sederoff and N. Wheeler. 2010. Genomics of forest and ecosystem health in the Fagaceae: meeting report. Tree Genet. Genomes 6(5): 815–820.

Kremer, A., A.G. Abbott, J.E. Carlson, P.S. Manos, C. Plomion, P. Sisco, M.E. Staton, S. Ueno and G.G. Vendramin. 2012. Genomics of Fagaceae. Tree Genet. Genomes 8(3): 583–610.

Kremer, A. 2016. Microevolution of European temperate oaks in response to environmental changes. C. R. Biol. 339(7-8): 263–267.

Kubitzki, K., J.G. Rohwer and V. Bittrich. 1993. Flowering plants · Dicotyledons - Magnoliid, Hamamelid and Caryophyllid families. 1st ed. Vol. II. VII vols. The Families and Genera of Vascular Plants. Berlin, Germany: Springer-Verlag Berlin Heidelberg.

Kumar, S. and A.K. Pandey. 2013. Chemistry and biological activities of flavonoids: an overview. Scientific World J. 2013: e162750.

Kuppusamy, S., P. Thavamani, M. Megharaj, R. Nirola, Y.B. Lee and R. Naidu. 2016. Assessment of antioxidant activity, minerals, phenols and flavonoid contents of common plant/tree waste extracts. Ind. Crops Prod. 83: 630–634.

Kwak, J.Y., J.K. Seok, H.-J. Suh, Y.-H. Choi, S.S. Hong, D.S. Kim and Y.C. Boo. 2016. Antimelanogenic effects of luteolin 7-sulfate isolated from *Phyllospadix iwatensis* Makino. Br. J. Dermatol. 175(3): 501–511.

Kwon, S.-H., Y.-J. Hwang, S.-K. Lee and K.-C. Park. 2016. Heterogeneous pathology of melasma and its clinical implications. Int. J. Mol. Sci. 17(6): 824.

Lande, R., E. Botti, C. Jandus, D. Dojcinovic, G. Fanelli, C. Conrad, G. Chamilos et al. 2014. The antimicrobial peptide LL37 is a T-cell autoantigen in psoriasis. Nat. Commun. 5: 5621.

Larios, C., A. Casas, M. Vallejo, A.I. Moreno-Calles and J. Blancas. 2013. Plant management and biodiversity conservation in Náhuatl homegardens of the Tehuacán Valley, Mexico. J. Ethnobiol. Ethnomed. 9: 74.

Lee, A.-Y. 2015. Recent progress in melasma pathogenesis. Pigm. Cell Melanoma R. 28(6): 648–660.

Lee, C.S., E.B. Jeong, Y.J. Kim, M.S. Lee, S.J. Seo, K.H. Park and M.W. Lee. 2013. Quercetin-3-O-(2″-galloyl)-α-L-rhamnopyranoside inhibits TNF-α-activated NF-κB-induced inflammatory mediator production by suppressing ERK activation. Int. Immunopharmacol. 16(4): 481–487.

Lee, D.H. and C.S. Lee. 2016. Flavonoid myricetin inhibits TNF-α-stimulated production of inflammatory mediators by suppressing the Akt, mTOR and NF-κB pathways in human keratinocytes. Eur. J. Pharmacol. 784: 164–172.

Lesur, I., G. Le Provost, P. Bento, C. Da Silva, J.-C. Leplé, F. Murat, S. Ueno et al. 2015. The oak gene expression atlas: insights into Fagaceae genome evolution and the discovery of genes regulated during bud dormancy release. BMC Genomics 16(1): 112.

Liang, Y., S. Kang, L. Deng, L. Xiang and X. Zheng. 2014. Inhibitory effects of (–)-epigallocatechin-3-gallate on melanogenesis in ultraviolet A-induced B16 murine melanoma cell. Trop. J. Pharm. Res. 13(11): 1825.

Lønnberg, A.S., L. Skov, A. Skytthe, K.O. Kyvik, O.B. Pedersen and S.F. Thomsen. 2016. Smoking and risk for psoriasis: a population-based twin study. Int. J. Dermatol. 55(2): e72–78.

Luna-José, A.L., L. Montalvo-Espinosa and B. Rendón-Aguilar. 2003. Los usos no leñosos de los encinos en México. Bol. Soc. Bot. México 72(1): 107–117.

Manos, P.S., J.J. Doyle and K.C. Nixon. 1999. Phylogeny, biogeography and processes of molecular differentiation in *Quercus* subgenus *Quercus* (Fagaceae). Mol. Phylogenet. Evol. 12(3): 333–349.

Martinez, R.M., F.A. Pinho-Ribeiro, V.S. Steffen, T.C.C. Silva, C.V. Caviglione, C. Bottura, M.J.V. Fonseca et al. 2016. Topical formulation containing naringenin: efficacy against ultraviolet B irradiation-induced skin inflammation and oxidative stress in mice. PLoS One 11(1): e0146296.

McDougal, K.M. and C.R. Parks. 1984. Elevational variation in foliar flavonoids of *Quercus rubra* L. (Fagaceae). Am. J. Bot. 71(3): 301–308.

Mehrbani, M., R. Choopani, A. Fekri, M. Mehrabani, M. Mosaddegh and M. Mehrabani. 2015. The efficacy of whey associated with dodder seed extract on moderate-to-severe atopic dermatitis in adults: a randomized, double-blind, placebo-controlled clinical trial. J. Ethnopharmacol. 172: 325–332.

Michalek, I.M., B. Loring, S.M. John and World Health Organization. 2016. Global report on psoriasis. 1st ed. Switzerland: World Health Organization.

Moin, A., Z. Jabery and N. Fallah. 2006. Prevalence and awareness of melasma during pregnancy. Int. J. Dermatol. 45(3): 285–288.

Nagata, H., S. Takekoshi, R. Takeyama, T. Homma and R. Yoshiyuki Osamura. 2004. Quercetin enhances melanogenesis by increasing the activity and synthesis of tyrosinase in human melanoma cells and in normal human melanocytes. Pigment. Cell Res. 17(1): 66–73.

Nasr Bouzaiene, N., F. Chaabane, A. Sassi, L. Chekir-Ghedira and K. Ghedira. 2016. Effect of apigenin-7-glucoside, genkwanin and naringenin on tyrosinase activity and melanin synthesis in B16F10 melanoma cells. Life Sci. 144: 80–85.

Nestle, F.O., D.H. Kaplan and J. Barker. 2009. Psoriasis. N. Engl. J. Med. 361(5): 496–509.

Ni, C. and M.W. Chiu. 2014. Psoriasis and comorbidities: links and risks. Clin. Cosmet. Investig. Dermatol. 7: 119–132.

Nixon, K.C. 1993. The genus Quercus in Mexico. pp. 447–458. *In*: Ramamoorthy, T.P. , R. Bye, A. Lot and J. Fa (eds.). Biological Diversity of Mexico: Origins and Distribution. New York: Oxford University Press.

Nixon, K.C. 2006. Global and neotropical distribution and diversity of oak (genus Quercus) and oak forests. pp. 3–13. *In*: Kappelle M. (eds.). Ecology and Conservation of Neotropical Montane oak Forests. Ecological Studies (Analysis and Synthesis), vol 185. Springer, Berlin: Heidelberg.

Ortonne, J.P., I. Arellano, M. Berneburg, T. Cestari, H. Chan, P. Grimes, D. Hexsel, S. Im, J. Lim, H. Lui, A. Pandya, M. Picardo, M. Rendon, S. Taylor, J. Van der Veen and W. Westerhof. 2009. A global survey of the role of ultraviolet radiation and hormonal influences in the development of melasma. J. Eur. Acad. Dermatol. Venereol. 23(11): 1254–1262.

Palombo, R., I. Savini, L. Avigliano, S. Madonna, A. Cavani, C. Albanesi, A. Mauriello, G. Melino and A. Terrinoni. 2016. Luteolin-7-glucoside inhibits IL-22/STAT3 pathway, reducing proliferation, acanthosis, and inflammation in keratinocytes and in mouse psoriatic model. Cell Death Dis. 7(8): e2344.

Parisi, R., D.P.M. Symmons, C.E.M. Griffiths and D.M. Ashcroft. 2013. Global epidemiology of psoriasis: a systematic review of incidence and prevalence. J. Invest. Dermatol. 133(2): 377–385.

Passeron, T. 2013. Melasma pathogenesis and influencing factors—an overview of the latest research. J. Eur. Acad. Dermatol. Venereol. 27(s1): 5–6.

Peairs, A., R. Dai, L. Gan, S. Shimp, M.N. Rylander, L. Li and C.M. Reilly. 2010. Epigallocatechin-3-gallate (EGCG) attenuates inflammation in MRL/Lpr mouse mesangial cells. Cell. Mol. Immunol. 7(2): 123–132.

Rakić, S., S. Petrović, J. Kukić, M. Jadranin, V. Tešević, D. Povrenović and S. Šiler-Marinković. 2007. Influence of thermal treatment on phenolic compounds and antioxidant properties of oak acorns from Serbia. Food Chem. 104(2): 830–834.

Randhawa, M., I. Seo, F. Liebel, M.D. Southall, N. Kollias and E. Ruvolo. 2015. Visible light induces melanogenesis in human skin through a photoadaptive response. PLoS One 10(6): e0130949.

Rellstab, C., S. Zoller, L. Walthert, I. Lesur, A.R. Pluess, R. Graf, C. Bodénès, C. Sperisen, A. Kremer and F. Gugerli. 2016. Signatures of local adaptation in candidate genes of oaks (*Quercus* spp.) with respect to present and future climatic conditions. Mol. Ecol. 25(23): 5907–5924.

Rho, H.S., A.K. Ghimeray, D.S. Yoo, S.M. Ahn, S.S. Kwon, K.H. Lee, D.H. Cho and J.Y. Cho. 2011. Kaempferol and kaempferol rhamnosides with depigmenting and anti-inflammatory properties. Molecules 16(12): 3338–3344.

Rhodes, L.E., G. Darby, K.A. Massey, K.A. Clarke, T.P. Dew, M.D. Farrar, S. Bennett, R.E.B. Watson, G. Williamson and A. Nicolaou. 2013. Oral green tea catechin metabolites are incorporated into human skin and protect against UV radiation-induced cutaneous inflammation in association with reduced production of pro-inflammatory eicosanoid 12-hydroxyeicosatetraenoic acid. Br. J. Nutr. 110(5): 891–900.

Rice-Evans, C.A., N.J. Miller and G. Paganga. 1996. Structure-antioxidant activity relationships of flavonoids and phenolic acids. Free Radic. Biol. Med. 20(7): 933–956.

Rivas-Ubach, A., A. Gargallo-Garriga, J. Sardans, M. Oravec, L. Mateu-Castell, M. Pérez-Trujillo, T. Parella, R. Ogaya, O. Urban and J. Peñuelas. 2014. Drought enhances folivory by shifting foliar metabolomes in *Quercus ilex* Trees. New Phytol. 202(3): 874–885.

Rocha-Guzmán, N.E., J.R. Medina-Medrano, J.A. Gallegos-Infante, R.F. González-Laredo, M. Ramos-Gómez, R. Reynoso-Camacho, H. Guzmán-Maldonado and S.M. González-Herrera. 2012. Chemical evaluation, antioxidant capacity, and consumer acceptance of several oak infusions. J. Food Sci. 77(2): C162–166.

Rodríguez-Arámbula, A., B. Torres-Álvarez, D. Cortés-García, C. Fuentes-Ahumada and J.P. Castanedo-Cázares. 2015. CD4, IL-17, and COX-2 are associated with subclinical inflammation in malar melasma. Am. J. Dermatopath. 37(10): 761–766.

Rokhsar, C.K. and R.E. Fitzpatrick. 2005. The treatment of melasma with fractional photothermolysis: a pilot study. Dermatol. Surg. 31(12): 1645–1650.

Rosales-Castro, M., R.F. González-Laredo, N.E. Rocha-Guzmán, J.A. Gallegos-Infante, J. Peralta-Cruz, J. Morré and J.J. Karchesy. 2011. Chromatographic analysis of bioactive proanthocyanidins from *Quercus durifolia* and *Quercus eduardii* barks. Acta Chromatogr. 23(3): 521–529.

Rosales-Castro, M., R.F. González-Laredo, N.E. Rocha-Guzmán, J.A. Gallegos-Infante, M.J. Rivas-Arreola and J.J. Karchesy. 2012. Antioxidant activity of fractions from *Quercus sideroxyla* bark and identification of proanthocyanidins by HPLC-DAD and HPLC-MS. Holzforschung 66(5): 577–584.

Routray, W. and V. Orsat. 2012. Microwave-assisted extraction of flavonoids: a review. Food Bioprocess Technol. 5(2): 409–424.

Sakar, M.K., D. Şöhretoğlu, M. Özalp, M. Ekizoğlu, S. Piacente and C. Pizza. 2005. Polyphenolic compounds and antimicrobial activity of *Quercus aucheri* Leaves. Turk. J. Chem. 29(5): 555–559.

Sallé, A., L.-M. Nageleisen and F. Lieutier. 2014. Bark and wood boring insects involved in oak declines in Europe: current knowledge and future prospects in a context of climate change. Forest Ecol. Manag. 328: 79–93.

Salminen, J.-P., T. Roslin, M. Karonen, J. Sinkkonen, K. Pihlaja and P. Pulkkinen. 2004. Seasonal variation in the content of hydrolyzable tannins, flavonoid glycosides, and proanthocyanidins in oak leaves. J. Chem. Ecol. 30(9): 1693–1711.

Sánchez-Burgos, J.A., M.V. Ramírez-Mares, M.M. Larrosa, J.A. Gallegos-Infante, R.F. González-Laredo, L. Medina-Torres and N.E. Rocha-Guzmán. 2013. Antioxidant, antimicrobial, antitopoisomerase and gastroprotective effect of herbal infusions from four *Quercus* species. Ind. Crops Prod. 42: 57–62.

Sarkar, R., S. Chugh and V.K. Garg. 2012. Newer and upcoming therapies for melasma. Indian J. Dermatol. Venereol. Leprol. 78(4): 417–428.

Sato, K. and M. Toriyama. 2009. Depigmenting effect of catechins. Molecules. 14(11): 4425–4432.

Schmidt, H.H.H.W., R. Stocker, C. Vollbracht, G. Paulsen, D. Riley, A. Daiber and A. Cuadrado. 2015. Antioxidants in translational medicine. Antioxid. Redox Signal. 23(14): 1130–1143.

Schwartz, J., A.W.M. Evers, C. Bundy and A.B. Kimball. 2016. Getting under the skin: report from the International Psoriasis Council Workshop on the role of stress in psoriasis. Front. Psychol. 7: 87.

Şen, A., I. Miranda, S. Santos, J. Graça and H. Pereira. 2010. The chemical composition of cork and phloem in the rhytidome of *Quercus cerris* bark. Ind. Crops Prod. 31(2): 417–22.

Şen, A., I. Miranda and H. Pereira. 2012. Temperature-induced structural and chemical changes in cork from *Quercus cerris*. Ind. Crops Prod. 37(1): 508–513.

Shin, S., L.-X. Wang, X.-Q. Zheng, L.-P. Xiang and Y.-R. Liang. 2014. Protective effect of (–)-epigallocatechin gallate against photo-damage induced by ultraviolet A in human skin fibroblasts. Trop. J. Pharm. Res. 13(7): 1079.

Singh Joshan, D. and S.K. Singh. 2013. Investigational study of *Juglans regia* extract and quercetin against photoaging. Biomedicine & Aging Pathology 3(4): 193–200.

[SIRE] Sistema de Información para la Reforestación. 2012. *Quercus laurina* Humb et Bonpl. SIRE-CONAFOR. Mexico. (Online: 19 January 2017) Available in: http://www.conafor.gob.mx:8080/documentos/docs/13/993Quercus%20laurina.pdf.

Skurić, J., N. Oršolić, D. Kolarić, D. Đikić, V. Benković, A. Horvat Knežević and D. Lisičić. 2011. Effectivity of flavonoids on animal model psoriasis—Thermographic evaluation. Periodicum Biologorum 113(4): 457–463.

Sofen, B., G. Prado and J. Emer. 2016. Melasma and post inflammatory hyperpigmentation: management update and expert opinion. Skin Therapy Lett. 21(1): 1–7.

Solano, F. 2014. Melanins: skin pigments and much more -types, structural models, biological functions, and formation routes. New J. Sci. 2014: 1–28.

Stalikas, C.D. 2007. Extraction, separation, and detection methods for phenolic acids and flavonoids. J. Sep. Sci. 30(18): 3268–3295.

Stevanato, R., M. Bertelle and S. Fabris. 2014. Photoprotective characteristics of natural antioxidant polyphenols. Regul. Toxicol. Pharmacol. 69(1): 71–77.

Stevanovic, T., P.N. Diouf and M.E. García-Pérez. 2009. Bioactive polyphenols from healthy diets and forest biomass. Curr. Nutr. Food Sci. 5(4): 264–295.

Sullivan, Alexis R., S.A. Owusu, J.A. Weber, Andrew L. Hipp and O. Gailing. 2016. Hybridization and divergence in multi-species oak (*Quercus*) communities. Bot. J. Linn. Soc. 181(1): 99–114.

Sun, L., J. Zhang, X. Lu, L. Zhang and Y. Zhang. 2011. Evaluation to the antioxidant activity of total flavonoids extract from persimmon (*Diospyros kaki* L.) leaves. Food Chem. Toxicol. 49(10): 2689–2696.

Taherkhani, N. and N. Gheibi. 2014. Inhibitory effects of quercetin and kaempferol as two propolis derived flavonoids on tyrosinase enzyme. Biotech. Health Sci. 1(2): e22242.

Tejerina, D., S. García-Torres, M. Cabeza de Vaca, F.M. Vázquez and R. Cava. 2011. Acorns (*Quercus rotundifolia* Lam.) and grass as natural sources of antioxidants and fatty acids in the 'montanera' feeding of iberian pig: intra- and inter-annual variations. Food Chem. 124(3): 997–1004.

Torres-Álvarez, B., I.G. Mesa-Garza, J.P. Castanedo-Cázares, C. Fuentes-Ahumada, C. Oros-Ovalle, J. Navarrete-Solis and B. Moncada. 2011. Histochemical and immunohistochemical study in melasma: evidence of damage in the basal membrane. Am. J. Dermatopath. 33(3): 291–295.

Tovar-Sánchez, E. and K. Oyama. 2004. Natural hybridization and hybrid zones between *Quercus crassifolia* and *Quercus crassipes* (Fagaceae) in Mexico: morphological and molecular evidence. Am. J. Bot. 91(9): 1352–1363.

Tovar-Sánchez, E., F. Rodríguez-Carmona, V. Aguilar-Mendiola, P. Mussali-Galante, A. López-Caamal and L. Valencia-Cuevas. 2012. Molecular evidence of hybridization in two native invasive species:

Tithonia tubaeformis and *T. rotundifolia* (Asteraceae) in Mexico. Plant Syst. Evol. 298(10): 1947–1959.

Tsai, M.-J., Y.-B. Huang, J.-W. Fang, Y.-S. Fu and P.-C. Wu. 2015. Preparation and characterization of naringenin-loaded elastic liposomes for topical application. PLoS One 10(7): e0131026.

Valencia-A, S. 2004. Diversidad del género *Quercus* (Fagaceae) en México. Bol. Soc. Bot. México 75: 33–53.

Valencia-Cuevas, L., D. Piñero, P. Mussali-Galante, S. Valencia-Ávalos and E. Tovar-Sánchez. 2014. Effect of a red oak species gradient on genetic structure and diversity of *Quercus castanea* (Fagaceae) in Mexico. Tree Genet. Genomes. 10(3): 641–652.

Vicentini, F.T.M.C., T. He, Y. Shao, M.J.V. Fonseca, W.A. Verri, G.J. Fisher and Y. Xu. 2011. Quercetin inhibits UV irradiation-induced inflammatory cytokine production in primary human keratinocytes by suppressing NF-κB pathway. J. Dermatol. Sci. 61(3): 162–168.

Vijayalakshmi, A., V. Ravichandiran, V. Malarkodi, S. Nirmala and S. Jayakumari. 2012. Screening of flavonoid 'quercetin' from the rhizome of *Smilax china* Linn. for anti-psoriatic activity. Asian Pac. J. Trop. Biomed. 2(4): 269–275.

Vijayalakshmi, A., V. Ravichandiran and K. Masilamani. 2014. Antipsoriatic and inhibitory effects of an oral dosage form containing bioflavonoids on inflammatory cytokines IL-1α, IL-1β, IL-6, IL-8, IL-17 and TNF-α. Indian J. Pharm. Educ. 48(Suppl): 139–148.

Warren, J.M., J.H. Bassman, D.S. Mattinson, J.K. Fellman, G.E. Edwards and R. Robberecht. 2002. Alteration of foliar flavonoid chemistry induced by enhanced UV-B radiation in field-grown *Pinus ponderosa*, *Quercus rubra* and *Pseudotsuga menziesii*. J. Photochem. Photobiol. B. 66(2): 125–133.

Weng, Z., A.B. Patel, M. Vasiadi, A. Therianou and T.C. Theoharides. 2014. Luteolin inhibits human keratinocyte activation and decreases NF-κB induction that is increased in psoriatic skin. PLoS One 9(2): e90739.

Wölfle, U., A. Heinemann, P.R. Esser, B. Haarhaus, S.F. Martin and C.M. Schempp. 2012. Luteolin prevents solar radiation-induced matrix metalloproteinase-1 activation in human fibroblasts: a role for p38 mitogen-activated protein kinase and interleukin-20 released from keratinocytes. Rejuvenation Res. 15(5): 466–475.

Wu, M., W.H. Zhang, C. Ma and J.Y. Zhou. 2013. Changes in morphological, physiological, and biochemical responses to different levels of drought stress in Chinese cork oak (*Quercus variabilis* Bl.) seedlings. Russ. J. Plant Physiol. 60(5): 681–692.

Yamauchi, K., T. Mitsunaga and I. Batubara. 2011. Isolation, identification and tyrosinase inhibitory activities of the extractives from *Allamanda cathartica*. Natural Resources 2(3): 167–172.

Yang, D.-J., S.-C. Liu, Y.-C. Chen, S.-H. Hsu, Y.-P. Chang and J.-T. Lin. 2015. Three pathways assess anti-inflammatory response of epicatechin with lipopolysaccharide-mediated macrophage RAW264.7 cells. J. Food. Biochem. 39(3): 334–343.

Yao, K., H. Chen, K. Liu, A. Langfald, G. Yang, Y. Zhang, D.H. Yu et al. 2014. Kaempferol targets RSK2 and MSK1 to suppress ultraviolet radiation-induced skin cancer. Cancer Prev. Res. (Phila) 7(9): 958–967.

Zhang, S., X. Liu, L. Mei, H. Wang and F. Fang. 2016. Epigallocatechin-3-gallate (EGCG) inhibits imiquimod-induced psoriasis-like inflammation of BALB/C mice. BMC Complement. Altern. Med. 16(1): 334.

The Newer Chemistry of Condensed Tannins and Its Foams Application

Antonio Pizzi

Introduction

Condensed tannin extracts consist of flavonoid units that have undergone varying degrees of condensation. They are invariably associated with their immediate precursors (flavan-3-ols, flavan-3,4-diols), other flavonoid analogs (Drewes and Roux 1963, Roux and Paulus 1961), carbohydrates and traces of amino- and imino-acids (Saayman and Roux 1965). Monoflavonoids and nitrogen-containing acids are present in concentrations which are too low to influence the chemical and physical characteristics of the extract as a whole. However, the simple carbohydrates (hexoses, pentoses and disaccharides) and complex glucuronates (hydrocolloid gums) as well as oligomers derived from hydrolysed hemicelluloses are often present in sufficient quantity. Equally, carbohydrate chains of various lengths (Drovou et al. 2015, Abdalla et al. 2014a,b) are also some times linked to flavonoid units in the tannin (Fig. 1).

All these materials are often present in sufficient quantities to decrease and/or increase viscosity, and excessive variation in their percentages alters the physical properties of the natural tannin extract independently of the contribution of the degree of condensation of the tannin.

LERMAB, University of Lorrain, Epinal, France.

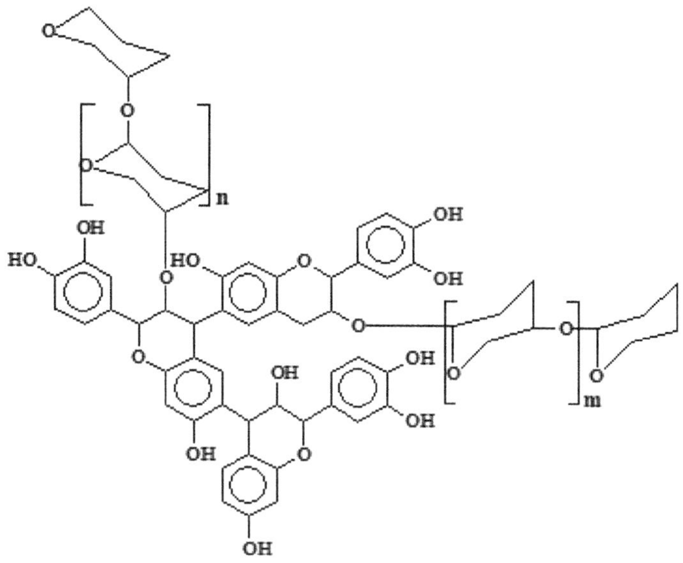

Fig. 1. Example of flavonoid trimer with long carbohydrate chains.

Monoflavonoids, also known as "phenolic non-tannins" (to which flavonoid dimers belongs too), represent the most studied group in the commercially important tannin extracts because of their relative simplicity. They comprise flavan-3,4-diols, flavan-3-ols, dihydrovlavonoids (flavonols), flavanones, chalcones and coumaran-3-ols, thus most of the known classes of flavonoid analogs (Drewes and Roux 1963, Roux and Paulus 1961, Roux 1972, Roux et al. 1975). Typical are those of the black mimosa bark extract (*Acacia mearnsii*, formerly *mollissima*, de Wildt) where the four possible combinations of resorcinol and phloroglucinol (A-rings) with catechol and pyrogallol (B-rings) coexist, although these monoflavonoids constitute a minor percentage (3%–5%) of the total phenolics of the tannin extract (Drewes and Roux 1963). In mimosa tannin extract only flavan-3,4-diols and certain flavan-3-ols (catechin) participate to tannin formation (Fig. 2).

In other tannins epicatechin, delphinidin and catechin gallate are present, always in minor percentages but with some exception. The main exception is cube gambier tannin extract where monoflavonoids, mainly catechin, can constitute up to 50% of the total extract. In mimosa bark tannin extract, each of the four combinations of resorcinol and phloroglucinol (A-rings) with catechol and pyrogallol (B-rings) are present. In this tannin the main polyphenolic pattern is represented by flavonoid analogs based on roninetinidin, thus on resorcinol A-ring and pyrogallol B-ring. This pattern is reproduced in approximately 70% of the phenolic part of the tannin. The secondary but parallel pattern is based on fisetinidin, thus on resorcinol A-rings and catechol B-rings. This represent about 25% of the total polyphenolic bark fraction. Superimposed on these two predominant patterns are two minor groups of A- and B-rings combinations. These are based on phloroglucinol (A-ring)-

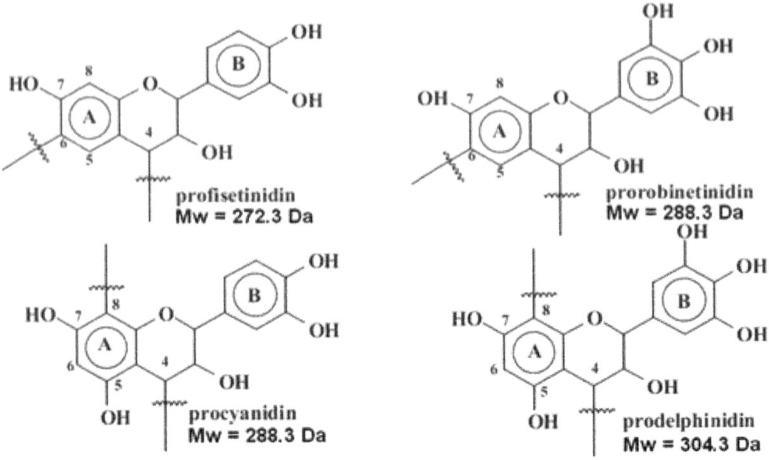

Fig. 2. The four basic flavonoid units present in commercial condensed tannins.

pyrogallol (B-ring) (gallocatechin/delphinidin) and on phloroglucinol (A-ring)-catechol (B-ring) flavonoids (catechin.epicatechin). These four patterns constitute 65%–84% of commercial mimosa bark extract. The remaining parts of the mimosa bark extract are the so-called "non-tannins", this definition coming from the leather industry where it is considered a "tannin" when polyphenolic oligomer is higher and comprising atrimer. It must be pointed out that the percentage of non-tannins varies considerable from tannin extract to tannin extract. For example, pecan nut tannin extract, a predominantly delphinidin tannin, contains no more than 5% of non-phenolic non-tannins (Pizzi 1983, 1994). The percentages are different even for different commercial tannins (Pizzi 1983, 1994). The non-phenolic non-tannins can be subdivided into carbohydrates, hydrocolloid gums, and some amino- and iminoacid fractions (Roux 1965, Pizzi 1983, Mitsunaga et al. 1998). For example, in commercial mimosa bark extract, the carbohydrates 1-pinitol and sucrose predominate, with glucose in a smaller proportion (Roux 1965ab, Pizzi 1983, Mitsunaga et al. 1998). The hydrocolloid gums contribute between 3% and 8% to the total weight of commercial mimosa bark extract and are a major contributor to its viscosity (Roux 1965, Pizzi 1983, Mitsunaga et al. 1998). Iminoacids such as L-pipecolic acid, L-4-hydroxy-*trans*-pipecolic acid and L-proline and traces of the aminoacids arginine, alanine, aspartic acid, glutamic acid and serine have also been reported (Roux 1965a,b, Pizzi 1983, Mitsunaga et al. 1998).

Similar flavonoid A- and B-ring patterns, although slightly different have been found for the other major commercial tannin extract, quebracho wood (*Schinopsis lorentzii* and *Schinopsis balansae*) wood tannin extract, where a predominance of fisetinidin rather than robinetinidin in the constituent flavonoid units has been determined (Mitsunaga et al. 1998). In quebracho extract there is apparent absence of quercetrin and myretrecin (Roux et al. 1975, Pizzi 1983, Mitsunaga et al. 1998, Clark-Lewis and Roux 1959) and much lower proportion or even absence of catechin and gallocatechin (delphinidin), thus practically lack of, or much mower level of phloroglucinol-like A-rings. This difference becomes very important from a structural point of view, as it will be indicated later, as while the quebracho tannin have mainly a linear structure, mimosa tannin presents a more branched structure, with very important effects on their viscosity, their stability and their non-leather use. Similar couplings of A- and B-ring types occur also in douglas fir and hemlock bark tannins.

However, different A- and B-ring couplings occurs in pine and other bark tannin species (Drovou et al. 2015, Pizzi 1983, Porter 1974, Navarrete et al. 2010, 2013, Ucar et al. 2013, Abdalla et al. 2014a,b, Saad et al. 2012, Vazquez et al. 2013). In this only two main patterns occur, predominantly phloroglucinol A-ring with catechol B-ring and two secondary patterns in much lower proportion of phloroglucinol A-ring with phenol B-ring (afzelechin) or of fisetinidin (resorcinol A-ring, catechol-B-ring) (Pizzi 1983, Porter 1974, Navarrete et al. 2010, 2013, Ucar et al. 2013, Abdalla et al. 2014a,b, Saad et al. 2012, Vazquez et al. 2013). In several procyanidin tannins such as some pine tannins and exotic African woods, catechin gallate or gallocatechin gallate does also occur as a constitutive unit of the tannin (Fig. 3).

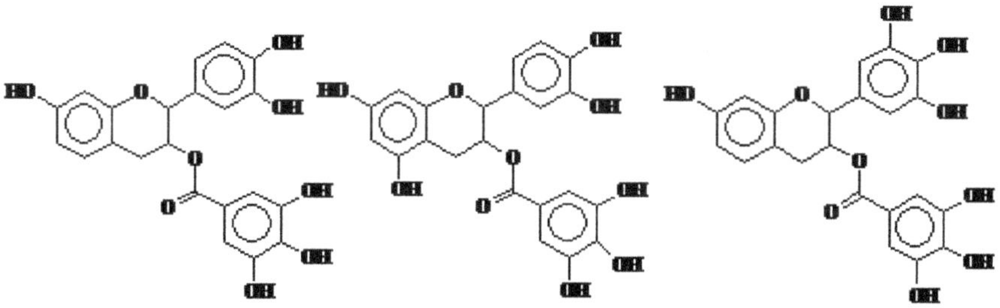

Fig. 3. Example of flavonoid gallate units present in some condensed tannins.

Roux et al. (Roux et al. 1975) have shown that condensation of robinetinidin and fisetinidin flavonoid units is based on a condensation forming a 4,6 interflavonoid link, following an initial 4,8 interflavonoid link formation between a phloroglucinol A-ring type unit, such as catechin, and a resorcinol A-ring type units. For a time it was thought that the positioning of the phloroglucinolic flavonoid unit was as the "lower" terminal unit of a 4,8 interflavonoid linkage. While extracts such mimosa show oligomers of this type (Pizzi 1983, Roux 1972, Hundt and Roux 1978) the latter finding of the so-called "angular" tannins indicated that the phloroglucinolic unit was not just the lower terminal unit of an oligomer but was a central units to which two resorcinol A-ring type units are linked (Pizzi 1983, Botha et al. 1978) (Fig. 4).

Moreover, an even later analysis derived from comparative applied results on the reactivity of aldehydes with different tannins (Pizzi 1983, 1994, 2003) showed that the phloroglucinol A-ring type units, be it catechin, epicatechin, or

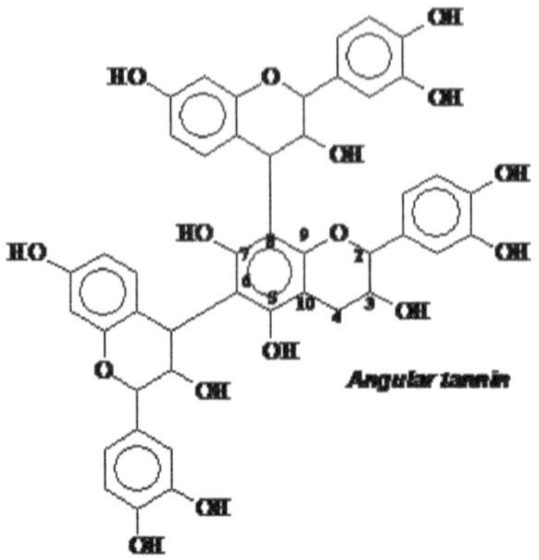

Fig. 4. An example of a so called "angular tannin" oligomer.

gallocatechin (delphinidin), were a branching point in the oligomers of some tannin-type (mimosa). This leads to the concept of such tannins been heavily "branched" rather than just "linear" (Fig. 5).

This was evident as two different rates of reaction with formaldehyde did not occur. Instead, the viscosity increase graphs as a function of time are smooth exponential curves indicating that only one reaction site type is present, the phloroglucinol unit A-ring reactive sites being blocked. This explained the particular practical differences in the use of quebracho and mimosa tannins, quebracho tannin sometimes being subject to a reaction of partial depolymerization while mimosa was not. Mimosa is heavily branched while quebracho is not, and thus this latter is more easily subject to cleavage of the interflavonoid linkage under some drastic conditions (Pizzi 1983, 1994, 2003).

In the case of procyanidin and prodelphinidin type tannins, these being composed exclusively of phloroglucinol A-ring type units, the interflavonoid link is always 4,8, and thus all these oligomers are "linear" (Fig. 6).

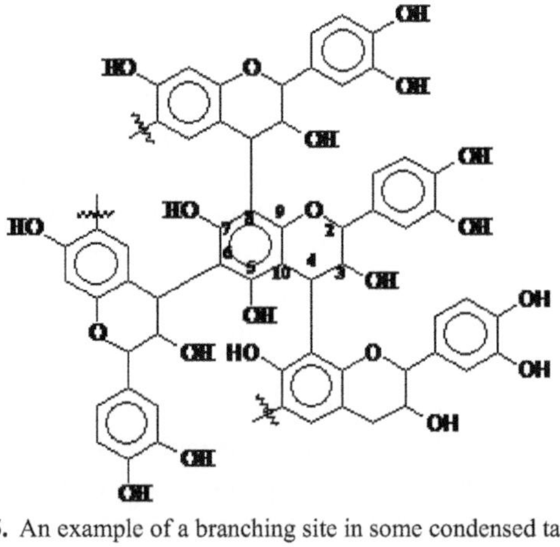

Fig. 5. An example of a branching site in some condensed tannins.

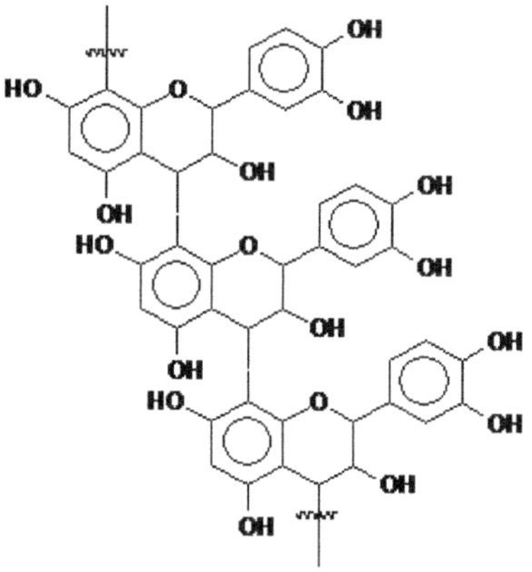

Fig. 6. A schematic representation of a 4,8-linked condensed procyanidin-type tannin.

Reactions of Condensed Flavonoid Tannins

Tannins are per se subject to a number of reactions that impinge on their adaptability to different uses. These basic reactions in tannin chemistry have been abundantly described in the relevant review literature and the reader is oriented towards these review for more complete information (Pizzi 1983, 1994). The basic reactions of tannin are their rearrangements by

  i) hydrolysis and acid and alkaline condensation (Roux et al. 1975, Pizzi 1983,1994, Sealy-Fisher and Pizzi 1992, Hemingway et al. 1986, Pizzi and Stephanou 1993a, McGraw et al. 1992), some leading to insoluble and unreactive compounds called "phlobaphenes" or "tanners red" (Sealy-Fisher and Pizzi 1992) (Fig. 7).

 ii) Sulphitation, this one of the older reactions used in tannin chemistry to decrease the tannins viscosity in water and improve their water solubility (Richtzenain and Alfredsson 1956, Roux 1965ab, Pizzi 1978a, 1979a, 1983), but the excess of which can be deleterious for some applications (Pizzi 1979a).

iii) The catechinic acid rearrangement: while this rearrangement is easily shown to occur in model compounds where the reaction is carried out in solution (Pizzi 1994, Ohara and Hemingway 1991), it is much less evident and more easily avoidable in tannin extracts where the colloidal nature of the extract limits markedly its occurrence. This is fortunate as otherwise some fast-reacting tannins such as pine, pecan, cube gambier, etc., could not be used to produce resins, adhesives and other thermosetting plastics as instead they have been successfully used for (McGraw et al. 1992, Pizzi 1982, Pizzi et al. 1993, Pizzi and Stephanou 1994, Valenzuela et al. 2012, Pizzi et al. 1994, Navarrete et al. 2011).

iv) Catalytic tannin autocondensation. Polyflavonoid tannins have been found to autocondense and harden when in the presence of particular compounds acting as catalysts. Foremost is the catalytic effect of small amounts (2%–3%)

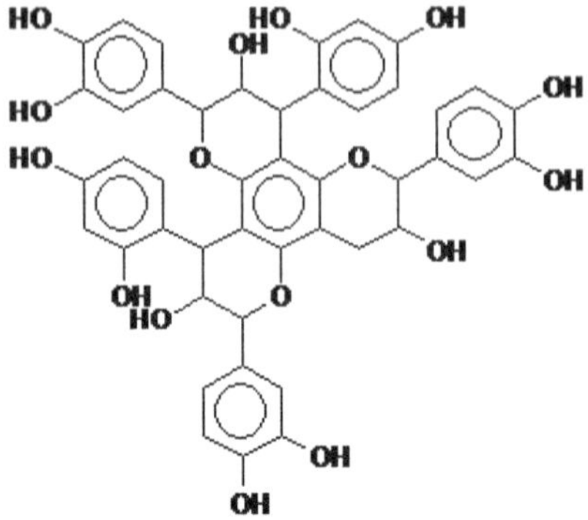

Fig. 7. An example of phlobaphene rearrangement ("tanners red") in a condensed tannin.

of silica smoke, or nanosilica or silicates at high pH (Meikleham et al. 1994). This reaction is rapid and is markedly exothermic, a concentrated solution of tannin at 40%–50% in water gelling and hardening at pH 12 and 25°C in 20–30 minutes. The strong exothermicity of the reaction leads to this result as the temperature increases several tens of degrees in a short time (Meikleham et al. 1994). Small amounts of boric acid and AlCl$_3$ were found to have the same effect (Meikleham et al. 1994) but are much less exothermic. Interestingly, even the presence of lignocellulosic material, such as placing the tannin on the wood surface has a catalytic effect on tannin autocondensation (Garcia et al. 1997, Garcia and Pizzi 1998a,b).

v) Tannin complexation of metals. Tannins readily complex metal ions (Slabbert 1992). This characteristic is at the base of a number of industrial applications. Thus, this characteristic is used to capture or precipitate toxic metals in water (Tondi et al. 2009a, Oo et al. 2009), to isolate a rare metal such as germanium from the copper matrix where it is mined, for paint primers for metal application, and for several other applications. An old example is the formation of Fe complexes used to prepare intensely black/violet inks by formation of ferric tannates. These coordination complexes are due to the ortho-diphenol hydroxyl groups on the tannin B-rings (Fig. 8).

vi) Reactivity and orientation of electrophilic substitutions of flavonoids. The relative accessibility and reactivity of flavonoid units is of interest for their use in resins and adhesives. The C8 site on the A-ring is the first one to react, for example with an aldehyde, and is the site, when free of highest reactivity (Roux et al. 1975, Pizzi 1983). The C6 site on the A-ring is also very reactive but less than the C8 site as the latter presents lower steric hindrance too (Roux et al. 1975, Pizzi 1983). The reactions involves in general only these two sites on the A-ring. The B-ring is particularly unreactive. A low degree of substitution at the 6' site of the B-ring can occur. In general at higher pHs such as pH 10 the B-ring start to react too contributing to cross-linking as well (Pizzi 1978b, Pizzi and Cameron 1986) (Fig. 9).

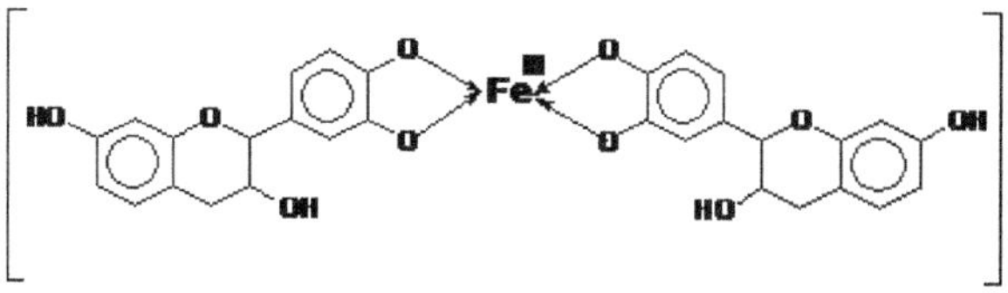

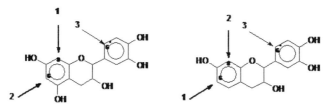

Fig. 8. An example of a ortho-diphenol Fe^{III} metal coordination complex.

Fig. 9. Preferred order of reactivity towards electrophilic substitution of different flavonoids.

Thus, for catechins and phlorogucinol A-ring type flavonoids, the reactivity sequence of the sites is C8 > C6 > C6' when these are free. For robinetinidin and fisetinidin, thus for resorcinol A-ring type flavonoids, the reactivity sequence is modified to C6 > C8 > C6' due to the greater accessibility and lower possibility of steric hindrance of the C6 site (Roux et al. 1975, Pizzi 1983). The curve of gel time of flavonoid tannins with aldehydes has always the shape of a bell curve. The longer gel time is at around pH 4 and fastest gel times at lower pHs and higher pHs. The curve reaches an almost asymptotic plateau of very high reactivity and short gel time at around pH 10 and higher and a fast reactivity too at pHs lower than 1–2 (Pizzi 1983, 1994). The shape of this curve is always the same, but the gel time value is different for different tannins, being slower for mimosa and quebracho, and much faster for procyanidin-type tannins (such as pine) (Pizzi 1994, Hillis and Urbach 1959a,b).

vii) Finally, the reactions of tannins with formaldehyde and other aldehydes, that has led to their extensive application as wood adhesives, will not be treated here as it has extensively been reviewed in depth already (Pizzi 1983, 1994), but other important, and novel reactions of tannin will be discussed.

Tridimensional Structure

While there is an abundant literature in chemical journals on the tridimensional structure of flavonoid monomers one point in which only scant literature exists is on the three-dimensional spacial configuration of flavonoid oligomers. Only one molecular mechanics study on this subject exists (Pizzi et al. 1986). This study shows the correlation that exists between the applicability of these materials and their 3D structure. For example, a tetraflavonoid of 4,8-linked catechins, all 3,4-cis is in helix configuration and when looked along the helix axis a characteristic structure presenting all the 4 B-rings pointing outwards appears (Fig. 10).

Fig. 10. Helix space conformation of a flavonoid tetramer as an example of configuration of unrestrained flavonoid chains in condensed tannins.

Such a structure, rendering particularly available the hydroxyl groups of the B-rings, obviously facilitates their use and reactions. Thus, adhesion to a lignocellulosic substrate, formation of metallic coordination complexes (Slabbert 1992, Tondi et al. 2009a, Oo et al. 2009), formation of polyurethanes with and without isocyanates (Thebault et al. 2015, Pizzi 1979b) are facilitated, and others where reaction of the B-ring is of interest (Pizzi 1978b), such as cross-linking at pH 10 and higher. Conversely, the "spring-like" structure contributes to some of the "resuscitation" behaviours of some plants by holding together the cellular walls and avoiding cellular walls cracking on drying (Pizzi and Cameron 1986).

Influence of Tannin Colloidal Behaviour on Reactions

Water solutions of 40%–50% polyflavonoid tannin extracts appear to be in a colloidal state as indicated by their zeta-potentials (Pizzi and Stephanou 1994). This is caused by both the presence of noticeable proportions of hydrocolloid gums (fragments of hemicelluloses) as well as the presence of higher molecular mass tannins. ^{13}C NMR has confirmed that, during chemical treatment of the tannin extracts, reactions occur in these colloidal solutions that would not be likely to occur in noncolloidal solutions such as used in model compounds experiments. These reactions centre on reactions occurring in the part of the tannin that is the non-aqueous environment within the colloidal micelles, away from water, within which reagents can migrate. An example of this is the role of an organic anhydride by addition of acetic or maleic anhydride to hot, concentrated water solutions of tannin, a reaction used to increase the reactivity of the tannin (Pizzi and Stephanou 1992, 1994a). While part of the anhydride is hydrolysed to acid in water, part of it does instead react within the colloidal micelles to give acetylation and maleation of some flavonoids of the tannin (Pizzi and Stephanou 1992, 1994a) contributing to the marked improvement of reactivity towards aldehydes of the tannin by allowing an alpha-set approach (Pizzi and Stephanou 1993b). Such reactions appear to be particularly beneficial to the quebracho and mimosa tannin extracts, and have some noticeable positive effects on the higher reactivity procyanidin-type tannin extracts, such as pine bark tannin, but less due to its already much higher reactivity. However, they may have deleterious effects on higher reactivity tannin extracts such as the pecan nut prodelphinidin tannin. The reason for the latter behaviour is the very low level, nearly absence of colloidal gums in the extract, and thus very low level of colloidal state, if any. These results are in line with the established zeta-potentials of the different tannin extracts measured (Pizzi and Stephanou 1992, 1994a).

New and Unusual Tannin Reactions

Recently, a number of reactions of tannins that could be useful for a number of different applications have come to light. The first of these is the reaction of flavonoid tannins with concentrated aqueous ammonia (Braghiroli et al. 2013). Catechin was also used as a model compound and treated under the same conditions as mimosa tannin extract was treated. Solid state ^{13}C NMR and MALDI ToF spectroscopy showed that unlike what was recently thought (Hashida et al. 2009), amination is not always regioselective and leading to the conversion of one single –OH group in C4' into an –NH$_2$ group. New reactions have been evidenced, clearly leading to multiamination of several phenolic hydroxygroups, heterocycle opening and oligomerisation and cross-linking through the formation of –N = bridges between flavonoid units, as shown in Fig. 11.

The amination reaction of condensed tannins was used, among others, to totally eliminate synthetic materials in the preparation of non-isocyanate polyurethanes derived from tannins (Thebault et al. 2017).

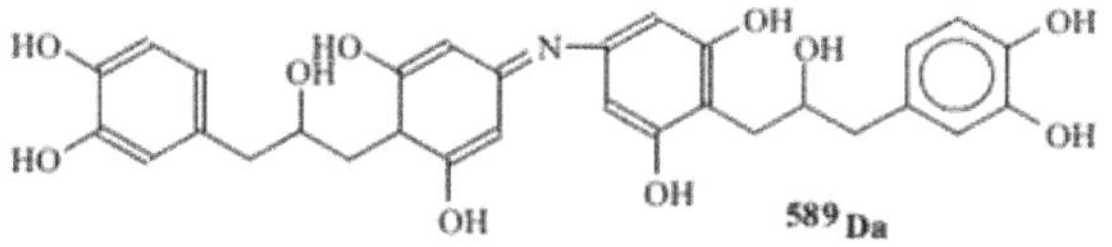

Fig. 11. An example of bridge forming in the reaction of ammonia with flavonoid monomers and oligomers.

Follow up of the reaction with ammonia was the development of rapid cross-linking by reaction of tannin extract with diamines and polyamines (Santiago et al. 2017a). Reaction of a condensed flavonoid tannin, namely mimosa tannin extract, with a hexamethylene diamine, has been investigated. Catechin was also used as a flavonoid model compound and treated under similar conditions. Solid-state CP-MAS ^{13}C NMR and MALDI-ToF mass spectroscopy showed that polycondensation compounds leading to resins were obtained by the reaction of the amines with the phenolic hydroxy groups of the tannin. Simultaneously, a second reaction leading to the formation of ionic bonds between the two groups occurred. These new reactions have been shown to clearly lead to the reaction of several phenolic hydroxyl groups, and flavonoid unit oligomerization, to form hardened resins. MALDI ToF analysis allowed us to observe the presence of compounds of the type shown in Fig. 12.

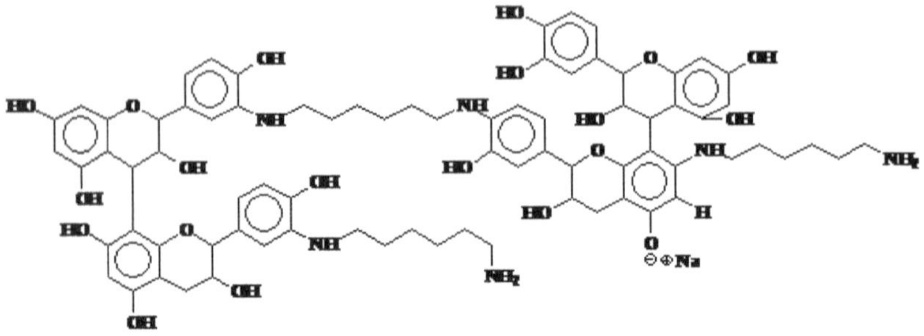

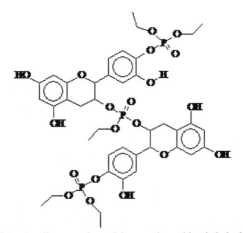

Fig. 12. An example of an oligomer formed by reaction of amines and polyamines with flavonoid monomers and oligomers.

All this clearly indicates how polymerization and cross-linking occurs.

The third very novel reaction is based on the reaction, oligomerization and cross-linking of tannins by triethyl phosphate (TEP) (Basso et al. 2017) in the presence or absence of ammonia (this latter being preferable to the yield). Reaction of condensation and cross-linking of catechin monomer as a model of condensed (flavonoid) tannin extracts and of mimosa tannin itself, as well as of resorcinol with triethyl phosphate (TEP) have been investigated. Solid state CP-MAS ^{13}C NMR, ^{31}P NMR and MALDI-ToF spectroscopy studies revealed that reaction occurs mainly on the C3 of the flavonoid heterocycle ring and on the aromatic C4' and C5' carbons of the flavonoids B-ring, while TEP does not appear to react on the A-ring. Structures of the type shown in Fig. 13 were obtained and the tannin cross-linked with or without being first reacted with ammonia (Fig. 13). The resin so obtained can produce hard thermoset plastics and films resistant up to temperatures in excess of 400°C.

A difference in the relative proportions of these two reaction sites for tannin and catechin has been noticed. The main reaction for the tannin appears to occur

Fig. 13. Example of tannin oligomers formed by reaction with triethyl phosphate showing the polymerisation based on this reaction.

on the C3 site of the heterocycle ring while catechin monomer reacts principally on the OH of the B-ring. This difference could be explained by the lower mobility of the tannin, due to its higher molecular weight and to its colloidal state. The reactions appear to be dependent on the temperature. The reaction appears to have a temperature of activation below which it does not appear to occur. Thus, it occurs readily at 185°C but does not at 100°C. This aspect needs further investigation. According to thermogravimetric analysis, materials obtained from the reaction of tannin with TEP showed high thermal stability. In this context, the potential of this reaction has been evaluated for the production of new heat resistant biomaterials and lacquers (Basso et al. 2014a, 2017, Polesel-Maris and Jutang 2017).

Lastly, among the new reactions of interest is the series of reactions leading to the preparation of non-isocyanate polyurethanes. At first condensed flavonoid tannins from maritime pine (*Pinus pinaster*), mimosa (*Acacia mearnsii*), and Radiata Pine barks, and quebracho (*Schinopsis lorentzii* and *balansae*) wood, were first reacted with dimethyl carbonate. Then hexamethylenediamine was added to these mixtures to form urethane linkages (Thebault et al. 2015). As these non-isocyanate polyurethanes (NIPU) still contained between 40 and 50% of non-biosourced materials, namely, diamine and the methyl carbonate, NIPU resins biosourced to a very high percentage level were prepared by the reaction of aminated mimosa tannin extract with commercial mimosa tannin extract prereacted with dimethyl carbonate (Thebault et al. 2017) (Fig. 14).

NIPU species of the type shown in Fig. 15 were between flavonoid oligomers were detected.

The reaction took place with ease at ambient temperature. Indications were that the polyurethanes obtained formed a hard film when cured at a temperature higher than 100°C. Furthermore, the carbohydrate fraction of the tannin extract appeared also to be carbonated and reacted to generate isocyanate free polyurethane linkages with the aminated tannins (Fig. 16).

This indicated that not only the polyphenolic fraction of the tannin extract but also its other major component can be used to prepare polyurethane resins.

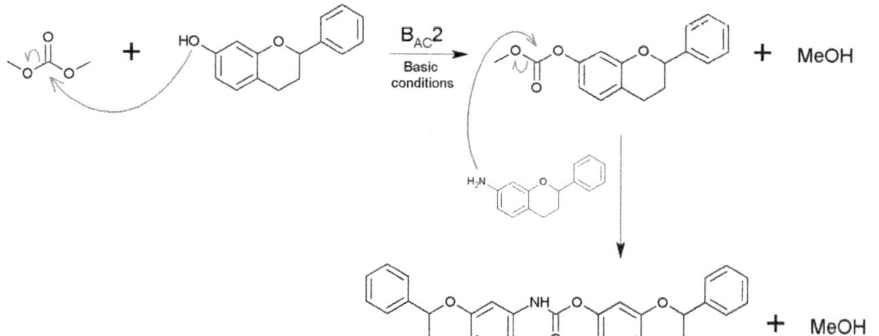

Fig. 14. Bimolecular nucleophilic substitution, acyl-cleaving, in basic catalysis (BAC2) mechanism between dimethyl carbonate and a flavonoid tannin molecule.

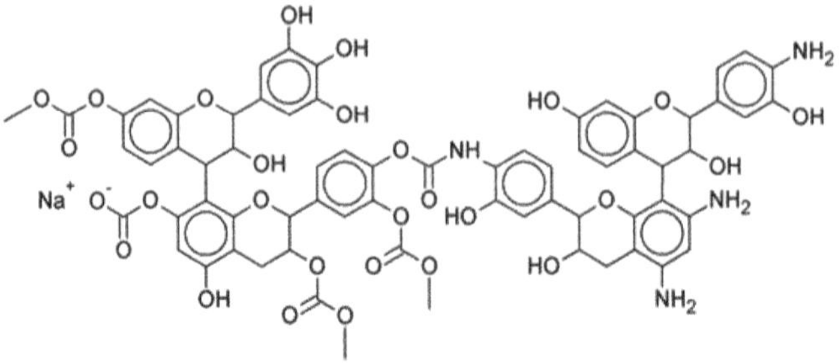

Fig. 15. An example of a non-isocyanate linked polyurethane involving condensed tannin flavonoid oligomers.

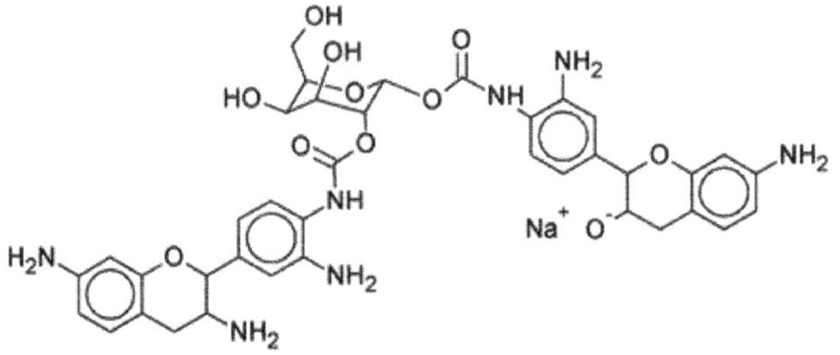

Fig. 16. Example of a non-isocyanate polyurethane formed by reaction of aminated and carbonated tannin units onto a carbohydrate monomers coming from the non-tannin fraction of the tannin extract used.

The chemistry and characteristic reactions of condensed flavonoid tannins have been the basis for their extended industrial utilization. It is on the basis of this chemistry that many heavily or totally biosourced materials have been developed. Among these are: industrialized wood panel adhesives (Pizzi 1978a, 1982, Valenzuela et al. 2012, Pizzi and Stephanou 1993c, 1994b, Santiago et al. 2016). Industrialized wood laminating adhesives (Pizzi and Roux 1978, Pizzi et al. 1982, Pizzi and Cameron 1984). Fire-resistant biosourced rigid and flexible foams for fire and acoustic isolation (Meikleham and Pizzi 1994, Tondi et al. 2008a,b, 2009a,b,c, Tondi and Pizzi 2009, Basso et al. 2011, 2013a,b,c, 2014a,b, Lacoste et al. 2013a,b, 2014a,b, 2015a). Hard plastics (Li et al. 2013) grinding disks for angle grinders (Lagel et al. 2015), and automotive brake pads (Lagel et al. 2016). Paper impregnating resins (Pizzi 1979b, Abdullah et al. 2013a,b), high pressure laminates (Abdullah et al. 2014), and impregnated composites (Pizzi et al. 2009, Sauget et al. 2013). Biosourced wood preservatives (Tondi et al. 2012a,b), polyurethanes by isocyanate reaction (Pizzi 1979c) and non-isocyanate polyurethane surface finishes and resins (Thebault et al. 2015, 2017), and others. There is to hope that on the basis of the same chemistry many other industrial products of progressively higher added value may also be developed in the future. In the next section the application to natural biobased foams is detailed as an example of a more recent development.

Applications

The reactions of tannins with formaldehyde and other aldehydes, that has led to their extensive application as wood adhesives, will not be treated here as extensively reviewed in depth already (Pizzi 1983, 1994), but another important and more recent application will be treated more in depth here.

Tannin-based fire resistant foams

The first records of tannin-based foams date from the early 1970s when a basic formulation was developed by a collaboration between D. Grey and D.G. Roux with A. Pizzi and A. Ryder in South Africa (Roux et al. 1975). These foams had severe problems and at that time, although a considerable effort was expended on them for many years by successive researchers, they did not achieve a good enough performance. Furthermore, they did not seem to interest industry then which was totally dominated by synthetic foams, also because their relative performance/cost structure seemed unfavourable at that time, in the public opinion, to totally biosourced materials. The first public record of a formulation that appeared to work well was published by Meikleham and Pizzi in 1994 (Meikleham and Pizzi 1994). This notwithstanding, interest in these materials came to the fore only in the late years 2000s (Tondi and Pizzi 2009).

There are several systems used to prepare foams, these varying from within foam chemical reactions in presence or not of a blowing agent, foaming under the impulse of the exotherm generated by a chemical reaction or by simply heating the mix, and foams created by simple mechanical agitation/stirring, as well as mixed systems. Tannin-furanic foams are no exception to this type of classification. The various systems developed will be briefly described.

Tannin-furanic foams by chemical reaction exotherm

Initially, and for quite some time afterwards, tannin-furanic foams powered by a reaction exotherm were the only one studied (Fig. 17). The early efforts used diethyl ether as a blowing agent and the exotherm allowing foaming was based on the acid autocondensation of furfuryl alcohol (Meikleham and Pizzi 1994). In this early work acid and alkali-catalysed rigid foams of comparable properties to synthetic phenolic rigid foams were prepared. The fluid polymer phase was based on a tannin-formaldehyde resin. Expansion (foaming) was brought about by a physical blowing agent, whereas dimensional stabilization was achieved through cross-linking to the desired density. In the case of the acid-catalysed foam, the heat-generating agent was the self-condensation of furfuryl alcohol. Tannin-furan copolymers were generated in the reaction. These foams moreover did not give any quantity of toxic gasses on carbonization (Tondi et al. 2008a). While this system worked well, when this research was taken up again in the late years 2000, the first problems that was necessary to solve were (i) the substitution of diethyl ether with

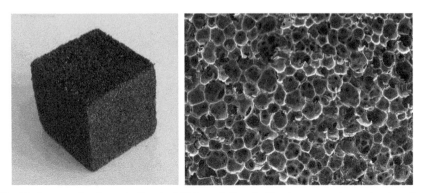

Fig. 17. An example of a tannin-furanic rigid foam (left) and of its structure as observed under a the scanning electron microscope.

a less dangerous and volatile solvent, and (ii) the elimination of formaldehyde. These foams were abundantly characterized. Mimosa tannin bark extract, pine bark tannin and quebracho tannin wood extract were used as building blocks together with furfuryl alcohol. Modifications by addition of up to 20% hydroxymethylated lignin, or smaller proportions of polyurethane, isocyanate (Li et al. 2012a) and an industrial surfactant were also characterized (Tondi and Pizzi 2009). Physical tests such as water absorption, compression resistance, direct flame behaviour and measure of foam cells dimensions were carried out (Tondi and Pizzi 2009). A ^{13}C-NMR analysis contributed to the chemical characterization of these foams.

Equally, resistance to fire and chemicals, absorption of various liquids, permeability, thermal conductivity and mechanical (compressive and tensile) strength were tested. Modifying the structure by addition of boric acids and/or phosphoric acid allowed a substantial increase in fire resistance. These early foam formulations produced rigid foams presenting good resistance to strong acid and bases, and to solvents (Tondi et al. 2009c). High affinity for water, but limited one for organic solvents was also put into evidence. Finally, slightly anisotropic mechanical properties were measured. The foams presented a brittle behaviour either in compression or in tension. However, their strength as well as their thermal conductivity were fully comparable with those of their synthetic phenolic counterparts. These early foams were also examined by X-ray microtomography (Tondi et al. 2009b, Zhao et al. 2010a,b). This gave by way of image treatment, additional useful informations on physical characteristics such as porosity, fraction of open cells, connectivity, tortuosity, and pore size distribution as a function of the foam's density.

To conclude, the great interest in these foams resided in their phenomenal fire resistance and potential thermal insulation (Celzard et al. 2011). They do not burn at all for whatever period when exposed at a flame of 1200°C or of higher temperatures. Only at 3000°C they start decomposing. The multiple uses for which they were tested will be discussed later, and will then become evident to a reader that just thinking of these materials only in term of thermal insulation would be limiting their much wider area of application.

The first modification implemented on these foams was to substitute the unacceptable diethyl ether with a much safer solvent, namely pentane, the same solvent as used for synthetic phenolic foams. This entailed a rebalancing of the formulation due to the higher boiling temperature of pentane.

The first fundamental foam reformulation was dictated by the need to eliminate formaldehyde from the formulation, formaldehyde having become unacceptable for health reasons (NIOSH 2000, NIAST 1977). Formaldehyde was eliminated and the new materials so obtained had, compared to the first generation of tannin-based foams, lower density, lower thermal conductivity, lower hydrophilicity but were also much less brittle due to a much higher flexibility. Such significantly improved characteristics could be obtained through the replacement of formaldehyde with furfuryl alcohol, also a renewable chemical, and a higher amount of blowing agent (Basso et al. 2011, Pizzi et al. 2016). This was followed by the further modification of eliminating both formaldehyde and solvent (pentane) to render the foams even greener, at 98% (Basso et al. 2013a). Comparison of kinetic curves describing the simultaneously-measured foams expansion, hardening, temperature and pressure variation as a function of time allowed to show the differences in process and foaming parameters as a function of time by the differences in formulation between the experimental and control foams and optimization of the foaming and hardening parameters involved (Basso et al. 2013a,b).

It was at this same time that the first successful try to prepare elastic tannin furanic foams occurred (Li et al. 2012b). In this first successful approach flexible tannin foams as opposed to the rigid tannin foams already prepared, were obtained by the addition of an external (non-reacted) plasticizer, namely glycerol. Glycerol was chosen for its high boiling temperature and the lack of evaporation, coupled to its lack of toxicity. Flexibility and spring-back of these experimental foams when subjected to a cyclic compression force followed by spring-back and compression again was quantified by both thermomechanical analysis at different temperatures as well as by compression/spring back hysteresis cycle tests in a universal testing machine. Tannin foams containing formaldehyde and without glycerol have been shown to reach a stress plateau indicative of structure crushing. Tannin foams without formaldehyde but without glycerol too, becomes very fragile, brittle and rigid just two months after their preparation, again presenting structure crushing with ageing. Instead, tannin foams without formaldehyde but with glycerol added do not show any change of flexibility with time and remain truly flexible (Li et al. 2012b). Their glass transition temperature was measured by thermomechanical analysis.

Furthermore, open cell foams obtained by the simultaneous coreaction of condensed flavonoid tannins with an alxoxylated fatty amine and polymeric diphenylmethane isocyanate yielded highly flexible/elastic polyurethane foams (Basso et al. 2014b). Copolymerised amine/isocyanate/tannin oligomers were identified by 13C NMR and MALDI-TOF spectroscopy. In general between 30% and 50% of natural tannins was added to the components used to obtain polymerization of the polyurethane. The characteristic of these new, partially biosourced polyurethanes is that the presence of tannin slows down burning, some of them can be made flame self-extinguishing and if burning they neither flow nor asperge flaming material around, contrary to what occurs with normal polyurethanes. This limits the possibility of transmitting fire to other materials in the same environment. Cyclic compression tests were carried out showing that after 50 cycles, foam recovery was in excess of 80%.

Tannin-based rigid foams were also modified with a hyperbranched synthetic polymer, namely hyperbranched poly(acylamide-ester) polyol, prepared by reaction of diethanolamine with succinic anhydride by a "one step" method (Li et al. 2012c). The hyperbranched poly(acylamide-ester) polyol prepared was acetalized by reaction with glutaraldehyde, and the product so obtained was used to modify the tannin-based foams. It was found that when 3.5 wt% acetalized poly(acylamide-ester) polyol was added, the compressive strength of the tannin-based foam improved by 36.6%. This occurred without any side effect to the other properties of the foam.

Pine bark tannins are much more reactive than mimosa and quebracho tannins experimented with up to 2012. Pine bark tannin/furanic foams were prepared for the first time in 2013 (Lacoste et al. 2013a,b, 2014a,b). The high reactivity of pine tannins in relation to other tannins induced fundamental changes in the tannin foam formulations in order to coordinate foam resin hardening, reaction exotherm and solvent blowing, allowing the formation of a rigid foam. For this work, an equipment named FOAMAT was used to record simultaneously temperature, pressure, velocity and dielectric polarization during foaming. The results highlight the role of surfactant (castor oil ethoxylate) and plasticizer (polyethylene glycol) during foam formation: polymerization, expansion, hardening, and shrinkage. In this work, foams density—and its physical properties—are either surfactant- or plasticizer-controlled. With polyethylene glycol and castor oil ethoxylate, homogeneous microstructure foams were obtained but polyethylene glycol made the foams more elastic and improved their shrinkage. Finally, pine tannin foams with and without formaldehyde were prepared. Their characteristics were tested as regards stress-strain curves, thermal conductivity, Young's modulus, compression strength, densification, densification rate and energy absorbed under compression. In most properties the pine tannin foams with formaldehyde have properties similar to mimosa tannin foams, in some cases slightly lower in others slightly higher. Pine tannin foams without formaldehyde showed lower mechanical strength and more elastic behaviour. This work permitted clearly the application of such type of tannin-furanic foam formulations to the whole class of very reactive procyanidin tannins, not only of different species of pine tannins (Lacoste et al. 2013a,b, 2014a,b) but also spruce tannins (Lacoste et al. 2015a, Cop et al. 2014, 2015a,b) and others. Formaldehyde-free, in reality any-aldehyde-free pine tannin foams, have also been developed but present lower mechanical resistance undergoing compression. Thus, formaldehyde-free pine tannin/furanic rigid foams were successfully prepared by a new approach, namely using alternative non-toxic (NIOSH 2000, NIAST 1977), non-volatile aldehydes as hardeners: glyoxal or glutaraldehyde (Lacoste et al. 2013b, Li et al. 2015). Furthermore, open cells structure tannin-furanic foams prepared using pine bark tannin and mimosa/quebracho type tannins have been shown to give good sound absorption/acoustic insulation characteristics at medium and high frequencies (1000–4000 Hz) with coefficient of acoustic absorption of 0.85–0.97 (Lacoste et al. 2015b). In this range they were superior to polyurethane foams, melamine foams, fibreglass and mineral wool acoustic insulations. Their acoustic absorption coefficient was lower—0.40–0.60 at lower frequencies (250–500 Hz). Compared with commercial foams, tannin-furanic foams have the same typical behaviour of light porous materials. The more open-cell is the foam the better is its sound absorption, with thicker samples absorbing better at medium frequencies.

One of the last problems to be solved has been the surface friability of these foams. One of their main drawbacks is the absorption of water within the foam itself. Another problem is the rather friable surface, which is a definite drawback for some potential applications. These two problems are minimized or eliminated by introducing a component of oil-grafted tannin in the foam formulation. The incorporation of fatty chains markedly decreased foam friability and increased water repellency in the body of the foams (Rangel et al. 2016).

Considering that procyanidin tannins are the greater repository and source of condensed tannins in the world, all these were very important results to allow any future diffuse utilization of tannin foams.

The considerable amount of research dedicated to these foams has allowed also to fairly clearly codify which are the essential parameters to consider when designing new tannin-furanic foams (Basso et al. 2015). It constitutes a guide for further progress in this field to anyone who cares to work on it.

Tannin foams by formation of mixed tannin-furanic and tannin-based polyurethanes

While phenolic foams can be clearly substituted to good effect with tannin-furanic foams, the market is particularly interested in the use of biobased polyurethane foams. This interesting situation came to the fore with an industrial plant trial for a plant where isocyanate had to be compulsorily used otherwise the plant could not run. This was furthermore quite a sizeable polyurethane foam panels line (approx. 18 thousand tons/year). Mixed phenolic-polyurethane-type rigid foams were developed using tannin-furfuryl alcohol natural materials co-reacted with polymeric isocyanate in the proportions imposed by the limitations inherent to the continuous industrial plants for polyurethane foams and used in a plant trial (Basso et al. 2014c). A variety of different copolymerization oligomers were formed. Urethanes appeared to have been formed with two flavonoid tannin sites, mainly at the flavonoid hydroxyl group at C3, but also, although less, on the phenolic hydroxyl groups of the flavonoid oligomers. Urethanes are also formed with (i) glyoxal in the formulation, be it pre-reacted or not with the tannin, (ii) with phenolsulfonic acid and (iii) with furfural. This latter one, however, greatly favours reaction with the A-ring of the flavonoids through a methylene bridge rather than reaction with the isocyanate groups to form urethanes (Basso et al. 2014c). All of the materials appeared to have co-reacted in a manner to form urethane and methylene bridges between all of the main components used. Thus, the tannin, the furfuryl alcohol, the isocyanate, the glyoxal and even the phenol sulfonic acid hardener formed a number of mixed species linked by the two bridge types. Several mixed species comprising 2, 3 and even 4 co-reacted different components were observed.

The more interesting result here, however, was an unusually different approach from the other interesting one taken to use tannins as polyols for polyurethanes that involve tannin oxypropylation (Garcia et al. 2015, 2016), thus an additional reaction step. The unusual result obtained was especially interesting because the results were obtained on an industrial plant line trial. Effectively what mainly occurred was that the glyoxal easily reacted with the tannin during the trial producing –OH groups much more easily approached by the isocyanate thus forming a glyoxalated tannin polyurethane in a single step, a remarkably useful outcome (Basso et al. 2014c). Thus species of the type shown in Fig. 18 were present.

The reaction of glyoxal with the tannin and then with isocyanates to form urethanes closely repeat the same reaction already used for wood adhesives but using the –CH$_2$OH groups formed by the reaction of formaldehyde with tannins, and also with synthetic phenolic and amino resins (Pizzi 1994, Pizzi and Walton 1992, Pizzi et al. 1993).

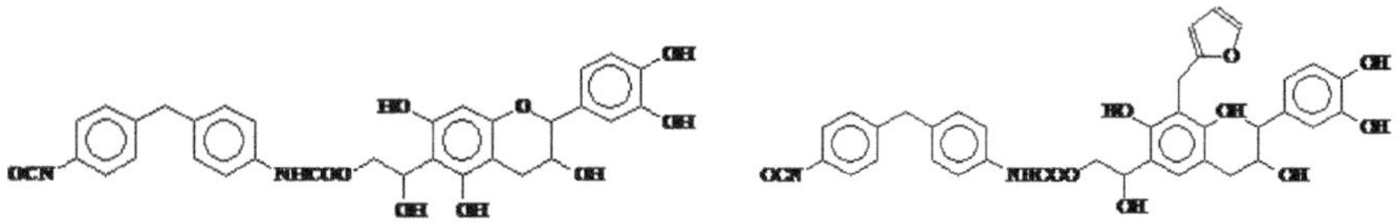

Fig. 18. Example of mixed tannin polyurethanes obtained by the reaction of the isocyanate group on the glyoxal groups pre-reacted with flavonoid tannin units. The reaction can be carried out simultaneously also, as used under industrial conditions.

Alkaline Tannin Foams by Application of External Heat

All the foam formulations decribed up to now, however, rely on the acid-catalyzed exothermal self-condensation reaction of furfuryl alcohol to provide blowing of the mixture to form the foam itself. They are all, then, without exception, tannin/furanic foams. The catalyst of all these foams is invariably a strong acid such as para-toluene sulphonic acid; hence the foams themselves are strongly acidic due to the strong acid catalyst used. Some self-neutralization systems originating from other tannin technologies (Pizzi et al. 1986) have been successfully tested. However, the fact remains that at some important stage of the process, the mix is strongly acidic and that acid can still be released in service. Such strong acidity, either permanent or transitory (as in self-neutralization) can be rather damaging in some applications where the loose acid might seep through and damage materials with which the foam is in contact, such as for instance wood.

Thus, the next major variations of condensed flavonoid tannin foams was the formulation of foams capable to be prepared under alkaline conditions (Basso et al. 2014d). This entailed the elimination of furfuryl alcohol from these formulations, as under alkaline conditions, foam preparation could not rely on the heat generated by the self-condensation of furfuryl alcohol, which occurs only under strongly acidic conditions. The approach used to formulate the alkaline foam then was: (i) the total elimination of furfuryl alcohol from the formulation, (ii) coupled with the use of an aldehyde hardener different from formaldehyde (for environmental reasons) and (iii) the application of moderate heat to allow foaming. These were not tannin/furanic foams as their acid-curing counterparts, but tannin only foams. The open cell foams were evaluated for bulk density, compressive strength, thermal conductivity and fire resistance. Their characteristics were similar to the acid-curing tannin/furanic foams.

Mechanically Blown and Mixed Mechanically/Chemically Blown Tannin Foams

Condensed flavonoid tannin-based foams from quebracho tannin extract have been developed using a new method of expansion based on what done for fire-fighting or tunneling foams, where a foam concentrate forms a stable liquid foam. This new mechanical method of expansion allows obtaining a solid porous material after curing and hardening at room temperature, through a liquid foam formed by a tannin resin and an aqueous solution of surfactant. The use of this new approach for preparing tannin-based rigid foams avoids the problem of shrinkage presented by many other formulations where physical or chemical foaming is employed. Non-ionic surfactants have also been used in the formulation to obtain smaller cells and improve the structure of the new material. Bulk density, mechanical and thermal properties and morphological appearance of the foams have been characterized and reported (Santiago et al. 2017b, Delgado et al. 2018). Equally a mixed mechanical-chemical blowing approach has been taken. These foams needed the reaction with polyamines (Santiago et al. 2017a) discussed in the new chemical reactions section to stiffen relatively quickly to avoid that the liquid projected foams pours/drips and breaks under the effect of gravity.

These latter foams are made using a combination of two expansion methods, one by mechanical expansion and the other method based on the release of water and other gases during the self-condensation of furfuryl alcohol (chemical expansion). The combination of both methods has allowed to overcome certain limitations found in the preparation of foams exclusively based on mechanical expansion, as the resulting density or the mechanical properties. The resin was characterized under stressed conditions by means of several rheological techniques, such as frequency and time sweeps and temperature ramps. The evolution of the linear viscoelastic functions during the reaction clearly showed the transition from viscous to strong gel-like behavior. The foaming and hardening process have been followed kinetically as well, and the resulting foams have been characterized in terms of density, mechanical properties and thermal conductivity (Santiago et al. 2017c).

Tannin-based Monolith Foams

Three types of monoliths-type foams can also be prepared with condensed tannins. The term monolith is preferred rather than foams in this case because their porous structure is not developed by traditional foaming. They can be obtained by emulsion-templating, or by whipping the emulsion until stiff, or by whipping an aqueous phase of resin in the presence of a surfactant, until a liquid, stable foam is obtained (Szczurek et al. 2014, Celzard et al. 2015).

Albumin/tannin-based monolith foams were also prepared under alkaline or acid conditions (Lacoste et al. 2015c). They were prepared by whipping up a water solution of a protein, egg albumin, mixed with a second water solution of condensed flavonoid tannins. Monolith foam formulations with four different condensed tannin bark extracts and different relative proportions of protein and tannin were developed with a rapid process where the protein has a double function: co-reagent and surfactant as well. Physical properties such as mechanical performance, thermal conductivity, and porous structure were characterized to identify the large potential of applications of these new cellular monolith foams, fully bio-based and probable of easier biodegradability. Moreover, some albumin-based monolith foams were found to be completely flexible under certain conditions.

Conclusions

Tannin based rigid foams appear suitable for a wide range of applications. Other than thermal insulation and acoustic insulation, obviously two of the more interesting areas, a third area is also of particular topical interest. This latter is one of the most promising, this being the carbonization of tannin foams to form carbon foams for a variety of applications, an area in which a considerable amount of research has been done in the last three years. A whole literature on the subject exists and the readers is invited to consult some of the initial basic work in this very interesting area as well as some of the more recent works on this subject (Tondi et al. 2008c, 2009d, 2010, Szczurek et al. 2011a,b, 2013, Celzard et al. 2012, Grishechko et al. 2013).

Another area of application of interest for the natural tannin-furanic foams is the field of hydroponics (Basso et al. 2016). Formaldehyde-free quebracho tannin foams were developed for horticultural/hydroponics and floral applications. These foams included in their composition a wetting agent and at least one compound able to neutralize the residual acidity derived from acid catalyst. This was necessary to bring the pH in line with what plants required. Their densities were between 0.048 and 0.066 g/cm^3 and compression strength between 0.07 and 0.09 MPa. Scanning electron microscopy (SEM) images showed open porosity and average cell size of 125–250 μm. Water absorption peaked at 98% (vol.) while residual pH value was 5. These new tannin foams do not result to be phytotoxic and are apt to the conservation of fresh cut flowers and as support matrices for horticulture and hydroponics. They have shown performances comparable or superior to commercial synthetic phenolic floral foam used as reference.

References

Abdalla, S., A. Pizzi, N. Ayed, F. Charrier, F. Bahabry and A. Ganash. 2014a. MALDI-TOF and 13C NMR analysis of Tunisian *Zyzyphus jubjuba* root bark tannins. Ind. Crops Prod. 59: 277–281.

Abdalla, S., A. Pizzi, N. Ayed, F. Charrier-El-Bouthoury, B. Charrier, F. Bahabry and A. Ganash. 2014b. MALDI-TOF analysis of Aleppo pine (*Pinus haleppensis*) bark tannin. BioResources. 9(2): 3396–3406.

Abdullah, U.H., A. Pizzi, X. Zhou, K. Rode, L. Delmotte and H.R. Mansouri. 2013a. Mimosa tannin resins for impregnated paper overlays. Eur. J. Wood Prod. 71: 153–162.

Abdullah, U.H., X. Zhou, A. Pizzi and A. Merlin. 2013b. A note on the surface quality of plywood overlaid with mimosa (*Acacia mearnsii*) tannin and melamine urea formaldehyde impregnated paper: Effects of moisture content of resin-impregnated papers before pressing on the physical properties of overlaid panels. Int. Wood Prod. J. 4: 253–256.

Abdullah, U.H., A. Pizzi and X. Zhou. 2014. High pressure paper laminates from mimosa tannin resin. Int. Wood Prod. J. 5(4): 224–227.

Basso, M.C., X. Li, S. Giovando, V. Fierro, A. Pizzi and A. Celzard. 2011. Green, formaldehyde-free foams for thermal insulation. Adv. Materials Lett. 2(6): 378–382.

Basso, M.C., S. Giovando, A. Pizzi, A. Celzard and V. Fierro. 2013a. Tannin/furanic foams without blowing agents and formaldehyde. Ind. Crops Prod. 49: 17–22.

Basso, M.C., A.Pizzi and A.Celzard. 2013b. Influence of formulation on the dynamics of preparation of tannin based foams. Ind. Crops Prod. 51 396–400.

Basso, M.C., A. Pizzi and A. Celzard. 2013c. Dynamic monitoring of tannin foams preparation: Surfactant effects. BioResources 8(4): 5807–5816.

Basso, M.C., C. Lacoste, A. Pizzi, E. Fredon and L. Delmotte. 2014a. Flexible tannin-furanic films and lacquers. Ind. Crops Prod. 61: 352–360.

Basso, M.C., S. Giovando, A. Pizzi, H. Pasch, N. Pretorius and L. Delmotte. 2014b. Flexible-elastic copolymerized polyurethane-tannin foams. J. Appl. Polym. Sci. 131: DOI 10.1002/app.40499.

Basso, M.C., A. Pizzi, C. Lacoste, L. Delmotte, F.A. Al-Marzouki, S. Abdalla and A. Celzard. 2014c. Tannin-furanic-polyurethane foams for industrial continuous plant lines. Polymers 6: 2985–3004.

Basso, M.C., S. Giovando, A. Pizzi, M.C. Lagel and A. Celzard. 2014d. Alkaline tannin rigid foams. J. Renew. Mat. 2: 182–185.

Basso, M.C., M.C. Lagel, A. Pizzi, A. Celzard and S. Abdalla. 2015. First tools for tannin-furanic foams design. Bioresources. 10(3): 5233–5241.

Basso, M.C., A. Pizzi, F. Al-Marzouki and S. Abdalla. 2016. Horticultural/hydroponics and floral foams from tannins. Ind. Crops Prod. 87: 177–181.

Basso, M.C., A. Pizzi, J. Polesel-Maris, L. Delmotte, B. Colin and Y. Rogaume. 2017. MALDI-TOF and 13C NMR analysis of the cross-linking reaction of condensed tannins by triethyl phosphate. Ind. Crops Prod. 95: 621–631.

Botha, J.J., D. Ferreira and D.G. Roux. 1978. Condensed tannins. Circular dichroism method of assessing the absolute configuration at C-4 of 4-arylflavan-3-ols, and stereochemistry of their formation from flavan-3,4-diols. J. Chem. Soc. Chem. Comm. 0: 698–700.

Braghiroli, F., V. Fierro, A. Pizzi, K. Rode, W. Radke, L. Delmotte, J. Parmentier and A. Celzard. 2013. Condensation reaction of flavonoid tannins with ammonia. Ind. Crops Prod. 44: 330–335.

Celzard, A., A. Szczurek, P. Jana, V. Fierro, M.C. Basso, S. Bourbigot, M. Stauber and A. Pizzi. 2015. Latest progresses in tannin-based cellular solids. J. Cellular Plastics. 51(1): 89–102.

Celzard, A., V. Fierro, G. Amaral-Labat, A. Pizzi and J. Torero. 2011. Flammability assessment of tannin-based cellular materials. Polym. Degrad. Stabil 96: 477–482.

Celzard, A., G. Tondi, D. Lacroix, G. Jeandel, B. Monod, V. Fierro and A. Pizzi. 2012. Radiative properties of tannin-based, glasslike, carbon foams. Carbon 50: 4102–4113.

Cop, M., M.-P. Laborie, A. Pizzi and M. Sernek. 2014. Curing characterisation of spruce tannin-based foams using the advances isoconversional method. Bioresources 9(3): 4643–4655.

Cop, M., C. Lacoste, M. Conradi, M.-P. Laborie, A. Pizzi and M. Sernek. 2015a. The effect of the composition of spruce and pine tannin-based foams on their physical, morphological and compression properties. Ind. Crops Prods. 74: 158–164.

Cop, M., B. Gospodaric, K. Kempainen, S. Giovando, M.-P. Laborie, A. Pizzi and M. Sernek. 2015b. Characterization of the curing process of pine and spruce tannin-based foams by different methods. Eur. Polym. J. 69: 29–37.

Clark-Lewis, J.W. and D.G. Roux. 1959. Natural occurrence of enantiomorphous leucoanthocyanidian: (+)-mollisacacidin (gleditsin) and quebracho(–)-leucofisetinidin. J. Chem. Soc. 0: 1402–1406.

Delgado, C., F.J. Santiago V. Fierro, A. Pizzi and A. Celzard. 2018. Optimisation of "green" tannin-furanic foams for thermal insulation by experimental design. Mater. Design 139: 7–15.

Drewes, S.E. and D.G. Roux. 1963. Condensed tannins. 15. Interrelationships of flavonoid components in wattle-bark extract. Biochem. J. 87: 167–172.

Drovou, S., A. Pizzi, C. Lacoste, J. Zhang, S. Abdalla and F.M. Al-Marzouki. 2015. Flavonoid tannins linked to long carbohydrate chains—MALDI ToF analysis of the tannin extract of the african locust bean. Ind. Crops Prod. 67: 25–32.

Garcia, D.E., W.G. Glasser, A. Pizzi, S. Paczkowski and M.-P. Laborie. 2015. Hydroxypropyl tannin from Pinus pinaster bark as polyol source in urethane chemistry. Eur. Polym. J. 67: 152–165.

Garcia, D.E., C.A. Fuentealba, J.P. Salazar, M.A. Perez, D. Escobar and A. Pizzi. 2016. Mild hydroxypropylation of polyflavonoids obtained under pilot-plant scale. Ind. Crops Prod. 87: 350–362.

Garcia, R., A. Pizzi and A. Merlin. 1997. Ionic polycondensation effects on the radical autocondensation of polyflavonoid tannins-An ESR study. J. Appl. Polym. Sci. 65: S2623–2632.

Garcia, R. and A. Pizzi. 1998a. Polycondensation and autocondensation networks in polyflavonoid tannins, Part 1: final networks. J. Appl. Polym. Sci. 70: 1083–1091.

Garcia, R. and A. Pizzi. 1998b. Polycondensation and autocondensation networks in polyflavonoid tannins, Part 2: polycondensation vs. autocondensation. J. Appl. Polym. Sci. 70: 1093–1110.

Grishechko, L., G. Aamaral-Labat, A. Szczurek, V. Fierro, A. Pizzi, V. Kuznetsov and A. Celzard. 2013. New tannin-lignin aerogels. Ind. Crops Prod. 41: 347–355.

Hashida, K., R. Makino and S. Ohara. 2009. Amination of pyrogallol nucleus of condensed tannins and related polyphenols by ammonia water treatment. Holzforschung 63: 319–326.

Hemingway, R.M., P.E. Laks, G.W. McGraw and R.E. Kreibich. 1986. Tannin cold set adhesives. *In*: Christiansen, A.W. and A.H. Conner (eds.). Wood Adhesives in 1985: Status and Needs, Forest Products Society, Madison, Wisconsin.

Hillis, W.E. and G. Urbach. 1959a. The reaction of (+)-catechin with formaldehyde. J. Chem. Technol. Biotechnol. 9(9): 474–482.

Hillis, W.E. and G. Urbach. 1959b. Reaction of polyphenols with formaldehyde. J. Chem. Technol. Biotechnol. 9(12): 665–673.

Hundt, H.K.L. and D.G. Roux. 1978. Condensed tannins: determination of the point of linkage in 'terminal'(+)-catechin units and degradative bromination of 4-flavanylflavan-3,4-diols. J. Chem. Soc., Chem. Comm. 0: 696–698.

Jahanshahi, S., A. Pizzi, A. Abdolkhani, K. Doosthoseini, A. Shakeri, M.C. Lagel and L. Delmotte. 2016. MALDI-TOF and 13C-NMR and FT-MIR and strength characterization of glycidyl ether tannin epoxy resins. Ind. Crops Prod. 83: 177–185.

Konai, N., D. Raidandi, A. Pizzi and L. Meva'a. 2017. Characterisation of *Ficus sycomorus* using ATR-FTMIR, MALDI-TOF MS and ^{13}C NMR methods. Eur. J. Wood Prod. 75: 807–815.

Lacoste, C., M.C. Basso, A. Pizzi, M.-P. Laborie, A. Celzard and V. Fierro. 2013a. Pine tannin-based rigid foams: mechanical and thermal properties. Ind. Crops Prod. 43: 245–250.

Lacoste, C., M.C. Basso, A. Pizzi, M.-P. Laborie, D. Garcia and A. Celzard. 2013b. Bioresourced pine tannin/furanic foams with glyoxal and glutaraldehyde. Ind. Crops Prod. 45: 401–405.

Lacoste, C., A. Pizzi, M.C. Basso, M.-P. Laborie and A. Celzard. 2014a. *Pinus pinaster* tannin/furanic foams: Part 1, Formulations. Ind. Crops Prod. 52: 450–456.

Lacoste, C., A. Pizzi, M.C. Basso, M.-P. Laborie and A. Celzard. 2014b. *Pinus pinaster* tannin/furanic foams: Part 2: Physical properties. Ind. Crops Prod. 61: 531–536.

Lacoste, C., M. Cop, K. Kampainen, S. Giovando, A. Pizzi, M.P. Laborie, M. Sernek and A. Celzard. 2015a. Biobased foams from condensed tannins from Norway spruce (*Picea abies*) bark. Ind. Crops Prod. 73: 144–153.

Lacoste, C., M.C. Basso, A. Pizzi, A. Celzard, E. Ella Bang, N. Gallon and B. Charrier. 2015b. Pine (*P. pinaster*) and quebracho (*Schinopsis lorentzii/balansae*) tannin based foams as green acoustic absorbers. Ind. Crops Prod. 67: 70–73.

Lacoste, C., M.C. Basso, A. Pizzi, M.-P. Laborie and A. Celzard. 2015c. Natural albumin/tannins cellular foams. Ind. Crops Prods. 73: 41–48.

Lagel, M.C., J. Zhang and A. Pizzi. 2015. Cutting and grinding wheels for angle grinders with a bioresin matrix. Ind. Crops Prod. 67: 264–269.

Lagel, M.C., L. Hai, A. Pizzi, M.C. Basso, L. Delmotte, S. Abdalla, A. Zahed and F.M. Al-Marzouki. 2016. Automotive brake pads made with a bioresin matrix. Ind. Crops Prod. 85: 372–381.

Li, X., M.C. Basso, V. Fierro, A. Pizzi and A. Celzard. 2012a. Chemical modification of tannin/furanic rigid foams by isocyanates and polyurethanes. Maderas 14: 257–265.

Li, X., A. Pizzi, M. Cangemi, V. Fierro and A. Celzard. 2012b. Flexible natural tannin-based and protein-based biosourced foams. Ind. Crops Prod. 37: 389–393.

Li, X., H. Essawy, A. Pizzi, L. Delmotte, K. Rode, D. Le Nouen, V. Fierro and A. Celzard. 2012c. Modification of tannin based rigid foams using oligomers of a hyperbranched poly(amine-ester). J. Polym. Res. 19: 21–29.

Li, X., A. Nicollin, A. Pizzi, X. Zhou, A. Sauget and L. Delmotte. 2013. Natural tannin/furanic thermosetting moulding plastics. RSC Advances 3: 17732–17740.

Li, X., A. Pizzi, X. Zhou, V. Fierro and A. Celzard. 2015. Formaldehyde-free prorobitenidin/profisetinidin tannin/furanic foams based on alternative aldehydes: glyoxal and glutaraldehyde. J. Renew. Mat. 3: 142–150.

McGraw, G.W., T.G. Rials, J.P. Steynberg and R.W. Hemingway. 1992. Chemistry of pecan tannins and analysis of cure of pecan tannin-based cold-setting adhesives with a DMS "Micro-Beam" test. pp. 979–990. *In*: Hemingway, R.W. and P.E. Laks (eds.). Plant Polyphenols. Plenum Press, New York, USA.

Meikleham, N. and A. Pizzi. 1994. Acid and alkali-setting tannin-based rigid foams. J. Appl. Polym. Sci. 53: 1547–1556.

Meikleham, N., A. Pizzi and A. Stephanou. 1994. Induced accelerated autocondensation of polyflavonoid tannins for phenolic polycondensates. Part 1: 13C NMR, 29Si NMR, X-ray and polarimetry studies and mechanism. J. Appl. Polym. Sci. 54: 1827–1845.

Mitsunaga, T., T. Doi, Y. Kondo and I. Abe. 1998. Color development of proanthocyanidins in vanillin-hydrochloric acid reaction. J. Wood Sci. 44: 125–130.

Navarrete, P., A. Pizzi, H. Pasch, K. Rode and L. Delmotte. 2010. MALDI-TOF and 13C NMR characterisation of maritime pine industrial tannin extract Ind. Crops Prod. 32: 105–110.

Navarrete, P., A. Pizzi, F. Bertaud and S. Rigolet. 2011. Condensed tannin reactivity inhibition by internal rearrangements: Detection by CP-MAS 13C NMR. Maderas 13: 59–68.

Navarrete, P., A. Pizzi, H. Pasch, K. Rode and L. Delmotte. 2013. Characterisation of two maritime pine tannins as wood adhesives. J. Adh. Sci. Technol. 27: 2462–2479.

[NIOSH] National Institute for Occupational Safety and Health. 2000. Registry of Toxic Effects of Chemical Substances. National Institute for Occupational Safety and Health, Washington, DC, USA.

[NTIS] National Technical Information Service. 1977. Publication AD-A125–539. National Technical Information Service, Washington, DC, USA.

Ohara, S. and R.W. Hemingway. 1991. Condensed tannins: The formation of a diarylpropanol-catechinic acid dimer from base-catalyzed reactions of (+)-catechin. J. Wood Chem. Technol. 11: 195–208.

Oo, C.W., M.J. Kassim and A. Pizzi. 2009. Characterization and performance of *Rhizophora apiculata* mangrove polyflavonoid tannins in the adsorption of copper (II) and lead (II). Ind. Crops Prod. 30: 152–161.

Pasch, H., A. Pizzi and K. Rode. 2001. MALDI-TOF mass spectrometry of polyflavonoid tannins. Polymer. 42: 7531–7539.

Pizzi, A. 1978a. Wattle-based adhesives for exterior grade particleboard. Forest Prod. J. 28(12): 42–47.

Pizzi, A. 1978b. Tannin formaldehyde exterior wood adhesives through flavonoid B-ring cross-linking. J. Appl. Polym. Sci. 22: 2397–2399.

Pizzi, A. and D.G. Roux. 1978. The chemistry and development of tannin-based weather- and boil-proof cold-setting and fast-setting adhesives for wood. J. Appl. Polym. Sci. 22: 1945–1954.

Pizzi, A. 1979a. Sulphited tannins for exterior wood adhesives. Colloid Polym. Sci. 257: 37–40.

Pizzi, A. 1979b. Tannin-based overlays for particleboard. Holzforsch. Holzverwert. 31: 59–61.

Pizzi, A. 1979c. Tannin-based polyurethane adhesives. J. Appl. Polym. Sci. 23: 1889–1990.

Pizzi, A., D. Du, T. Rossouw, W. Knuffel and M. Singmin. 1980. "Honeymoon" phenolic and tannin-based fast setting adhesive systems for exterior grade fingerjoints. Holzforsch. Holzverwert. 32(6): 140–151.

Pizzi, A. 1982. Pine tannin adhesives for particleboard. Holz Roh Werkst. 40: 293–301.

Pizzi, A. 1983. Tannin based adhesives. pp. 178–246. *In:* Pizzi, A. (ed.). Wood Adhesives Chemistry and Technology. Marcel Dekker, New York, USA.

Pizzi, A. and F.A. Cameron. 1984. Fast-set adhesives for glulam. Forest Prod. J. 34: 61–65.

Pizzi, A. and F.A. Cameron. 1986. Flavonoid tannins—structural wood components for draught resistance mechanism of plants. Wood Sci. Technol. 20: 119–124.

Pizzi, A., F.A. Cameron and N.J. Eaton. 1986. The tridimensional structure of polyflavonoids by conformational analysis. J. Macromol. Sci. Chem. A23: 515–538.

Pizzi, A., R. Vosloo, F.A. Cameron and E. Orovan. 1986. Self-neutralizing acid-set PF wood adhesives. Holz Roh Werkst. 44: 229–234.

Pizzi, A. and T. Walton. 1992. Non-emulsifiable, water-based diisocyanate adhesives for exterior plywood, Part 1: novel reaction mechanisms and their chemical evidence. Holzforschung. 46: 541–547.

Pizzi, A. and A. Stephanou. 1992. Theory and practice of non-fortified tannin adhesives for particleboard. Holzforsch. Holzverwert. 44: 62–68.

Pizzi, A., E.P. Von Leyser, J. Valenzuela and J.G. Clark. 1993. The chemistry and development of pine tannin adhesives for exterior particleboard. Holzforschung 47: 164–172.

Pizzi, A., J. Valenzuela and C. Westermeyer. 1993. Non-emulsifiables, water-based, diisocyanate adhesives for exterior plywood, Part 2: industrial application. Holzforschung. 47: 69–72.

Pizzi, A. and A. Stephanou. 1993a. A comparative 13C NMR study of polyflavonoid tannin extracts for phenolic polycondensates. J. Appl. Polym. Sci. 50: 2105–2113.

Pizzi, A. and A. Stephanou. 1993b. On the chemistry, behaviour and cure acceleration of phenol-formaldehyde resins under very alkaline conditions. J. Appl. Polym. Sci. 49: 2157–2160.

Pizzi, A. and A. Stephanou. 1993c. Comparative and differential behaviour of pine vs. pecan nut tannin adhesives for particleboard. Holzforsch. Holzverwert. 45: 30–33.

Pizzi, A. 1994. Advanced Wood Adhesives Technology. Marcel Dekker, New York, USA.

Pizzi, A., J. Valenzuela and C. Westermeyer. 1994. Low-formaldehyde emission, fast pressing, pine and pecan tannin adhesives for exterior particleboard. Holz Roh Werkst. 52: 311–315.

Pizzi, A. and A. Stephanou. 1994a. A ^{13}C NMR study of polyflavonoid tannin adhesives intermediates, Part 2: colloidal state reactions. J. Appl. Polym. Sci. 51: 2125–2130.

Pizzi, A. and A. Stephanou. 1994b. Fast vs. slow-reacting non-modified tannins extracts for exterior particleboard adhesives, Holz Roh Werkst. 52: 218–222.

Pizzi, A., N. Meikleham, B. Dombo and W. Roll. 1995. Autocondensation-based, zero-emission, tannin adhesives for particleboard. Holz Roh Werkst. 53: 201–204.

Pizzi, A. 2003. Natural phenolic adhesives 1: Tannin. pp. 573–598. *In*: 2nd edition A. Pizzi and K.L. Mittal (eds.). Handbook of Adhesive Technology 2nd edition. Marcel Dekker, New York, USA.

Pizzi, A., R. Kueny, F. Lecoanet, B. Masseteau, D. Carpentier, A. Krebs, F. Loiseau, S. Molina and M. Ragoubi. 2009. High resin content natural matrix-natural fibre biocomposites. Ind. Crops Prod. 30: 235–240.

Pizzi, A., M.C. Basso, A. Celzard, V. Fierro and S. Giovando. 2016. Composition for manufacturing a tannin-based foam material, foam material obtainable from it, and manufacturing process thereof. US 9,302,413. Assigned to Silva Chimica and University of Lorraine.

Polesel-Maris, J. and I. Jutang. 2017. Antiadhesives coatings based on condensed tannins. Patent WO2017/037393 A1. Assigned to SEB Development.

Porter, L.J. 1974. Extractives of *Pinus radiata* bark. 2. Procyanidin constituents. New Zealand J. Sci. 17: 213.

Radebe, N., K. Rode, A. Pizzi and H. Pasch. 2013. Microstructure elucidation of polyflavonoid tannins by MALDI-TOF-CID. J. Appl. Polym. Sci. 127: 1937–1950.

Rangel, G., H. Chapuis, M.C. Basso, A. Pizzi, C. Delgado-Sanchez, V. Fierro, A. Celzard and C. Gerardin. 2016. Improving water repellency and friability of tannin-furanic foams by oil-grafted flavonoid tannins. BioResources 11: 7754–7768.

Ricci, A., G.P. Parpinello, A. Scghwertner-Palma, N. Teslic, C. Brilli, A. Pizzi and A. Versari. 2017. Analytical profiling of food-grade extracts from grape (*Vitis vinifera* sp.) seeds and skins, green tea (*Camellia sinensis*) leaves and Limousin oak (*Quercus robur*) heartwood using MALDI-TOF-MS, ICP-MS and spectrophotometric methods. J. Food Comp. Analysis 59: 95–104.

Richtzenhain, H. and B. Alfredsson. 1956. Uber Ligninmodellsubstanzen. Chem. Ber. 89: 378.

Roux, D.G. and E. Paulus. 1961. Condensed tannins. 8. The isolation and distribution of interrelated heartwood components of *Schinopsis* spp. Biochem. J. 78: 785–789.

Roux, D.G. 1965a. Modern applications of mimosa extract. Leather Industries Research Institute, Grahamstown, South Africa.

Roux, D.G. 1965b. Wattle Tannin and Mimosa Extract, Leather Industries Research Institute, Rhodes University Press, Grahamstown, South Africa.

Roux, D.G. 1972. Recent advances in the chemistry and chemical utilization of the natural condensed tannins. Phytochemistry 11: 1219–1230.

Roux, D.G., D. Ferreira, H.K.L. Hundt and E. Malan. 1975. Structure, stereochemistry, and reactivity of natural condensed tannins as basis for their extended industrial application. J. Appl. Polym. Sci. Appl. Polym. Symp. 28: 335–353.

Saad, H., F. Charrier-El-Bouthoury, A. Pizzi, K. Rode, B. Charrier and N. Ayed. 2012. Characterization of pomegranate peel tannin extractives. Ind. Crops Prod. 40: 239–246.

Saayman, H.M. and D.G. Roux. 1965. The origins of tannins and flavonoids in black-wattle barks and heartwoods, and their associated 'non-tannin' components. Biochem. J. 97: 794–801.

Santiago, F.J., G. Foyer, A. Pizzi, S. Caillol and L. Delmotte. 2016. Lignin-derived non-toxic aldehydes for ecofriendly tannin adhesives for wood panels. Int. J. Adhesion Adhesives. 70: 239–248.

Santiago, F.J., A. Pizzi, M.C. Basso, L. Delmotte and A. Celzard. 2017a. Polycondensation resins by flavonoid tannins reaction with amines. Polymers 9(2): 37, 1–16.

Santiago, F.J., C. Delgado, M.C. Basso, A. Pizzi, V. Fierro and A. Celzard. 2017b. Mechanically blown wall-projected tannin-based foams. Ind. Crops Prod. (in press).

Santiago, F.J., A. Tenorio-Alfonso, C. Delgado, M.C. Basso, A. Pizzi, A. Celzard, V. Fierro, M.C. Sanchez and J.M. Franco. 2017c. Projectable tannin foams by nechanical and chemical expansion. Ind. Crops Prod. (in press).

Sealy-Fisher, V.J. and A. Pizzi. 1992. Increased pine tannins extraction and wood adhesives development by phlobaphenes minimization. Holz Roh Werkst. 50: 212–220.

Slabbert, N. 1992. Complexation of condensed tannins with metal ions. pp. 421–429. *In*: Hemingway, R.W. and P.E. Laks (eds.). Plant Polyphenols: Biogenesis, Chemical Properties, and Significance. Plenum Press, New York.

Sauget, A., A. Nicollin and A. Pizzi. 2013. Fabrication and mechanical analysis of mimosa tannin and commercial flax fibers biocomposites. J. Adhesion Sci. Technol. 27: 2204–2218.

Szczurek, A., G. Amaral-Labat, V. Fierro, A. Pizzi, E. Masson and A. Celzard. 2011a. The use of tannin to prepare carbon gels: part 1. Carbon aerogels. Carbon. 49: 2773–2784.

Szczurek, A., G. Amaral-Labat, V. Fierro, A. Pizzi and A. Celzard. 2011b. The use of tannin to prepare carbon gels: part 2. Carbon cryogels. Carbon. 49: 2785–2794.

Szczurek, A., V. Fierro, A. Pizzi, M. Stauber and A. Celzard. 2013. Carbon meringues derived from flavonoid tannins. Carbon. 65: 214–237.

Szczurek, A., V. Fierro, A. Pizzi, M. Stauber and A. Celzard. 2014. A new method for preparing tannin-based foams. Ind. Crops Prod. 54: 40–53.

Thebault, M., A. Pizzi, H. Essawy, A. Baroum and G. Van Assche. 2015. Isocyanate free condensed tannin-based polyurethanes. Eur. Polym. J. 67: 513–523.

Thebault, M., A. Pizzi, F.J. Santiago, F.M. Al-Marzouki and S. Abdalla. 2017. Isocyanate-free polyurethanes by coreaction of condensed tannins with aminated tannins. J. Renew. Mat. 5: 21–29.

Tondi, G., A. Pizzi, E. Masson and A. Celzard. 2008a. Analysis of gasses emitted during carbonization degradation of polyflavonoid tannin/furanic rigid foams. Polym. Degrad. Stabil. 93: 1539–1543.

Tondi, G., A. Pizzi and R. Olives. 2008b. Natural tannin-based rigid foams as insulation in wood construction. Maderas 10: 219–227.

Tondi, G., A. Pizzi, H. Pasch and A. Celzard. 2008c. Structure degradation, conservation and rearrangement in the carbonization of polyflavonoid tannin/furanic rigid foams—a MALDITOF investigation. Polym. Degrad. Stabil. 93: 968–975.

Tondi, G. and A. Pizzi. 2009. Tannin based rigid foams: characterisation and modification. Ind. Crops Prod. 29: 356–363.

Tondi, G., C.W. Oo, A. Pizzi, A. Trosa and M.-F. Thevenon. 2009a. Metal adsorption of tannin-based rigid foams. Ind. Crops Prod. 29: 336–340.

Tondi, G., S. Blacher, A. Leonard, A. Pizzi, V. Fierro, J.M. Leban and A. Celzard. 2009b. X-ray microtomography studies of tanninderived organic and carbon foams. Microscopy & Microanalysis 15: 395–402.

Tondi, G., W. Zhao, A. Pizzi, V. Fierro and A. Celzard. 2009c. Tannin-based rigid foams: a survey of chemical and physical properties. Bioresource Techn. 100: 5162–5169.

Tondi, G., V. Fierro, A. Pizzi and A. Celzard. 2009d. Tannin-based carbon foams. Carbon 47: 1480–1492.

Tondi, G., A. Pizzi, L. Delmotte, J. Parmentier and R. Gadiou. 2010. Chemical activation of tannin derived carbon foams. Ind. Crops Prod. 31: 327–334.

Tondi, G., S. Wieland, N. Lemenager, A. Petutschnigg, A. Pizzi and M.-F. Thevenon. 2012a. Efficacy of tannin in fixing boron in wood: fungal and termite resistance. Bioresources 7: 1238–1252.

Tondi, G., S. Wieland, T. Wimmer, M.F. Thevenon, A. Pizzi and A. Petutschnigg. 2012b. Tannin-boron preservatives for wood buildings: mechanical and fire properties. Eur. J. Wood Prod. 70: 689–696.

Ucar, M.M., G. Ucar, A. Pizzi and O. Gonultas. 2013. Characterisation of *Pinus brutia* bark tannin by MALDI-TOF and ¹³C NMR. Ind. Crops Prod. 49: 679–704.

Valenzuela, J., E. Von Leyser, A. Pizzi, C. Westermeyer and B. Gorrini. 2012. Industrial production of pine tannin-bonded particleboard and MDF. Eur. J. Wood Prod. 70: 735–740.

Vazquez, G., A. Pizzi, M.S. Freire, J.S. Antos, G. Antorrena and J. Gonzalez-Alvarez. 2013. MALDI-TOF, HPLC-ESI-TOF, and 13C NMR characterisation of chestnut (*Castanea sativa*) shell tannins. Wood Sci.Technol. 47: 523–535.

Zhao, W., A. Pizzi, V. Fierro, G. Du and A. Celzard. 2010a. Effect of composition and processing parameters on the characteristics of tannin-based rigid foams. Part 1: Cell structure. Mater. Chem. Phys. 122: 175–182. Zhao, W., V. Fierro, A. Pizzi, G. Du and A. Celzard. 2010b. Effect of composition and processing parameters on the characteristics of tannin-based rigid foams. Part 2: Physical properties. Mater. Chem. Phys. 123: 210–217.

Index

A

Acid hydrolysis 172, 176, 177, 183, 198
Acoustic cavitation 153
Amino acids coupling 155
anti-inflammatory 265, 269
applications of floss fibers 35
Asclepias syriaca 23, 24, 26–28, 31, 32, 35, 37
Attenuated total reflection Fourier transform
    infrared spectroscopy (ATR-FTIR) 100,
    102

B

Bacterial nanocellulose 172, 173, 202
bio-based wood impregnation systems 124
Biocarbon 3, 5–11, 13–19
Biomass 69–80, 83, 85–94
Biosorbent 3–11, 13–19

C

Calotropis gigantean 23, 26, 27, 31, 36, 37
Calotropis procera 23, 25, 26, 31, 35
Carbohydrate solicitation 102, 113
carbon fiber 223–228
Cellulose nanocrystals 168, 172–177, 179, 183,
    184, 194
Cellulose nanofibers 168, 201
Chemical composition 3–7, 9, 10, 18
chemical composition of hollow floss fibers 29
chemometrics for NIR 45
Condensation index 102
condensation reaction resins 120
conifer resin 231, 232, 234, 240, 242, 245–247
cooperage industry 42, 55
Cpressaceae 232, 234, 235, 237, 247, 248

D

densification with glycerol 125, 127
Drying of nanocellulose 180, 186
Dynamic-mechanical properties 169, 170, 198,
    204

E

Elastic properties 174
electrospinning 223, 224, 228
Esterification 169, 180, 181, 183–186, 204
Eucalyptus pilularis 100, 102–106, 110, 112,
    113
Eucalyptus saligna 100, 102, 103, 107–109,
    111–114
Extractive compounds 104, 114

F

Fatty acids 104, 109, 113, 114
flavonoids 253, 254, 257–260, 265, 266,
    268–272
foam types 293, 296, 297, 300
Food packaging composites 163

G

Guayacyl 103, 110

H

Hemp fibers 5–13, 16–19
high purity lignin 221, 228
hollow floss fibers 22–24, 27–30
hydrodistillation 232, 235, 239, 240

I

impregnation processes 117, 118

K

kapok 23–25, 27–34, 36–38

L

Lead ions 6, 8, 9, 13–18

M

melasma 253, 254, 259, 261–266, 271, 272
melt-spinning 225, 226, 228

milkweed 23–26, 28–32, 36–38
monoterpenes 232, 235, 237, 239, 241

N

Nanocellulose chemical modification 178
Nanocellulose polyester composites 167, 170, 203, 204
NIR spectroscopy 42, 43, 45–47, 50–55, 57, 59, 60

O

Organic conductive composites 159
organosolv processes 214, 215, 217, 219

P

perfumer's palette 244
physical properties of floss fibers 31
Pinaceae 232, 235, 237, 241, 245, 248
polylactic acid (PLA) 130, 131, 138
Polypyrrole 149, 150, 158–164
psoriasis 253, 254, 259, 261, 262, 264, 265, 269–272
Pyrolysis (Py-GC/MS) 100

Q

Quercus 253–257, 259, 260, 266, 267, 269–271

R

Reinforcement agent 149
resin acids 232–234, 245

S

sol-gel densification 132, 133
Sonochemistry 147, 151

structure of floss fibers 29
Structures 283, 287, 288, 290, 293–297, 299, 300
Sulfuric acid 176–179
Surface hydroxyl groups 169, 178, 179, 184, 186, 204
Syringyl 103

T

tannins chemistry 280, 285, 292
TEMPO-Oxidized Cellulose Nanofibres 147–151, 153–155, 157–159, 161
therapeutic properties 240
thermal gasification 69, 71, 72, 80, 85, 88, 91, 92, 95
Thermochemical changes 99, 100, 102, 113
Thermogravimetric analysis (TGA) 100, 101, 112, 113

U

Ultrasound 147–149, 151–156, 163

V

vinyl polymers 123, 138
volatile oil 234, 235

W

Welding 99, 100, 102–104, 109–114
wood impregnation 117, 120, 121, 124, 127, 133, 135

X

X-ray photoelectron spectroscopy (XPS) 100, 101, 109–113

Printed and bound by CPI Group (UK) Ltd, Croydon, CR0 4YY

24/10/2024

01778304-0006